5, 6, 7, 8, 11, 13

Preliminary Chemical Process Design and Economics

D. William Tedder
School of Chemical Engineering
Georgia Institute of Technology
Atlanta, GA 30332-0100

Hickory Mountain
Marietta, GA

December 5, 2005

ISBN 0-9776308-0-3

Contents

Preface

These notes introduce chemical process design. They are intended for a senior level design course. It is assumed that students have already studied and understand calculus, linear algebra, thermodynamics, mass transfer, stagewise, and other unit operations. Some of the chapters briefly review material students should have learned in earlier courses. Others relate these topics to the synthesis, design, economics, and optimization of chemical processes. The author's intention is to encourage students to review information already learned and to think about it in new ways.

D. W. Tedder
Georgia Institute of Technology
School of Chemical & Biomolecular Engineering
Atlanta, Georgia 30332-0100
December, 2004

Glossary

Adjusted Purchased Equipment Cost Purchased cost adjusted for operating pressures and materials of construction.

Algorithm A list of steps that describes the desired manipulation of data.

Annuity Annual income derived from an initial investment or principal. The term "annuity" refers to any periodic payment (usually annual) used to repay a loan, accumulate a sum or deplete a principal.

Approach Temperature The difference in temperature between entering and exiting streams in a unit operation where heat transfer occurs (typically a countercurrent heat exchanger).

Auxiliary Investment Items that are of common use, e.g., cooling towers. Auxiliary equipment may be more easily liquidated that the fixed investment.

Base Case The first process flow diagram that is designed to produce a product. The estimated cost of the Base Case is a reference that is used to evaluate the attractiveness of alternative cases (i.e., other process flow diagrams). As a design project evolves, the original Base Case may be discarded in favor of another, more attractive process that is then treated as the new Base Case.

Bare Module Cost includes the following costs: (1) the purchased fob equipment cost, (2) other direct materials cost, (3) the direct construction labor cost, and (4) the indirect cost.

Battery limits The chemical manufacturing area of a plant site. Usually, the battery limits includes all processing equipment, but it excludes administrative buildings, utilities, storage, and transportation facilities, and other auxiliary facilities that are not directly involved in manufacturing. Additional support or auxiliary facilities are referred to as the Offsites Services although they are located on the plant site.

Capitalized Cost The initial principal required to guarantee that funds will always be available to replace an item of equipment that has worn out.

CNDM Composition Node Design Method - A design model which requires that a node of fixed composition be defined at each point in a unit operation where a process stream enters or exits.

CDS Coefficient of Difficulty of Separation - A heuristic which favors easy splits, distributed components and balanced columns.

Complex Towers A tower consisting of more than two stripping or rectifying sections. A complex tower produces more than two products.

Composition Node Each point in a unit operation where a stream composition is either fixed or estimated using simplified design equations.

Cooling Requirements, Minimum That minimum cooling utility which is required by a heat exchanger network even when the best possible stream matches are made. It is computed using the TI Problem Table.

Cost Indices An index value for a given point in time showing the cost at that time relative to a certain base time.

Design Equations An approximate set of equations which can be used to easily design and optimize the system of interest. Design equations are used to compute equipment specifications from the given design goals.

Decision Variables Those decisions which can be made to alter the structure of a process flowsheet. For example, using distillation versus solvent extraction to achieve a separation may be a decision variable.

Degrees of Freedom The number of design variables minus the number of linearly independent equations.

Design Model A model concerned with the specification of equipment configurations that can yield product streams for a given feed.

Design Parameters Values for fixed conditions or constraints as required for a particular process.

Design Variables Those state or structural decision variables that may be adjusted within a given process flowsheet to minimize costs or achieve other goals.

Direct Search Any variable search technique that adjusts the variables based upon a pattern (*e.g.* secant, golden search).

Direct Sequence A sequence of towers consisting of two or more columns where the bottoms product from the first column is distilled in the second tower, *etc.* The most volatile product is removed first.

Dynamic Programming A stepwise optimization procedure that can be used to synthesize chemical networks starting with the downstream unit operations. Its guarantee of optimality is limited to acyclic systems.

Economic Objective Function The function describing a process which may be either minimized or maximized in order to synthesize and optimize the most economical process. For example, venture cost may be minimized as an objective function whereas venture profit and net present worth are usually maximized when used as objective functions.

ESI Ease of Separation Index - The ratio of relative volatilities of components between which two different splits take place.

ESM Equilibrium Stage Model - A rating model which assumes that the vapor and liquid streams leaving a stage are in thermodynamic equilibrium. Equilibrium stage models are also referred to as plate-to-plate models and they require extensive trial and error calculations to compute energy and material balances for countercurrent unit operations.

Enumeration Technique A method of synthesis in which all possible unit operation combinations are evaluated and listed. It is only practical when there are relatively few possible combinations.

Evolutionary Search consists of examining variations and/or combinations of flowsheets by relaxing or discounting some heuristics to generate modified networks. Alternatively, an evolutionary search may be based upon examining a rank ordered list of alternative structures.

Fixed Capital Investment The capital needed to construct the necessary manufacturing and plant facilities.

Flowrate Heat Capacity The product of a process stream's mass flowrate and its heat capacity.

Flowsheet A simplified diagram of a process showing the major process streams and unit operations.

Flowsheet Integration The improvement of a flowsheet through such techniques as heat exchanger matching or vapor recompression to minimize processing costs (usually refers to reductions in utility consumption).

fob Free on Board - The purchased equipment cost excluding transportation costs. The purchased equipment cost after it is loaded onto a common carrier at the vendor's site.

Golden Search A single variable optimization search technique that adjusts the bounds by rejecting the lower or upper point and continues until a specified tolerance has been reached in the optimization variable.

Gradient Search A multivariate search method that adjusts the variables based upon gradient information (i.e. the response surface characteristics)

Grass-Roots Plant A complete plant that is constructed on a new site. The cost of a grass-roots plant includes the costs of land purchases, site development, both auxiliary or offsites facilities, and battery limits facilities where manufacturing occurs.

Gross Profit The difference between total annual sales and annual operating costs.

Heat Exchanger Matching The procedure of matching hot and cold process streams to cool and heat each other such that the utility requirements and their respective costs may be minimized.

Heat Requirement, Minimum When hot and cold process streams are matched in the best possible manner, the network may still require some minimum heating utility. This requirement may be calculated for a given network using the TI Problem Table.

Hessian The n x n matrix of second partial derivatives for a function.

Heuristic A rule of thumb or design rule which simplifies the analysis in some way.

Incremental Return on Investment The rate of return on a small amount of money invested in addition to some base amount.

Inexact Line Search A search which interpolates between an exact line search and the method of steepest descent and therefore is not strictly along the search direction indicated by either.

Information Flow The direction in which calculations are made based on previous results from previous calculations. This direction can be deliberately chosen to minimize trial and error calculations.

Internal Pricing Utility charges which are assigned to a process as an operating cost only. These charges include both the fixed and operating expenses associated with the utility, but appear as an operating expense for the battery limits process.

Inverted Sequence The sequence consisting of two or more distillation columns where the distillate product from the first column is distilled in the second tower, *etc.* The least volatile product is recovered first.

Jacobian Matrix The gradient of a function, or that function's set of partial first derivatives.

Labor Costs Those costs that are proportional to annual labor requirements including direct operating labor, supervision and overhead.

Material Costs MAT, $/yr The product of the annual rate at which raw materials are consumed by a process, times their respective costs (e.g., in $/lb).

Method of Steepest Descent A search method in which the direction of search is the negative of the Jacobian vector or the local gradient.

Model A preliminary pattern for a process.

MNU The minimum number of heat exchanger units for matching a network of hot and cold process streams. The MNU is equal to the number of process streams transferring heat above the pinch temperature plus the number of streams transferring heat below the pinch temperature.

Neighbor A flowsheet which results from interchanging the relative positions of two adjacent subproblems.

Net Profit The gross profit minus the capital recovery and annual taxes.

Network A system of process components or unit operations that are interconnected.

Operating Cost The sum of the direct production costs, fixed charges and plant overhead costs.

Ordered Branching A search which begins with the upstream tower and applies the heuristic to do the cheapest separation next. This heuristic with forward branching and backtracking may locate the optimal sequence.

Overspecification When process variables are specified to an extent which exceeds the available degrees of freedom. In this case, it is not possible to satisfy all of the fundamental equations exactly, but over- specification can often be done in design work so as to lead to a small and acceptable level of overdesign.

Pattern Search A search which adjusts the variables in a specified pattern.

Perpetuity An annuity paid out over an indefinite length of time. It refers to periodic payments generated by a principal amount sufficiently large so that the periodic payments will continue indefinitely.

Pinch Point The point in the column at which the vapor rate first becomes equal to the minimum rate for that section containing the pinch point as the reboiler rate is reduced. That is, the operating line touches the equilibrium curve at the pinch point in binary distillation.

Pinch Temperature In a heat exchanger network, that temperature at which the composite energy transfer curve for the cold streams first intersects the composite energy transfer curve for the hot streams.

Positive Definite The quadratic form $\boldsymbol{x}^t \boldsymbol{A}\boldsymbol{x}$ is positive definite if all of its eigenvalues are less than zero at a stationary point.

Precedence Order Precedence Ordering is a method for analyzing a set of equations so that they may be solved sequentially, one-at-a-time, rather than simultaneously. It is not always possible to solve an equation set sequentially, but it is usually helpful whenever a Precedence Order can be established.

Problem Table A systematic listing of data used in the temperature interval method of heat exchanger matching.

Process Integration The combination of two or more processes or subprocesses to achieve a design goal.

Process Optimization The systematic adjustment of design variables to achieve either reductions in cost or to maximize profitability. Optimization may use gradient or pattern searches or adjustments to the process structure.

Process Synthesis The operation of assembling pieces of process equipment together into optimal networks that minimize the cost when the resulting system is operated at the best possible state.

Production Cost The production cost includes operating costs which are proportional to the annual production rate. It includes costs for raw materials, utilities, maintenance and distribution.

Profit Upper Bound (PUB) An approximate measure of the profitability of a reaction path, defined as the annual sales (S) minus the annual cost of raw materials (MAT)(i.e., PUB = S - MAT). Since there are other costs besides the annual MAT costs, the true profitability is less than the computed PUB. If a PUB is negative, then it is impossible to make a profit using that reaction path.

Quadratic Interpolation A one-dimensional search technique which is applicable to multivariable optimization problems. A response is modeled as a quadratic curve in r by evaluating three points in the region where the answer may be found. (There is no requirement for upper and lower bounds.) Improved values of the optimization vector X may be found by interpolation in r.

Quadratic Programming A nonlinear programming technique, the goal of which is to find extreme points for a quadratic objective function with linear constraints.

Rating Model A mathematical model that evaluates the performance of a processing unit based upon fundamental considerations such as equilibrium. In a rating model, the feeds and equipment specifications are given information while the products are calculated as dependent variables.

Reaction Path Selection (or Screening) An analysis method that establishes criteria for determining if a new chemical process has any chance of being economical and realizing a profit. Usually the analysis consists of a production/consumption analysis and determines an upper bound on profitability which must be positive for the reaction to be economical. The gross profit can be used as the upper bound.

Recovery Factor The fraction of input feed that is recovered in a separation step in a chemical process. The recovered portion is recycled to the initial reactor.

Relative Volatility The volatility of the *ith* component relative to the *jth* component is K_i/K_j, where $K_i = y_i/x_i$ is the vapor-liquid equilibrium constant.

Return on Investment (ROI) The net profit realized divided by the capital investment. Usually an investment must be studied to determine if it will satisfy a minimum return criterion before it is made.

Risk The uncertainty associated with whether or not a capital investment will be recovered by the venture.

Sales (S, \$/yr) The product of the sales price per unit of product and the annual number of units actually sold.

Search A method of optimizing a process by determining the values of design variables that will either minimize the process costs or maximize the profitability.

Shortcut Design Method Design methods which provide a rapid procedure for estimating the operating state of any unit operation.

Simple Tower A distillation tower in which the feed stream is split into two streams, the distillate and the bottoms products (*i.e.* there are no side streams in a simple tower).

Systematic Biases These are usually assumptions that result in systematic errors. Say, for example, incorrect assumptions regarding materials of construction are made. Then the resulting cost estimate will be systematically biased, as opposed to random errors in cost. Suppose an inappropriate thermodynamic package is used to simulate a chemical process. The resulting simulation model output will have systematic errors (or biases) that invalidate the findings.

Task Solution Heuristics A set of *rules of thumb* that are used as guidelines so that economically feasible alternative flowsheets can be quickly identified.

Ticking Off In heat exchanger matching, ticking off refers to the marking of a stream that has had its heating or cooling requirements met.

Total Capital Investment The sum of the fixed capital investment and the working capital.

Tower Section A collection of trays that are located between two adjacent composition nodes.

Tree Search Branches on the tree represent logical connections or paths between the nodes on the decision tree. Each node represents a decision point and each branch represents a different decision. A tree search is simply any method that can be used to evaluate the relative merits of the alternative decisions.

Vapor and Liquid Flows The vapor and liquid flows passing each other counter currently in a distillation tower.

Venture Cost A measure of the cost of a particular operation. It takes into account the net profit and the investment risk. It enables one to specify a minimum rate of return on the investment.

Venture Profit A measure of the profits realized from a particular operation. It takes into account the net profit and the investment risks. It combines the net profit and rate of return criteria into a single objective function.

Working Capital The funds required to finance the day-to-day expenses primarily in the form of inventories. Working Capital also includes cash on hand for payment of operating expenses and accounts payable.

Nomenclature

a power rule for size dependence of equipment cost, eqn 4.4

A Annuity (annual payment), eqn 3.32, eqn 3.33

$APUB$ Adjusted Profit Upper Bound, eqn 2.44

C total cost in eqn 2.4; free energy of consuption, eqn 2.36; annual capacity in eqn 2.43; total manufacturing cost, eqn 3.14

C^a_{bm} actual bare module cost, Table 4.3

CD cash dividends, eqn 3.7

C_F complexity factor, eqn 2.42

CF, CF_i cash flow, eqn 3.6, cash flow in ith year, eqn 3.19, eqn 7.27, eqn 7.31

C_I annual costs proportional to total capital investment, eqn 2.4 and eqn 3.14

C_L annual labor costs, eqn 2.4 and eqn 3.14

C^a_p fob purchased cost of equipment fabricated of required materials of construction, Table 4.3

C^{cs}_p fob purchased cost of equipment fabricated of carbon steel, eqn 4.4, Table 4.3

C_Q annual product costs (e.g., raw materials, utilities, etc.), eqn 2.4

CR capital recovery, eqn 3.5, eqn 3.24, eqn 3.24

$C(t)$ cost at a given gime, eqn 4.2

D pipe diameter, eqn 11.1, eqn 11.6, eqn 11.8, eqn 11.9, eqn 11.16, eqn 11.17, eqn 11.18, eqn 11.19

DEP_i annual depreciation in ith year, eqn 3.24

E, E_o fob purchase cost, at required and references sizes, eqn 4.4; fob purchase cost with carbon steel construction, Table 4.3

E' fob purchased equipment cost when manufactured of special materials of construction, Table 4.3

E_c capacity elasticity defined in eqn 2.2

E_d demand elasticity defined in eqn 2.1

f friction factor, eqn 11.1, eqn 11.7, eqn 11.8, eqn 11.9

F value of a future sum, eqn 7.1, eqn 7.2, eqn 7.3

F^a_{bm}, ${}^{cs}_{bm}$ actual and carbon steel bare module factors, Table 4.3

f_M materials factor, eqn 2.42

F_c complexity factor, eqn 11.16, eqn 11.17, eqn 11.18

F_m materials cost factor, Table 4.3

F_n future sum in nth year, eqn 7.2, 7.7

F_p pressure cost factor, Table 4.3, eqn 4.7

f_P pressure factor in eqn 2.41

f_T temperature factor, eqn 2.39

f_t tubing factor, Table 4.3

g capital recovery, annuity, or sinking fund factor, eqn 3.21; gravity factor, eqn 11.1

GP gross profits, eqn 3.4, eqn 3.16, eqn 3.18

H_p polytropic head, eqn 11.2

i_a discrete compound annual interest rate, eqn 7.2

i_e effective annual earnings rate for an entity, eqn 8.9

i_{min} minimum acceptable annual earnings rate for an entity, eqn 8.9

i_s simple interest rate, eqn 7.1

I_A installed cost of auxiliary equipment, eqn 3.8

I_C current assets, eqn 3.10

I_{CE} CE Plant Cost index, eqn 2.43

I_D depreciable investment, eqn 3.13, eqn 3.16

I_F fixed assets, eqn 3.9

I_L land investment, eqn 3.8, eqn 3.16

I_o original purchase cost, eqn 3.31

I_P installed cost of process battery limits, eqn 3.8

I_S start-up costs, eqn 3.8, eqn 3.16

I_W working capital, eqn 3.8

k heat capacity at constant pressure divided by heat capacity at constant volume, eqn 11.4

K_i frictional resistance for ith fitting, eqn 11.1

L path length, eqn 11.1; length of straight pipe, eqn 11.18, eqn 11.19

L_e equivalent length of pipe, per foot of straight pipe, L, eqn 11.18, eqn 11.19

LEL lower explosion limit, example 10.1, eqn 10.2

MAT annual raw material costs, eqn 2.6, eqn 2.44

$\dot{m}$ mass flow rate, eqn 11.1

n years, eqn 7.1

NC_i net cost in ith year, eqn 3.25

N_f estimated number of fittings per foot of straight pipe, eqn 11.16

N_F number of major units (or processing steps) in a flowsheet, predesign estimates, see Section 2.5.2

N_v estimated number of valves per foot of stright pipe, eqn 11.17

NP net profits, eqn 3.5, eqn 3.21, eqn 3.24, eqn 8.1

NPC net present cost, eqn 8.7

NTI net taxable income, eqn 3.16

OTH all costs except raw material costs, eqn 2.5

p price in eqn 2.1 and 2.2

p_i system pressure at ith state, eqn 11.1

P present value, eqn 7.3, eqn 3.32,

P_i^{sat} pure component vapor pressure for ith species eqn 3.33, eqn 7.1, eqn 7.12, present worth or net present worth, eqn 7.27, eqn 7.31; estimated number of plant employees, eqn 6.1

P^* reference pressure, eqn 2.40, eqn 4.7

P_l,P_b laboratory and bulk quantity prices, eqn 2.6

P_E, P_S price per unit of energy consumed, and selling price, eqn 2.37

P_v present value or book value, eqn 3.30

PI profitability index, eqn 7.34

PUB Profit Upper Bound, eqn 2.6

q volumetric flowrate, eqn 11.6

Q demand in eqn 2.1 and 2.2

Q_l, Q_b laboratory and bulk quantities, eqn 2.34

Q_P annual product production rate, eqn 3.15

r relative risk penalty factor, eqn 8.9, eqn 8.10

r_p compression ration per stage, eqn 11.2

R ideal gas constant, eqn 11.2

RC replacement cost, eqn 3.28, eqn 3.29, eqn 3.31

Re Reynolds number, eqn 11.7, eqn 11.8

RE retained earnings, eqn 3.7

ROI return on investment, eqn 8.2

S annual sales, eqn 2.6 and eqn 3.15, eqn 3.18

S, S_o required and reference sizes, eqn 4.4

S_n value of a future sum in the nth year, eqn 7.12

t_i tax rate in ith year, eqn 3.19, eqn 3.25

T temperature, absolute scale, eqn 11.2

TCI total capital investment, eqn 3.8

TDH total developed head, eqn 11.14, eqn 11.15

TIC total installed cost for entire plant, eqn 4.3

TIC_{BL} total installed cost for a plant battery limits

u_i stream velocity at ith state, eqn 11.1

UEL upper explosion limit, eqn 10.1, eqn 10.3

VC annual venture cost, eqn 8.3, eqn 8.8

VP annual venture profit, eqn 8.1

$\dot{w}_f$ shaft work, eqn 11.1

X plant capacity in tons/day, Wessel correlation, eqn 6.3

Y estimated man-hours/ton-processing-step, Wessel correlation, eqn 6.3

z_i elevation at ith state, eqn 11.1

Z compressibility factor, eqn 11.2

ΔG_f^o Gibbs energy of formation, eqn 2.36

ρ stream density, eqn 11.1, eqn 11.6

ϕ compression parameter, eqn 11.2, eqn 11.3, eqn 11.5

μ liquid viscosity, eqn 11.6

ϵ pipe roughness factor, eqn 11.8, eqn 11.9

ϵ_i pump efficiency factor, eqn 11.14

Chapter 1

Introduction

The dictionary definition of *design* means *to make, draw, or prepare preliminary plans or sketches of, ...to form or make (plans, schemes, etc.) in the mind; conceive, invent* [29]. This broad definition applies equally well both to the design and to synthesis or invention, and it can be applied to either process or product design. Design is both a left- and right-brain activity. It involves analytical steps, but also cognitive steps such as pattern recognition and the ability to assemble processing steps into meaningful structures. In the case of product design, it involves the recognition of useful combinations of products, often the selection of a few more desirable products from among a large number of alternatives.

The Accreditation Board for Engineering and Technology (ABET)[1] offers a somewhat broader definition of engineering design:

> *...the process of devising a system, component, or process to meet desired needs. It is a decision-making process (often iterative), in which the basic sciences, mathematics, and engineering sciences are applied to convert resources optimally to meet a stated objective. Among the fundamental elements of the design process are the establishment of objectives and criteria, synthesis, analysis, construction, testing, and evaluation.*

This second definition also includes optimization and synthesis.

In chemical process design, the goal is to conceive, invent, draw, size, and test chemical processes, and to eventually estimate their capital and operating costs. When chemical engineers *design* processes, they usually generate cost estimates too, so chemical process design relates closely to *cost engineering*. The American Association of Cost Engineers (AACE) defines [21] the latter activity as:

> *...that area of engineering practice where engineering judgment and experiences are utilized in the application of scientific principles and techniques to the problems of cost estimation, cost control, and profitability.*

It therefore makes sense to study process synthesis and chemical engineering cost estimation together.

The traditional approach to design follows the first dictionary definition more closely. Practically speaking, one is often pleased just to have a feasible design completed (i.e., a base case) because of the complexity involved in generating it. Even it if is not optimal, a feasible base case that is reasonably detailed and provides a sensible economic analysis is an important milestone for any design project.

Because of the complexity of many chemical processes, process optimization and the invention (or synthesis) of alternative flowsheets often receive secondary emphases. To a certain extent, synthesis heuristics can help, but the natural tendency is to become embroiled in the details of analyzing individual unit operations and to overlook the interactions between them. In chemical process design is is usually very easy to lose the forest for the trees, but network interactions are often very important. In fact, it is very important to generate an initial estimate of the entire process, considering all of the economic factors, before focusing too much on any one component or subsystem.

Heuristic: 1 *Initially focus on the overall process, not individual components. It is more important to determine those processing steps that are economically most important than to optimize individual steps. Don't lose sight of the forest for the trees.*

Frequently, the costs of reliably designing a reasonable base case and its components are high, so the detailed evaluation of alternative flowsheets only becomes practical if the engineer resorts to approximations such as described here.

Optimization comes from the verb *to optimize* which means *to plan or prepare plans in order to secure maximum efficiency,...to make the most of,...to work toward obtaining an optimum*, that is, *the condition or degree producing the best result* [29].

In chemical process design, two factors are often in competition and need optimization — the operating costs and the capital investment. Those operating states which minimize operating costs (e.g., minimum reflux in distillation) often increase capital investment. In the case of distillation, for example, operating at minimum reflux means that an infinite number of theoretical trays are required. Thus, the capital investment becomes infinite. For a given separation, increasing the reflux ratio in a distillation column moves the operating line away from the equilibrium curve and results in fewer required theoretical trays, but also increases the operating costs (i.e., the reboiler and condenser duties). So these two factors are in competition. Optimum design implies that a balance has been achieved between these two competing costs.

The best alternative, however, is one which reduces both the annual operating costs and the required capital investment to build the facility. In the case of distillation, a optimum reflux ratio exists for most columns which balances the annual operating costs with the capital investment, but a different unit operation may exist that would have both lower operating cost and capital requirements, compared to an equivalent distillation system, even when the latter has been optimized. Process synthesis is concerned with finding such alternatives in an efficient manner.

Synthesis means *the assembling of separate or subordinate parts into a new form, ...the combination of separate elements into a whole,...a process of reasoning*

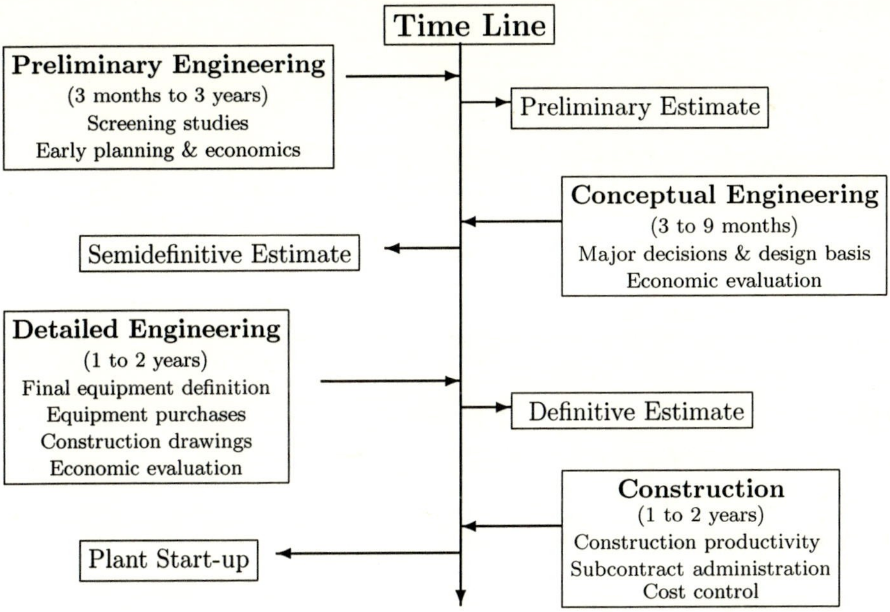

Figure 1.1: Events and Phases in a design Project (adapted with permission from Chemical Engineering, McGraw-Hill, July 1, 1975).

from the whole to a part, from the general to the particular, by deductive reasoning. The synthesis of chemical processes usually refers to the invention or discover of new flowsheets with improved features. It is one of the higher forms of design and optimization, and is often the most exciting and can provide the most dramatic increases in profitability. The synthesis process is difficult to describe and quantify because of the many processing combinations that usually exist, and because of the difficulties in describing how to approach an inherently creative activity. However, it often yields the greatest benefits and improvements. Benefits may include reduced operating and capital costs, better controllability, greater processing flexibility, increased product yield, and lower waste management costs.

1.1 The Design Project

In chemical engineering, a design project is a course or plan of action leading to the detailed description of a manufacturing process that achieves specified goals. Projects may occur over in brief periods like one or two days. More commonly, they extend over several years and involve several phases.

Table 1.1: Typical steps during Preliminary Engineering Phase

1. Draw Block Diagrams and develop concepts.
2. Draw initial Process Flow Diagrams (PFDs) for Base Case.
3. Develop Major Equipment List for Base Case.
4. Compute preliminary Energy & Material Balances for Base Case.
5. Determine Major Equipment Specifications for Base Case.
6. Complete preliminary safety, environmental, and waste management analyses.
7. Estimate Total Installed Cost of Battery Limits and the Entire Plant.
8. Estimate Total Product Cost and Project Profitability.
9. Write up results and submit Preliminary Estimate Report.

1.1.1 The Phases in a Project Life

Major projects are typically completed in phases. Figure 1.1, for example, illustrates four major phases:

1. Preliminary Engineering,
2. Conceptual Engineering,
3. Detailed Engineering, and
4. Construction.

Each of these phases results in a major event:

1. Completion of the Preliminary Estimate,
2. Completion of the Semidefinitive Estimate,
3. Completion of the Definitive Estimate, and
4. Plant Construction and Start-up.

The first three phases are partly repetitive. They involve similar steps, but with increasing levels of detail. During Preliminary Engineering, it is often necessary to make assumptions that have not been verified by experimental measurements. The design engineer is sometimes forced to make approximations and her goal is to model reality as accurately as possible. During Conceptual Engineering, these assumptions are evaluated and confirmed. The flowsheet is modified as needed to accommodate the more detailed understanding of its behavior, based upon understanding during Phase II. Detailed Engineering includes final equipment definition, purchases, the development of a critical path to complete the project. Plant Start-up is the natural climax to the successful design project.

In the Preliminary Engineering Phase, only a small group, perhaps only one or two engineers, are involved. Typical steps and characteristics of the Preliminary Engineering Phase are summarized in Tables 1.1 and 1.2. At that point, engineers are often working closely with several chemists and technicians to devise a base case.

Table 1.2: Typical characteristics of Preliminary Engineering Phase

1. Small working group
2. Engineers work closely with chemists
3. Consider many alternatives besides the Base Case
4. Complete synthesis studies to develop a list of alternatives
5. Define the best Base Case
6. Limited experimental studies
7. Often includes assumptions that need experimental verification
8. Total Product Cost and Project Profitability are less certain

Table 1.3: Typical steps during Conceptual Engineering Phase

1. Revise Process Flow Diagrams.
2. Experimentally verify energy & material balances.
3. Complete pilot studies.
4. Complete corrosion tests.
5. Complete detailed safety, environmental, and waste management studies.
6. Experimentally confirm key assumptions.
7. Revise and Expand Major Equipment List.
8. Complete optimization studies.
9. Select Architect/Engineering (A/E) firm.
10. Complete Semidefinite cost estimates.

This base case is then evaluated using simplified methods to yield a Preliminary Estimate.

There are several important objectives during Preliminary Engineering. The greatest flexibility exists during this phase of the project and one objective is to consider the widest range of structural alternatives (synthesis) that will achieve the desired result. During Preliminary Engineering, different unit operations are considered as well as different chemistry to achieve goals, and a multiple cases are often defined. The flowsheet revisions can be quite substantial during this phase.

A second goal during Preliminary Engineering is to obtain an objective measure of the economic problem areas. In order to do so, simplified methods are used to evaluate the steps in the process and to arrive at conclusions. Because simplified design methods are used, there is greater uncertainty in the cost estimates. On the other hand, major problem areas (e.g., processing steps that are particularly expensive) are more readily identified so they can be subjected to synthesis and analysis to resolve them.

Bench testing is completed as part of Preliminary Engineering, but parts of the flowsheet may not have been experimentally evaluated. The best available material and energy balances, together with simplified design methods, are used to estimate costs. Generalized cost correlations may be used to estimate capital costs with some quotes of the more expensive items in the flowsheet.

Table 1.4: Typical characteristics of Conceptual Engineering Phase

1. Involves larger working group
2. Use vendor pilot studies; obtain equipment guarantees
3. Obtain preliminary bids for most expensive equipment items
4. Extensive experimental testing
5. Eliminate problem areas
6. Initiate licensing petitions
7. Select plant site
8. Prepare critical path diagram

Table 1.5: Typical steps during Detailed Engineering Phase

1. Complete final Process Flow Diagrams.
2. Complete final, verified Energy & Material Balances.
3. Complete Piping & Instrumentation Diagrams (P&IDs).
4. Finalize equipment specifications.
5. Complete detailed plant layout.
6. Complete all licensing requirements.
7. Complete utility line diagrams (ULDs).
8. Complete isometric diagrams.
9. Complete scale models & virtual reality models.
10. Complete Definitive Cost Estimate.
11. Begin Construction

In the second phase, Conceptual Engineering, all of the assumptions made in the preliminary design are reviewed and modified as needed based upon better experimental data with fewer assumptions. Energy and material balances, for example, are usually recalculated in greater detail. Also, the cost estimate is more detailed and more accurate. Typical steps in and characteristics of the Conceptual Engineering Phase are summarized in Tables 1.3 and 1.4 Fewer equipment items are omitted. Essential details (e.g., materials of construction and required vessel wall thicknesses) are reviewed. Quotes are obtained from vendors for major equipment items. Pilot studies are completed. Assumptions made during Preliminary Engineering are experimentally verified and revised as required. The flowsheet may be modified somewhat, but the changes are generally less drastic than those made during Preliminary Engineering when alternative structures were considered and compared to each other. Safety analysis is completed in more detail. Negotiations with appropriate licensing authorities are initiated. Potential locations for the new facility are identified. A Semidefinitive Cost Estimate is prepared and submitted to management for review and approval.

During the Detailed Engineering Phase (also see Tables 1.5 and 1.6), the flowsheet is essentially frozen. Process flow diagrams are finalized, along with energy and material balances. All equipment items have been sent out to vendors for bids

Table 1.6: Typical characteristics of Detailed Engineering Phase

1. Often involves a large working group
2. Process Flow Diagrams are frozen
3. Costs based on final bids for all equipment and site development work
4. All experimental testing completed
5. Safety protocol and training programs in place
6. Operator training programs initiated
7. Plant site purchased
8. Neighborhood information programs underway

and firm cost quotes are in hand, along with estimated lead time to provide each item when required during construction. All pilot studies have been completed. Detailed diagrams are available describing plant layout. Piping and instrumentation diagrams, utility line diagrams, and isometric diagrams are completed, usually by an architect/engineering (A/E) firm that has been contracted to complete the construction. Formerly, detailed scale models of the plant were built and used at the construction site to guide the effort. More commonly now, virtual reality computer systems are provided by the A/E as part of the computer aided design (CAD) system to enable workers to view the finished areas of the plant during construction. Finally, a Definitive Cost Estimate is prepared and submitted to management for review and approval.

The costs of each phase in a design project are successively higher. If the results of the Preliminary Estimate are favorable, then management may choose to commit funds to complete the second project phase and generate a Semidefinitive Estimate. Again, if the economic results are favorable, then management may choose to approve the remaining project phases as shown in Figure 1.1 and build the plant. At the end of each of the first three phases, cost estimates are generated and are used to determine whether or not the project should be continued. During successively phases, the cost estimates become increasingly accurate and have less uncertainties. Ideally, the Preliminary Estimate accurately projects the final costs, but Preliminary Estimates are commonly low and, therefore, contingency is added to compensate. All calculations are checked and rechecked during the second and third phase to make sure that computed results are accurate.

Usually only one or two engineers are involved in Preliminary Engineering, but the numbers involved in a project grow considerably during the second and third phases, especially when an architect/engineering firm and equipment vendors become involved. Thus problems in communication become more challenging as the project progresses, but are simplified if effective techniques for process representation and analysis are adopted during Preliminary Engineering.

1.1.2 The Steps in a Preliminary Estimate

As suggested in Figure 1.1, the accepted approach to process design involves several phases in which a single process is evaluated at least two or three times before the

Table 1.7: The accepted steps in the Preliminary Design of chemical processes.

1. Complete a literature review and identify similar processes.
2. Complete a Process Flow Diagram (PFD).
3. Complete all energy and material balance calculations, and define all processing streams.
4. Complete all equipment and facility sizing and layout calculations using simplified methods.
5. Complete all equipment, building, and other facility cost estimates.
6. Complete an economic evaluation of the project costs (e.g., estimate cost/lb of product or feed, total installed cost, annual operating costs, labor requirements, etc.).
7. Compile Material Safety Data Sheets (MSDSs) and identify safety issues.
8. Complete preliminary environmental and waste management analyses.
9. List major assumptions and recommend areas where research work is needed to confirm these assumptions.
10. Identify problem areas in the process as designed.

plant is actually built. Each evaluation results in a cost estimate that is more accurate and detailed than its predecessor. Usually the cost estimates become higher as more is known about the process and its requirements.

It is usually necessary to complete several steps sequentially in order to generate a cost estimate for either a single component (or unit operation) in a process or an entire processing network. The basic steps leading to a Preliminary Estimate are summarized in Table 1.7. As shown, one of the first steps is the completion of a process flow diagram (or PFD) which indicates the major unit operations in the flowsheet and their interconnections. This drawing describes the key structural features of the design.

Energy and material balance calculations are necessary in order to size equipment items and to estimate utility costs. This step is usually one of the more difficult because: (1) material balances often require trial-and-error calculations and (2) they often involve hundreds of process streams that must be defined. Energy balances are usually more straightforward, if given an accurate and complete set of material balances. Of course, with some unit operations (e.g., distillation) neither is known very well until both are known.

Equipment sizing calculations are usually simplified during Preliminary Engineering. Overall heat transfer coefficients for heat exchangers, for example, may be assumed rather than calculated. Frictional resistances in pipelines are approximated rather than calculated exactly because piping diagrams and layout are not defined (and because time is short and it is more important to consider alternatives during that phase). However, any approximations that are used should lead to equipment sizes that are slightly oversized (10% overdesign is common). A degree of overdesign is needed initially to insure that the process design criteria are eventually met. Overdesign offers some insurance that new equipment will function properly and will not cause undue delays because it has to be rebuilt after installation.

There is also a tendency to underestimate equipment requirements, especially when less is known about a process. In such instances the PFD initially does not contain all of the unit operations required to enable it to function as desired. This common problem is the motivation for the heuristic:

Heuristic: 2 *The less that is known about a process, the better it looks. Chemical processes when constructed are commonly more complex and involved than when first conceived because one or more essential steps were omitted during Preliminary Engineering or the underlying chemistry was not well understood.*

Engineers should guard against the euphoria that commonly surrounds a newly-conceived process, especially when few experimental data are available.

Initially, generalized cost correlations may be used for estimation. In later phases, more detailed and precise estimates are needed. They should be based upon vendor quotes. However, reasonable preliminary cost estimates can often be obtained fairly easily from published correlations. Some of these are reproduced in subsequent chapters.

It is extremely important that Preliminary Cost Estimates be based upon the correct selection of materials of construction. A failure to correctly estimate the materials of construction can lead to low estimates by a factor of two or three or more. Preliminary Engineering should identify corrosion issues and describe appropriate tests to address them. When in doubt, it is best to specify materials of construction that are more resistant and expensive than may actually be required. This approach will emphasis the potential economic impacts of operating as planned, and will highlight the need to complete experimental testing (e.g., coupon corrosion tests) to determine whether or not less expensive materials of construction can be safely employed.

Step 6 in Table 1.7 consists primarily of simple accounting. The major problem encountered is a failure to include significant cost elements in the initial analysis. Because major equipment items are easily overlooked in a preliminary estimate, early estimates can be low.

The diligent cost engineer is careful to avoid two types of errors. First, errors of *commission* occur from either grossly underestimating the correct sizes of major equipment items, or else from failing to correctly select materials of construction. Second, errors of *omission* occur from leaving out major equipment items. For example, a laboratory chemist working on a reactor design may neglect the cost of solvent recycle to the reactor. Consequently, the cost engineer must either include a solvent recycle system in the flowsheet, or justify solvent recycle through a second party on economic grounds, or use some other means to compensate.

Safety, environmental, and waste management issues can be used to eliminate an alternative process from further consideration. Because of their importance, problems in any of these three areas must be identified early as suggested in Table 1.7. The identification of problem areas (i.e., Step 10 in Table 1.7), may not occur until experimental measurements have been completed. For this reason, design must be based upon consistent and thorough experimental evaluations of the process, followed by an in-depth analysis of the experimental data and its reproducibility.

1.2 Chemical Process Synthesis

Design studies in chemical engineering typically involve several steps:

1. Design a base case,
2. Optimize the base case,
3. Synthesize alternative processing structures, and
4. Verify the best structure using detailed simulation and cost analysis.

Process synthesis is often the most challenging step, but it is also most likely to identify significant improvements compared to the base case.

Process synthesis also has the greatest effect on the overall design problem. Through it the flowsheet structure is established. Thus, it largely determines the nature and extent of subsequent problems in the design. It also implicitly fixes the maximum potential for improvement.

Consider Table 1.7. Process synthesis is most closely related to the completion of a PFD since it entails the aggregation of various unit operations to define the overall process, and it must occur before a PFD can be drawn. Since there are usually many different unit operations that can perform the same function, and since there are usually many different ways in which these unit operations can be interconnected to meet the overall process flowsheet goals, synthesis is open-ended (i.e., it doesn't have a unique solution).

Once a flowsheet has been synthesized in some fashion, then improvements may result from state optimization. If a flowsheet is chosen that does not use the best available technology, then a significantly better process may exist that cannot be found from optimization alone, but requires additional synthesis. Although this improved process will not be discovered through optimizing the old flowsheet because it uses different technology, it may be discovered by repeating the process synthesis.

In many ways synthesis is the most important step in process design. It is also the most interesting because there are relatively few guidelines for synthesizing networks, but there are many alternatives. Therefore, the engineer has relatively few tools to guide the synthesis. Process synthesis can be highly creative and individualistic. It is is not easily reduced to a set of rules. Instead the rules of synthesis tend to be defined by the nature of the problem. The ones now available are simple and worth understanding, but the most important tool is the engineer's creativity and imagination.

The three major steps in chemical process design can also be summarized in terms of their information flow as shown in Fig. 1.2.

1.3 Problems in Process Synthesis

There are several problems in process synthesis that must be overcome in order for a design engineer to be effective. They include:

1. Representation of the alternatives,
2. Evaluation of the alternatives, and
3. Development of strategies to select among the alternatives.

It is important to have methods for representing flowsheets that are simple and compact, but that contain enough information so that each alternative can be identified visually. For example, the synthesis activity can be treated as a tree search where each node in the tree represents either an entire alternative flowsheet or else a portion of an alternative flowsheet. A tree search is then an organized search for optimality which considers all of the permutations of feasible alternatives as members of a hierarchy of paths (*i.e.* the tree). The branches in the tree represent logical connections or paths between the nodes. In this instance, having a compact representation permits a visual identification of each node that facilitates synthesis by inspection.

Each alternative needs evaluation of some type in order to establish the most attractive system. There are several heuristics, or rules-of-thumb, that can be used to perform preliminary screenings of alternative flowsheets. Ultimately, however, the alternatives, or some subset of the alternatives, must be examined more carefully using economic analysis. For chemical processes, this evaluation typically involves the estimation of annual operating costs and capital investment requirements as a minimum. In some instances the economic analysis will be confined to estimates for grass roots plants, but retrofitting costs are also important in others.

Since the calculation of costs for chemical processes is usually rather complicated, and the design equations may consist of sets of nonlinear coupled equations that require simultaneous solution, the estimation of cost at a single operating point can be challenging. Moreover, optimization to minimize processing costs for an equipment ensemble may require that these design equations be solved for hundreds or even thousands of different operating conditions. Consequently, economic evaluations of alternatives can become very tedious even on a high speed computer.

The strategy to select among alternatives is essential in most cases. Even if a particular flowsheet can be economically evaluated and optimized, there may be millions of alternative flowsheets that could be evaluated. Since it is often not practical to perform exhaustive searches and evaluation of these alternatives while also performing a detailed optimization of each one, the Task Solution Heuristics that

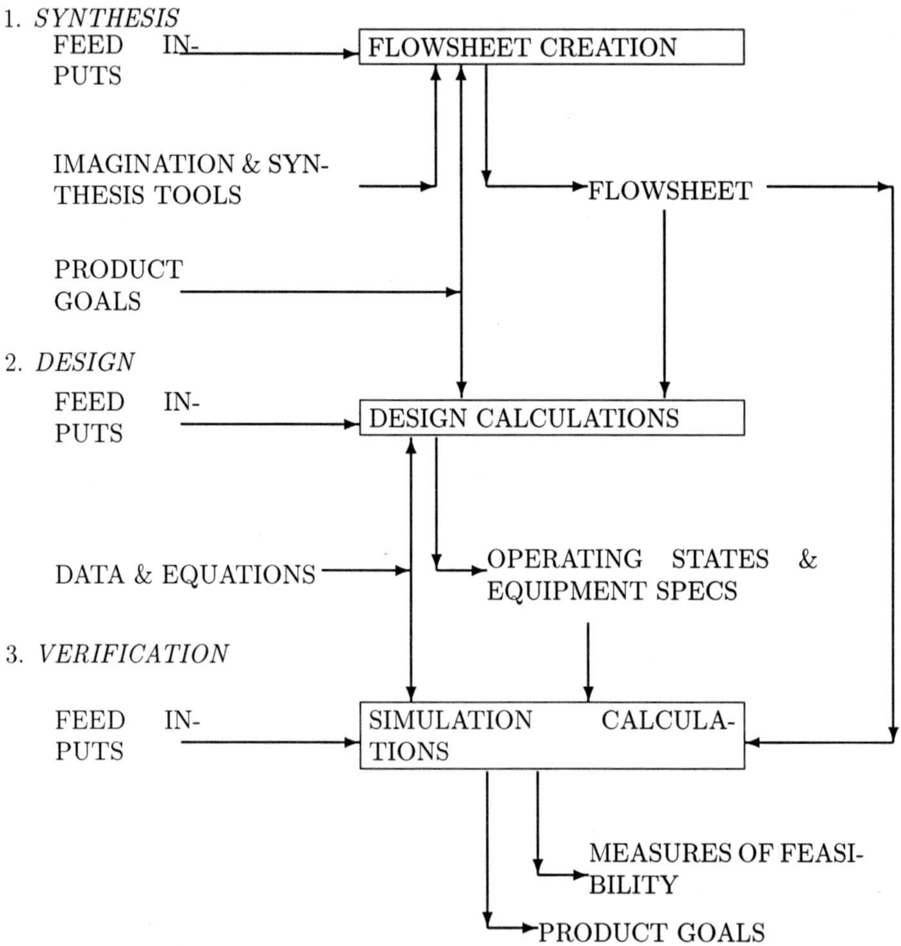

Figure 1.2: Process Synthesis defines the subsequent design and simulation calculations and their difficulty.

are discussed in latter chapters are useful tools to help make intelligent selections among the alternatives so that relatively few alternatives can be evaluated in great detail.

1.4 Approaches to Process Synthesis

There are at least three major approaches to process synthesis. These approaches include:

1. process decomposition,
2. evolutionary search, and
3. synthesis by optimization.

Using process decomposition a flowsheet is broken down into subtasks, or subsystems, that are suboptimized individually. Subsequently, the optimized subsystems are used as building blocks to synthesize the flowsheet. One effective method that is discussed further utilizes dynamic programming to assemble the building blocks into an integrated flowsheet.

With evolutionary search, a good flowsheet is chosen as a starting point. Changes are made in that flowsheet in an attempt to further reduce costs. Several investigators have developed rules for making these searches to improve certain types of flowsheets. In addition, one can also utilize evolutionary search techniques to improve the energy management schemes in sequences of distillation towers. Some examples of this latter application include heat exchanger integration, vapor recompression, and refrigeration cascade synthesis.

When synthesis by optimization is utilized, an integrated flowsheet is developed that contains all of the possible alternative flowsheets as subsystems that are embedded in the integrated flowsheet. In other words, a single stream may initially be connected to several parallel processing systems when, in fact, only one processing system would be used to treat that stream. The optimization, or synthesis in this case, operates by manipulating dummy variables that adjust the split fractions for streams that report to each of the parallel processing systems. The optimization terminates when the integrated flowsheet has been reduced to a single alternative by setting the flowrates to near zero values for the feed streams to the other alternatives under consideration. So this method strives to minimize processing costs by adjusting flowrates until the less desirable alternatives have only very small flowrates into and out of them. The residual flowsheet, processing the bulk of the streams that are split between alternatives, is presumed to be the optimal network.

1.5 Types of Synthesis Reasoning

There are several types of synthesis reasoning:

1. heuristic and

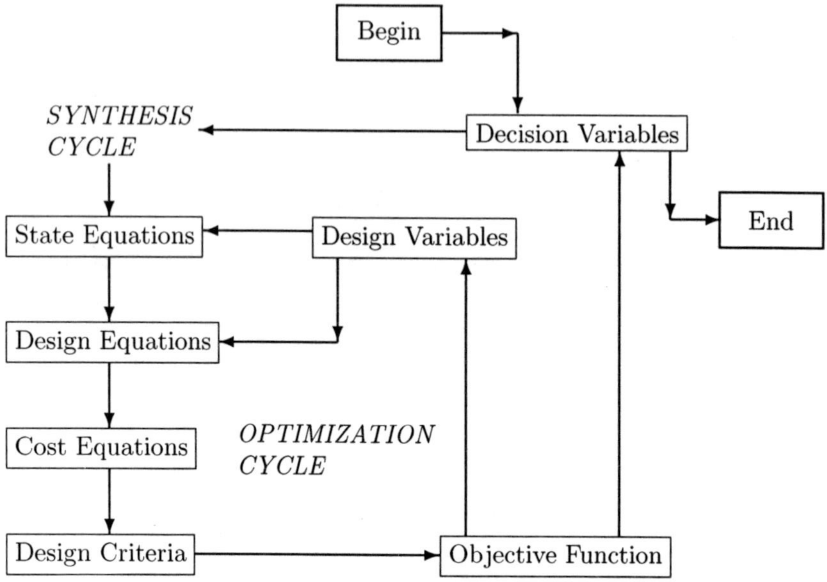

Figure 1.3: The information flow in Process Synthesis and Optimization are nested cycles with strong interactions. Both the decision and design variables affect the computed value of the objective function.

2. algorithmic or deterministic.

Heuristics are simply rules-of-thumb, or general observations, that are empirically true, but that are not necessarily based upon fundamental engineering principles. There are many heuristics that are common to everyday life. For example, the statements *a bird in the hand is better that two in the bush* or *everything is a trade-off* may be considered heuristics. Similarly, Murphy's famous laws are largely heuristic. Additionally, more specific heuristics have been developed which are useful for process synthesis, especially for the identification of attractive processing alternatives.

Algorithmic and deterministic syntheses utilize some type of algorithm or fundamental principle to perform the flowsheet synthesis. Two important examples of algorithmic and deterministic synthesis are dynamic programming and branch-and-bound tree searching.

Process synthesis and optimization are tightly bound together in that one cannot be performed well without the other. This condition is illustrated in Fig. 1.3 and this condition suggests why the computer can be useful in these types of preliminary design studies.

Table 1.8: The Process Synthesis Creed.

1. The network is *more important* than the components.
2. The *ultimate* performance targets are *attainable.* (i.e. a better design may be found by a more careful search among the alternatives.)
3. The *best* design reduces both operating costs *and* capital investment requirements.
4. *First* think in *simple* terms.

1.6 The Process Synthesis Heuristics

Dr. Rudy Motard [68] has developed a *Process Synthesis Creed* that appears as Table 1.8. This creed summarizes the attitude that must generally exist in the engineer's mind in order to successfully synthesize a new chemical process. Attitude plays a very important role in the synthesis process. In this regard, Dr. Motard's second point in Table 1.8 is particularly relevant, and it is clear that process improvements will not be found without engineering motivation.

Actually, synthesis requires a large portion of faith because it is necessary to believe that a new process can be developed before it actually exists. For example, one of the most difficult steps of faith is the formulation of new components in a process before the other downstream components have been conceptualized. The problem that arises is: *Suppose I can get from step A to step B, then what have I actually accomplished? I have no idea what to do after step B. Why should I go on with this? Faith*, however, *is the substance of things hoped for, the evidence of things not seen*, and faith insists that it is possible to achieve step C after completing step B. Faith simply believes that the synthesis will be useful and desirable when it is completed.

Often it's the simple things that count the most in design. Table 1.9 summarizes several general design rules that should be obvious, but seldom are when one is challenged to quickly generate accurate cost estimates for a complex chemical process. It's surprising how often large organizations will attempt to "engineer" bad chemistry. The factors weighing most heavily against the use of a particular technology may not be understood initially, but become clear with time as the process develops and experimental data are generated. Frequently, it's the larger organizations that experience greater difficulties in abandoning bad chemistry, particularly government organizations. Perhaps because of their visibility they have greater difficulty in admitting defeat. In any event, it is the underlying chemistry that makes or breaks a process, not the engineering. Engineering can improve processing characteristics, but if the precipitate resulting from a wash is a gooey mess, rather than nice, clean flowing crystals, changing the filter design will not

Table 1.9: Heuristic Design Rules

1. Never try to "engineer" bad chemistry. Don't try to solve chemical problems with equipment. Find better chemistry.
2. Talk to the chemists. Spend time understanding the chemistry and its limitations.
3. Try to find the material balance "loose ends" early in the design.
4. Be realistic. Remember that the less you know about a process, the better it looks.
5. The most obvious process solution is not necessarily the best.

eliminate the problem. The precipitate will still be gooey.

It is particularly important for engineers to communicate effectively with chemists during the early stages of a design, before a firm commitment to any particular chemistry has been made. Asking questions, spending time in the lab, and just observing often pay big dividends. Frequently, creative alternatives to problems can be found most readily just from a few hours working with technicians and getting a "feel" for what is going on, chemically speaking. A small amount of greasy, sticky stuff in the bottom of a round-bottom flask may not look much on a a small scale, but it may represent an expensive hazardous waste disposal cost for the full-scale plant. Often, side reactions are not mentioned during the preliminary analysis, but if the yield is only 50% of the reactants, a good design engineer is deeply concerned with determining the fate of the other 50% early in the analysis. It would be much better, for example, to find chemistry giving an 80 or 90% yield, rather than having to recycle large amounts of reagents or dispose of hazardous, polymeric wastes.

1.7 Process Synthesis Examples

Invariably, separation technologies will comprise part of a larger processing system that may include production and waste management operations. The optimal waste management strategy necessarily reflects the whole operation, rather than the most favorable conditions for any one unit, and the most attractive waste management schemes may involve hybrid separation systems. Stream recycle clearly must play an important role. The following examples briefly illustrate how separations, process synthesis, and waste management can be used to create better systems.

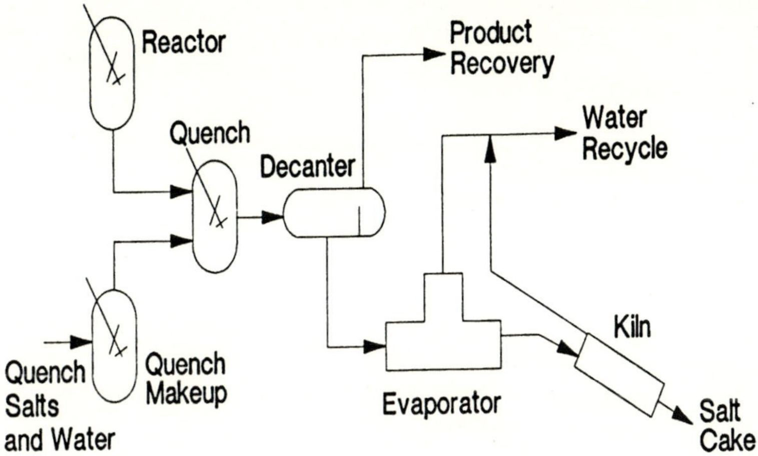

Figure 1.4: Conceptual flowsheet for brine waste management in specialty chemicals. Brine wastes are treated by evaporation and calcination to produce a solid waste form.

1.7.1 Brine Recycle

The first example is illustrated in Figures 1.4 and 1.5. The process manufactures an organometallic which, as a final step, undergoes hydrolysis to generate an organic-soluble product and an aqueous brine waste. This step is completed by mixing an alkaline salt with an intermediate organometallic and then decanting the two-phase system. As illustrated in Figure 1.4, the organic product is recovered in a solvent phase while the aqueous brine waste reports to an evaporator and rotary kiln dryer where a salt cake is produced for landfill disposal. Used water is recovered and recycled to the quench tank and other in-plant applications.

A more effective waste management strategy uses brine recycle as illustrated in Figure 1.5. In this case, both the expensive evaporator and kiln are eliminated from the flowsheet. They are replaced with a much less expensive carbon-steel plate-and-frame filter press. During plant operation, aqueous brine wastes are continuously recycled to the quench makeup tank where fresh quench salts are added to adjust the pH to the correct alkalinity. This brine mixture is then reused by recycle through the quench tank, decanter and filter press.

No salt cake is recovered until the brine waste concentrations reach their saturation levels, but the flowsheet in Figure 1.5 is preferable to that in Figure 1.4 for several reasons. In addition to reducing capital costs by eliminating an expensive evaporator and kiln, brine recycle reduces energy costs by eliminating the need to evaporate water. More importantly, the additional salting of the aqueous phase from brine recycle increases product recovery and reduces solvent losses to the aqueous phase. Since this brine recycle strategy makes it possible to process the waste

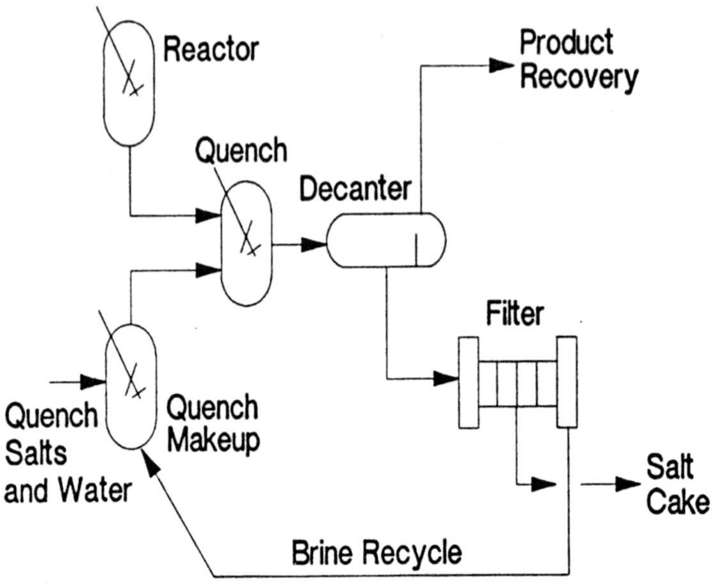

Figure 1.5: Conceptual flowsheet for brine waste management in specialty chemicals. Brine wastes are treated by waste recycle and filtration to recover a salt cake.

streams at ambient temperatures, severe corrosion problems are greatly reduced, and the waste systems can be fabricated out of carbon steel rather than hastalloy or titanium. Thus, there are multiple benefits from this simple, but effective, waste recycle scheme.

This example in improved waste management reduces operating costs, reduces capital investment, and increases product recovery. It achieves the goals listed in Table 1.8. The disadvantage of the flowsheet in Figure 1.5 is that it requires brine recycle and, particularly, the careful control of water additions to the brine recycle stream. Water is removed from this process in the solvent phase shown as the Product Recovery stream in Figure 1.5 and in the moisture present in the salt cake. In addition, a brine bleed may be required to control the buildup of impurities, but still the addition of water to the Quench Tank in Figure 1.5 is much less than in Figure 1.4.

1.7.2 Used Oil Reuse

Miller presents another, more striking, example of how waste recycle can yield large benefits [65]. His base flowsheet is shown in Figure 1.6. In the original process, used motor oils are distilled and recycled using a sulfuric acid wash and treatment with activated clay. The first distillation column in Figure 1.6 liberates water and a light fuel fraction. The bottoms product is treated with strong H_2SO_4 (e.g.,

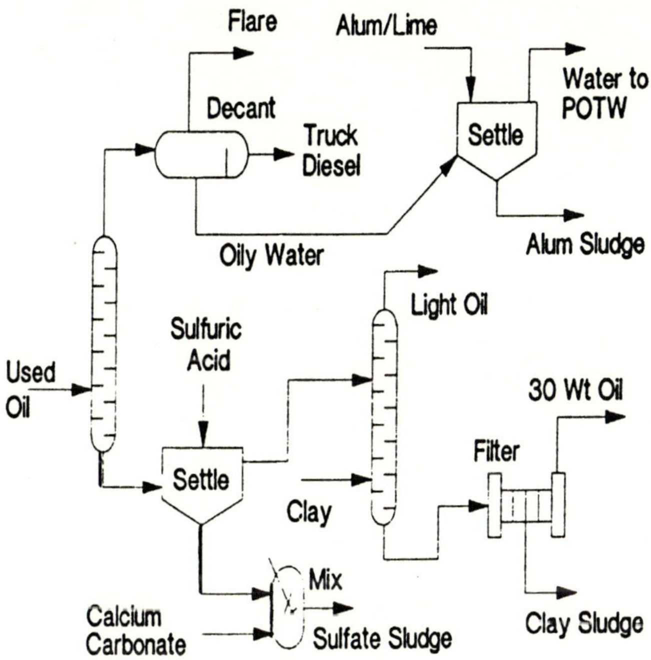

Figure 1.6: Acid-clay process for used motor oil reuse and recycle. (Adapted from Miller, courtesy of the Noyes Data Corporation [65]).

98%) to sulfate and precipitate all sludge components except stable lube oils. The acid sludge is then neutralized with $CaCO_3$ to yield an oily, $CaSO_4$ sludge that is landfilled.

The first column distillate in Figure 1.6 is decanted to yield an oily wastewater, truck diesel, and noncondensibles that are flared. The oily wastewater is treated using alum and lime precipitation to produce a second sludge waste that is landfilled. The aqueous wastewater overflow is sent to a public wastewater treatment facility.

The flowsheet in Figure 1.6 produces a third waste sludge as the acid-washed bottoms product is treated by a second distillation using activated clay to remove color bodies. The distillate from the second column is a light oil product. The bottoms product is treated by filtration to produce an oily spent-clay waste while the filtrate yields a 30 weight motor oil.

The original acid-clay recycle scheme in Figure 1.6 produces three sludge wastes, all of which are hazardous. Miller's recycle scheme in Figure 1.7 eliminates all of these wastes while still producing useful oil products for reuse. The key features are sludge recycle between the oily-water settler and the first distillation column, and

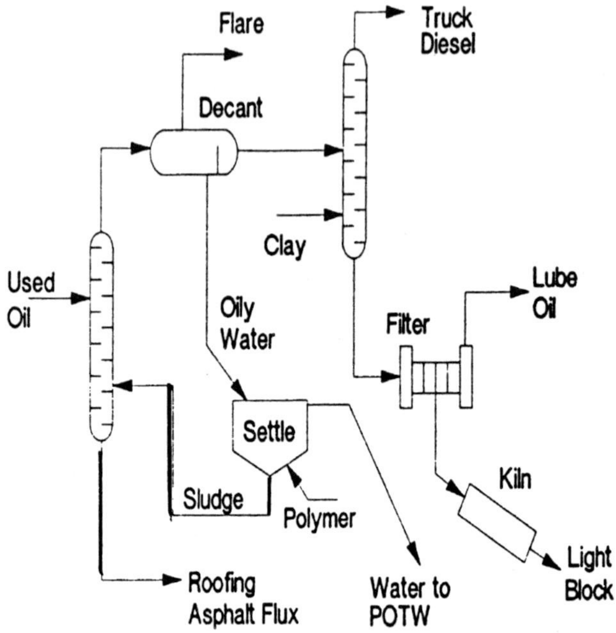

Figure 1.7: Miller process for used motor oil reuse and recycle. (Adapted from Miller, courtesy of the Noyes Data Corporation [65]). Note that all solid wastes are eliminated or used beneficially.

the use of a polymeric coagulant instead of alum and lime to produce a wastewater suitable for treatment by a public utility.

Figure 1.7 illustrates several other features that give this process superior systems properties. First, the $CaSO_4$ sludge is eliminated because sulfuric acid (and subsequent $CaCO_3$ neutralization) are no longer required. Second, the metals present in the used motor oil are incorporated into a new product, a roofing asphalt flux, instead of the sulfate sludge. Thus, the polymeric coagulant serves two purposes. It enhances the oil water separation by producing a sludge that settles faster, and it enables the metals in the sludge to become incorporated into the asphalt flux while in the distillation column.

The second tower and filter are virtually identical to those in Figure 1.6. The kiln in Figure 1.7 is new, but it could also be applied to the spent clay sludge in Figure 1.6. The concept is simple, but effective. Instead of landfilling the spent clay, it is used to manufacture lightweight aggregate for concrete block (e.g., cinder block). This strategy converts the third sludge waste to a usable by-product.

The remaining products generated from used motor oil in Figure 1.7 are similar to those generated in Figure 1.6. In addition, the flowsheet in Figure 1.7 eliminates two essential unit operations in Figure 1.6 (a settler and a mix tank to neutralize

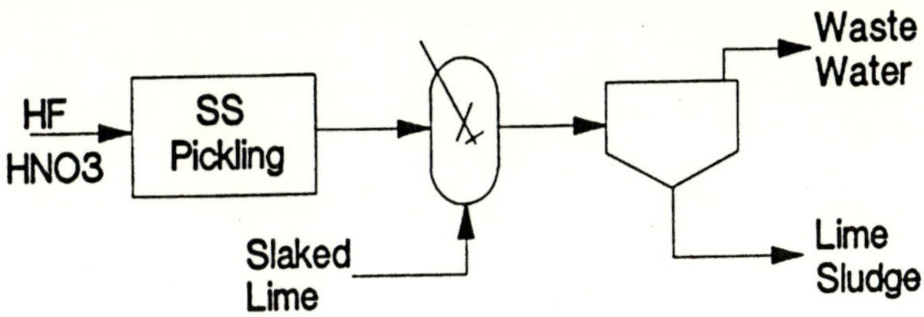

Figure 1.8: Conventional pickling acid waste management.

the sulfuric acid). In addition to eliminating much of the wastes, Miller's concept is also attractive because it otherwise utilizes the same separation operations as in Figure 1.6. Thus, the retrofit problems are minimal.

1.7.3 Acid Recycle

The most attractive technologies for recycle also offer reagent regeneration. In this way, the additional waste management costs can be partially offset by reduced reagent requirements. Thornburg et al. provide one example for recovering nitric and hydrofluoric acids from spent pickling liquors [105]. In a conventional pickling operation, acids are neutralized using slaked lime to produce a sludge waste and wastewaters as shown in Figure 1.8.

Acid recovery is implemented by using KOH rather than $Ca(OH)_2$ to neutralize the acids. Metal sludges are still precipitated as hydroxide wastes which are recovered as solids using filtration. Brine filtrate is then recycled using electrodialysis to split water and regenerate usable acids:

$$KF + H_2O \rightleftharpoons KOH + HF \tag{1.1}$$

$$KNO_3 + H_2O \rightleftharpoons KOH + HNO_3 \tag{1.2}$$

Figure 1.9 illustrates the entire recycle concept.

1.7.4 Dual-pH Precipitation

Batachelor et al. [6] discuss an advanced lime precipitation concept that uses two pHs to achieve a higher degree of cooling water recycle. A conventional process operates as indicated in Figure 1.10. A single pH, typically 9.5–10.5, is used to maintain water quality. A cooling water bleed is used to control the buildup of dissolved salts.

The dual-pH (or ultra-high lime) concept is illustrated in Figure 1.11. Water is still treated at a pH of 9.5–10.5 prior to use, but the bleed stream is recycled by additional (ultra-high) liming to a pH of 11–12. This strategy precipitates silicates

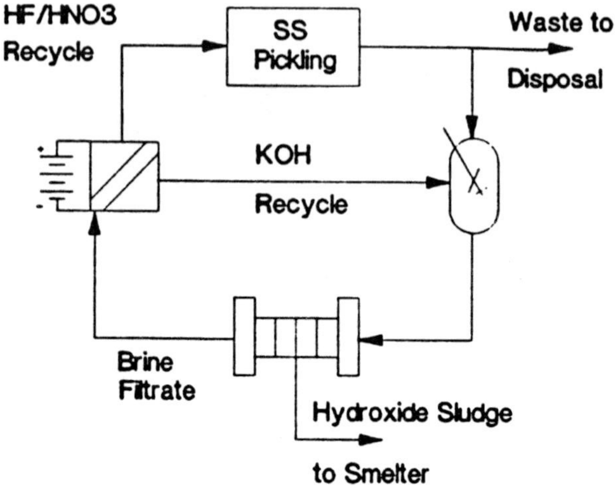

Figure 1.9: Pickling acid waste management and recycle using electrodialysis. (From Thornburg [105], courtesy of the American Institute of Chemical Engineers.)

and magnesium salts. Lime use is controlled by recycling the ultra-high lime effluent to the high-lime mix tank. This concept illustrates an advanced precipitation concept that is significantly enhanced by process integration. By eliminating (or at least reducing) the cooling water bleed stream, the potential exists for reducing chemical usage somewhat. The dual-pH scheme also reduces the production of secondary wastewaters.

The above examples illustrate the creative possibilities that exist and the growing importance of systems analysis to separations design and waste management. Benforado et al. present an intriguing specialty-chemicals example that achieves similar benefits for a batch process [9]. They substitute a reactive distillation step for a pressure reactor to achieve higher product yields. This modification together with improved cleanup and reagent recycle separations yields an almost 90% reduction in waste production.

1.8 Summary

Process design, optimization, and synthesis involve multiple steps. In many cases, a hierarchy of approaches is required in order to achieve optimal systems that are also reliable and meet all of the required processing constraints. Although the initial synthesis step is frequently the most important, it is also ill-defined and the engineer must resort to creative and highly individualistic approaches because of a lack of formal guidance and rules.

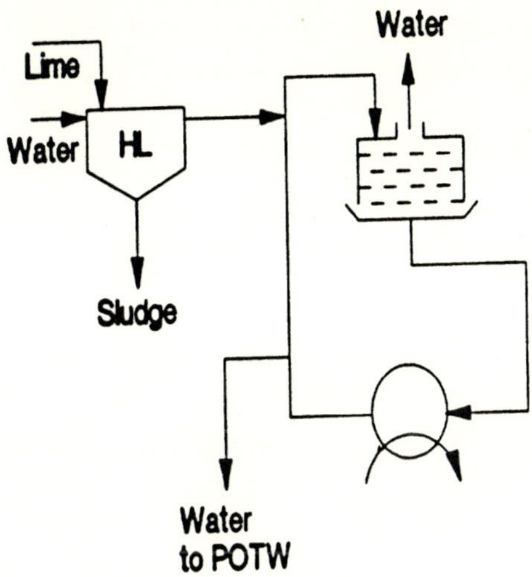

Figure 1.10: Conventional cooling water treatment with high lime treatment at a single pH.

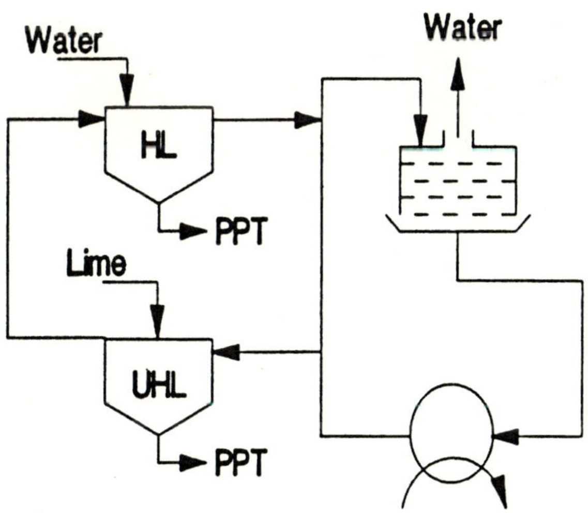

Figure 1.11: Cooling water recycle using ultra-high lime treatment (dual pH) concept. (Adapted from Batchelor et al. [6], courtesy of the Water Pollution Control Federation).

1.9 Problems

1.1 Suggest two alternative distillation sequences for separating a ternary mixture into three essentially pure product streams. What qualitative differences exist between these two sequences in terms of reboiler duties, condenser loads, and tower size requirements?

1.2 Absolute ethanol may be recovered from dilute fermentation liquors using a sequence of distillation towers that includes a beer still (a single stage which removes the ethanol from the beer feed), fractionator, and an azeotropic distillation system to recover the anhydrous product. Draw a conceptual flowsheet showing these processing steps. What strategies might exist for conserving energy in this type of chemical process?

1.3 Liquid-liquid solvent extraction may be used to recover absolute ethanol from beers and wines. What advantages might exist using this approach to ethanol purification, rather than a distillation sequence consisting of a beer still, a fractionator, and an azeotropic distillation system? How could the ethanol product be recovered if the solvent is liquefied carbon dioxide? How could ethanol be recovered if the solvent is an organic mixture whose normal boiling point is much higher than that of ethanol? What disadvantages does ethanol recovery by solvent extraction have relative to recovery by distillation?

1.4 Are individual components more or less important than the network in chemical process design? Why?

1.5 In your own words, explain how chemical process synthesis, design and optimization interact with each other.

1.6 What differences exist between process synthesis by decomposition versus by integration? What are some of the advantages and disadvantages of each?

1.7 Which of the following phrases best completes the next statement? Murphy's famous laws are: (1) proven algorithms for linearization, (2) always true, (3) usually confusing and obscure, (4) largely heuristic, or (5) based on gradient search methods.

1.8 Which of the following phrases best completes the statement below? Chemical process design: (1) requires only a single level of effort, (2) never results in flowsheet invention, (3) cannot involve synthesis, (4) usually involves several levels of detail, or (5) usually cannot be optimized.

Chapter 2

Product and Reaction Path Selection

2.1 Product Selection

A market study is often required to justify the initiation of a Preliminary Design Project. There are several key questions that this study should answer regarding any product under consideration:

1. **Demand Questions**
   (a) What has been the demand for this product over the past 10 years?
   (b) What is the current demand for this product?
   (c) Is market demand expected to increase or to decrease over the next 10 years?
2. **Capacity Questions**
   (a) How has industrial capacity for producing this material changed over the past 10 years?
   (b) What is the current industrial capacity for producing this product?
   (c) What is the projected capacity to produce this product over the next 10 years?
   (d) Have other chemical suppliers announced plans to expand capacity or to build new facilities to produce this product?
3. **Decision Questions**
   (a) What is the current Industry Operating Rate? (The Industry Operating Rate is the ratio of the demand to industrial capacity.)
   (b) Is there room in the market for a new facility (i.e., your proposed plant) to produce this chemical?

Careful literature and market studies are required to answer these questions. Current production capabilities of known competitors should be tabulated. Typically, three years are required from the time a decision is made to build new capacity until plant start-up, so demand projections over a 10 year horizon are not unreasonable.

For specialty chemicals, existing or prospective buyers may be contacted to determine whether or not they have an interest in buying larger quantities. It can be helpful to negotiate contractual arrangement to guarantee that the product will be sold at an agreed price, particularly for high-value added products. Such arrangements are less attractive for commodities (since they are typically low-value added products). In either case, such agreements should be tied to the costs of raw materials and/or energy prices which may fluctuate considerably over a three year period.

Initially, it is helpful to review the Current Periodicals listed below to find business articles that provide partial answers to the questions listed above. These periodicals are most conveniently reviewed using the Business and Management Databases that are available at many libraries online. Having found annual demand (i.e., sales) and production capability data, they may be plotted as shown in Figure 2.1. In this example, projections into the future are based upon a projected average annual growth in demand and nearly constant production capability. Other considerations include such things as: (1) Are there known external factors that are likely to affect demand in the future, but are not represented in the past 10 year history?, (2) Have other companies announced plans to increase capacity?, (3) Are there other social issues that may impact on the future use of this commodity (e.g., greenhouse gases, other environmental or potential health effects)?

If the projected demand exceeds the projected capacity, then room for new production capacity is expected to exist. As shown in Figure 2.1, an expansion opportunity then exists to introduce new capacity and to sell additional amounts of product.

The observed time-dependent behaviors of capacity and demand are seldom as simple as shown in Figure 2.1. In reality, these parameters oscillate over time as does the market price. Typically, the market price for a commodity exhibits behavior as shown in Figure 2.2. Cyclic behavior is often observed where the lower bound on the price is set primarily by cash costs (e.g., raw material costs, utilities, and labor) without any opportunity for the manufacturer to recover the cost of production facilities or depreciation. When the market price hits these lower bounds, a surplus of the chemical generally exists and capacity exceeds demand.

Heuristic: 3 *If production capacity exceeds demand, the sales price for a commodity tends toward actual cash costs of production (i.e., raw materials, utility costs, and labor). With technical improvements, this Price Lower Bound may become progressively lower over time. If commodity surpluses exist and a commodity price is at its Price Lower Bound, then bringing additional capacity online will not reduce the price further, but it will further increase the surplus.*

The Price Upper Bound in Figure 2.2 is determined by market conditions where capacity and demand are essentially in balance. Under these conditions, there is

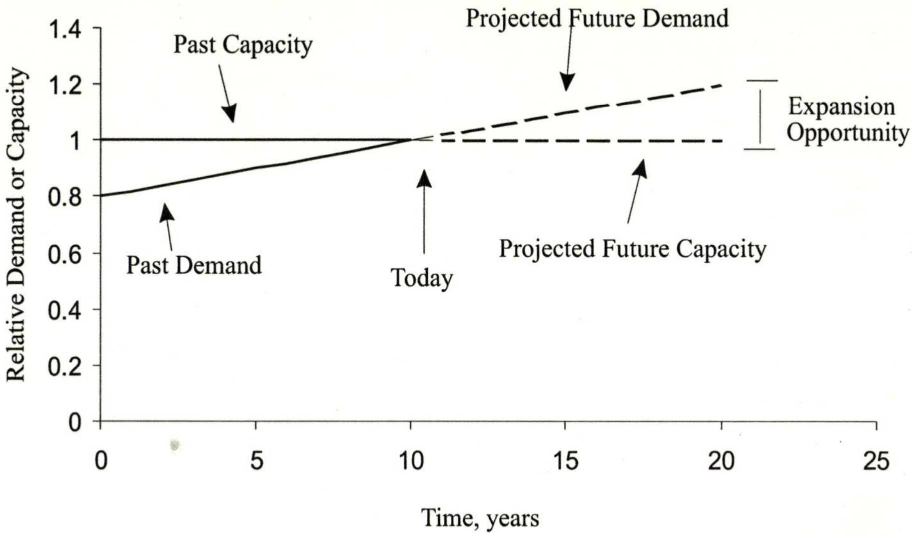

Figure 2.1: Projected growth in demand and anticipated constant capacity, based upon observed behavior over past 10 years.

no commodity surplus and all demand is satisfied (see Figure 2.2). Revenues are generated from product sales that cover the cost of facilities and other expenses, and the manufacturing operation is profitable. However, because of fluctuations in the commodity market, this "happy state" for the manufacturer does not continue indefinitely. At some point additional capacity comes on stream, triggered by the profitability of the "happy state" and price starts descending toward its lower bound. Over time, the Price Lower Bounds may become progressively lower due to the economies of scale and technological improvements, but the Price Upper Bound tends to be more constant.

Heuristic: 4 *The Price Upper Bound on commodity chemicals is set by market conditions at which the industrial capacity matches demand and neither surpluses nor shortages exist. Under these conditions, the market is at steady state (i.e., no commodity accumulation or deficit). This is also referred to as "equilibrium".*

2.1.1 Transition Prices

During periods of change (either from the Price Lower Bound to the Price Upper Bound or *vice versa*), the capacity and demand curves are determining factors that affect the Transition Prices. When the price is nearly constant, the market is either nearly at equilibrium (Price Upper Bound) or at a steady-state surplus condition (Price Lower Bound).

Transition Prices during times of increasing demand or capacity surplus can be

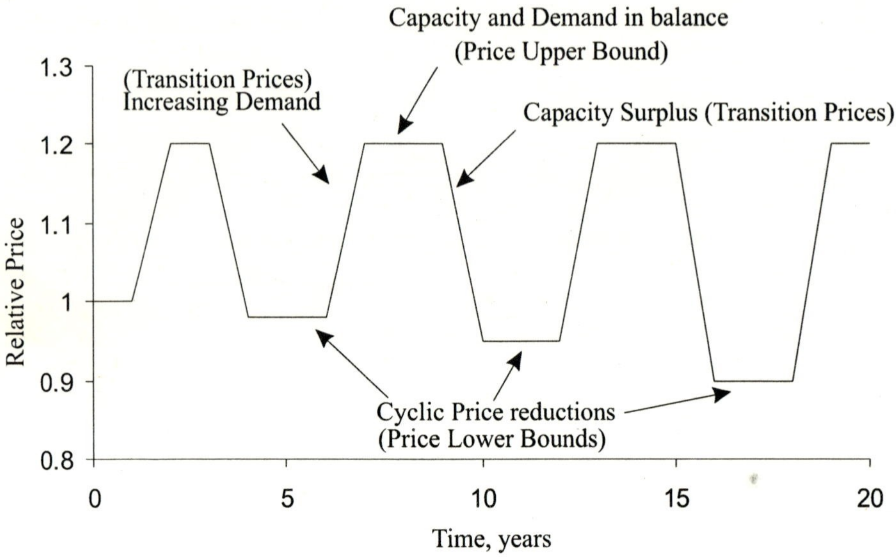

Figure 2.2: Typical cyclic price behavior for commodity chemicals. Cyclic price reductions occur due to economy of scale and technology improvements over time. Demand and supply elasticity determine the shape of the curve during periods of increasing demand and capacity surplus.

further analyzed using the concepts of demand and capacity elasticities. Demand and Capacity curves can be estimated using the concepts of Demand elasticity, E_d, and Capacity elasticity, E_c, and by assuming relatively small deviations from the equilibrium market operating point as shown in Figure 2.3.

Demand elasticity relates the percentage change in demand, dQ/Q, to the percentage change in price, dp/p.

$$E_d = -\frac{dQ/Q}{dp/p} = -\frac{pdQ}{Qdp} \tag{2.1}$$

Using this definition, E_d is usually a positive number. If the function, $Q = ap^{-n}$, reasonably relates p and Q, at least for small changes in these two variables, then the Demand elasticity is constant and $E_d = n$.

Suppose that $E_d = n$. Then $dQ/Q = -ndp/p$ and a decrease in price results in more demand. As n becomes larger, the demand increases more with smaller decreases in price. If $E_d > 1$, the the demand is said to be elastic. If $E_d < 1$, then the demand is said to be inelastic. Qualitatively, demand is more elastic whenever substitutes for a chemical or product exist and less so when they do not.

The Capacity elasticity, E_c, is defined in a similar manner, but with a change in sign.

$$E_c = \frac{dQ/Q}{dp/p} = \frac{pdQ}{Qdp} \tag{2.2}$$

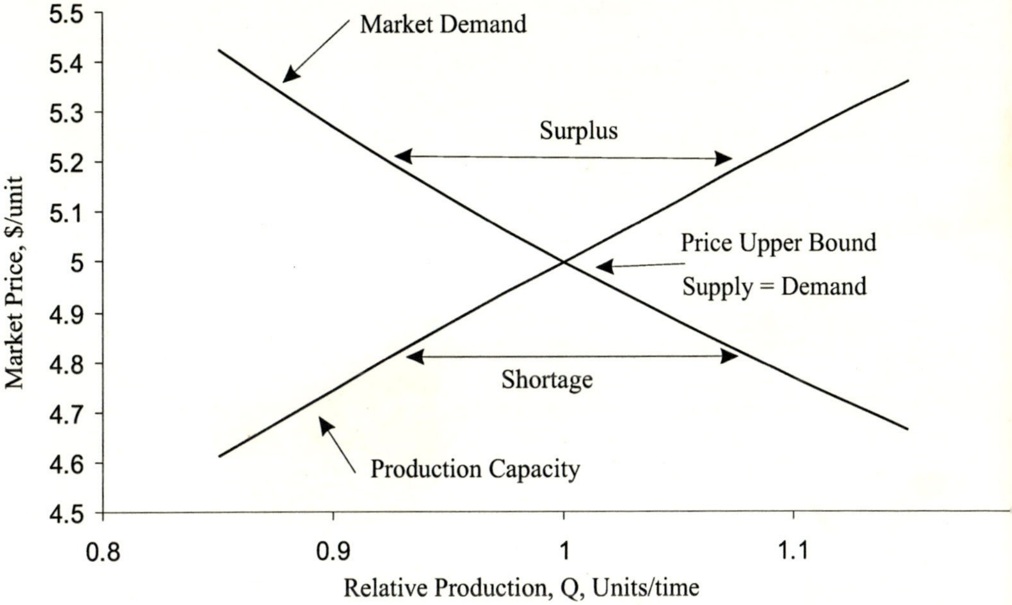

Figure 2.3: Supply and demand curve projections assuming constant elasticities with $E_d = E_c = 2$.

If $E_c = n$, then $dQ/Q = ndp/p$ and $Q = ap^n$, and an increase in price increases the capacity. As n becomes larger, the capacity increases more with smaller increases in price. If $E_c > 1$, the capacity is said to be elastic. If $E_c < 1$, then the capacity is said to be inelastic. Qualitatively, capacity is more elastic whenever it can be readily increased using existing facilities, and less so whenever capacity increases require the construction of new facilities.

Wei *et al* [116] studied 1972 data (prices and annual productions) for a range of products manufactured by the chemical processing industries. They found the approximate relationship

$$p = 5000Q^{-0.5} \quad \text{or} \quad Q = (25 \times 10^6)p^{-2} \tag{2.3}$$

which suggests $E_d = 2$. Making the assumption that $E_d = E_c = 2$ leads to the results plotted in Figure 2.3.

Heuristic: 5 *When completing Profit Upper Bound analysis, assume that $E_d = E_c = 2$ to adjust commodity prices for changes in raw material and product demand and capacity, if the anticipated new plant capacity significantly affects current capacity or demand conditions in the market.*

Example: 2.1 *A new chemical plant will increase the capacity for producing a chemical product by 20%. In so doing, it will increase the demand on a raw material* *by 10%. Estimate the decrease in product price and the increase in raw material price when this new plant goes online. Assume* $E_d = E_c = 2$.

Solution: 2.1

For the product price decrease: $p_2/p_1 = (Q_1/Q_2)^{0.5} = (1/1.2)^{0.5} = 0.913$
So a 8-9% reduction in selling price is expected.
For the raw material price increase: $p_2/p_1 = (Q_2/Q_1)^{0.5} = (1.1/1)^{0.5} = 1.049$
So a 4-5% increase in purchase price is expected.
The new plant will increase the product capacity, but it will also increase demand for its raw material feed. Increasing capacity tends to cause surpluses and depress the price. Increasing demand tends to inflate the price.

2.2 Economic References

The following citations are adapted from J. Wei, T.W.F. Russell, M.W. Swartzlander, The Structure of the Chemical Processing Industries: Function and Economics, McGraw-Hill, New York (1979).

2.2.1 Current Periodicals

1. Chemical and Engineering News
2. Chemical Week
3. Chemical Engineering Progress
4. CHEMTECH
5. Fortune
6. Forbes
7. Chemical Marketing Reporter
8. Chemical Engineering
9. Hydrocarbon Processing
10. Oil and Gas Journal
11. Economic Indicators
12. Survey of Current Business
13. The Wall Street Journal
14. Chemical Age
15. Economic Report of the President
16. Harvard Business Review

17. Foreign Affairs
18. The Economist
19. Far Eastern Economic Review

2.2.2 References

1. Paul A. Samuelson, Econmics, 10th ed, McGraw-Hill, New York, 1976
2. C.E. Ferguson, Microeconomic Theory, Irwin, Homewood, Ill
3. G.J. Stigler, The Theory of Price Macmillan, New York, 1976
4. R.N. Anthony, Management Accounting, Irwin, Homewood, Ill, 1964
5. M.J. Gordon and G. Shillinglaw, Accounting: A Management Approach, Irwim, Homewood, Ill 1974
6. S. Davidson (ed), Standard Handbook for Accountants, McGraw-Hill, New York, 1970
7. J. Backman, The Economics of the Chemical Industry, Manufacturing Chemists Association, Washington, 1970
8. Chemistry in the Economy, American Chemical Society, Washington, 1974
9. U.S. Bureau of the Census, Statistical Abstract of the United States, Washington (published each year)
10. SRI Internation, Chemical Economics Handbook, Menlo Park, CA

2.2.3 Abstracting Services and Indexes

1. Chemical Market Abstracts, predicasts, Inc., Cleveland
2. Chemical Industry Notes, Chemical Abstract Service, Columbus, Ohio
3. F and S Index of Corporations and Industries, Predicasts, Inc., Cleveland

2.2.4 On-Line Computer Retrieval Services

1. Lockheed Retrieval Services
2. Business Index (BUSI) on GTEL

2.2.5 Internet References

1. http://www.the-innovation-group.com/chemprofile.htm
2. http://www.chemicalmarketreporter.com/home/frameset.htm
3. http://www.cmr.com

2.3 Reaction Path Selection

Initially, it is useful to develop simple criteria for establishing whether or not a new chemical process has any chance of being economical and realizing a profit. One simple method for doing this analysis consists of production/consumption or input/output analysis and permits the calculation of an upper bound on profitability. If this calculated upper bound on profits is negative, then there is clearly not an incentive to continue the design analysis. On the other hand, if the upper bound on profits is positive, the plant may still not be economical. That is, a positive upper bound which will be described is a necessary, but not sufficient, measure of profitability.

The criterion is based upon the following heuristic suggested by Rudd, Powers and Siirola [85]:

> All other things being equal, the reaction path that has the largest difference between the value of the products and the cost of the raw materials is the most promising.

In other words, initially look only at the cost of raw materials and the value of the product while ignoring the cost of the chemical processing equipment itself.

In Chapters 3 and 6, the operating costs, C_Q, are defined to include raw material costs. The total cost, C, is defined as:

$$C = C_I + C_L + C_Q \tag{2.4}$$

These costs can be regrouped as:

$$C = MAT + OTH \tag{2.5}$$

where MAT = the annual costs of raw materials
OTH = all other annual costs
C_I = annual costs that are proportional to total capital investment (e.g., maintenance, property taxes, interest on capital loans, payments on principal)
C_L = annual labor costs (e.g., operating labor, supervision, overhead)
C_Q = annual product costs (e.e., raw materials, utilities, royalties, shipping costs)

Some measures of profitability are relatively difficult to calculate, especially during the early part of a Preliminary Design, because they require detailed information about the chemical process. However, one can define a Profit Upper Bound (PUB) which is easy to calculate and which also measures profitability:

$$PUB = S - MAT \tag{2.6}$$

where the term, S, is simply the annual sales as defined in Chapter 3. As defined, the PUB is an upper bound on actual profitability because it does not include all true costs (i.e., there are other costs besides MAT), and the calculated PUB is therefore greater than the true profitability. So it is an upper bound on profits or profitability.

Since the *PUB* is easy to calculate, it can be used to screen reaction paths. Moreover, if the calculated *PUB* for a reaction path is negative, then that manufacturing route cannot be profitable and should be dropped from further consideration. On the other hand, if the *PUB* is positive, then that reaction path should be given further consideration. Reaction paths with the largest *PUB* values are the most likely to be attractive, *all other things being equal.*

Example: 2.2 *Several different commodities may be produced in different quantities as indicated in the table below, which also gives the estimate Profit Upper Bound (PUB) for each commodity. If a chemical plant will be built that can produce a total of up to 400,000 lb/yr, then which combination of products A-F (or subset of A-F) should it produce to maximize its PUB expressed as \$/yr total for all products manufactured? Indicate the products, the annual amount in lb/yr of each that is produced, and the \$/yr in revenues from that product. Each product may be produced and sold in amounts that are less than or equal to the Demand amounts indicated in the table, but production for any product should not exceed its demand.*

Product	Demand, lb/yr	PUB, \$/lb
A	100,000	3.50
B	250,000	1.50
C	50,000	7.50
D	100,000	4.25
E	150,000	5.50
F	300,000	2.25

Solution: 2.2 *Calculate the maximum \$/yr for each product. Then begin with that product offering the highest PUB, then the next highest PUB, etc. Continue until the total lb/yr equals the maximum, 400,000. Producing the maximum amounts*

Product	Demand, klb/yr	PUB, \$/lb	k\$/yr	Choice	Accum klb/yr
A	100	3.50	350	4	400
B	250	1.50	375		
C	50	7.50	375	1	50
D	100	4.25	425	3	300
E	150	5.50	825	2	200
F	300	2.25	675		

of C, E, D, and A will generate $375 + 825 + 425 + 350 = \$1,975$ *k\$/yr.*

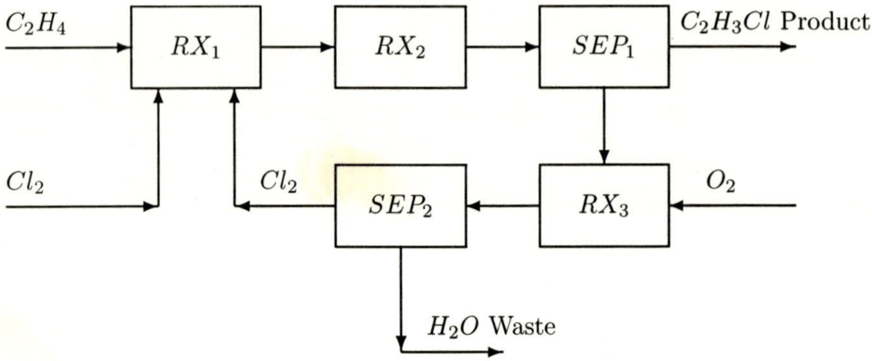

Figure 2.4: Flow diagram for ethylene chlorination example.

2.3.1 Vinyl Chloride Production Example

In this example, chlorine is reacted with ethylene to produce vinyl chloride. Hydrochloric acid is a side product, but it can be reacted with oxygen to produce chlorine and water. Water is discharged as a waste, but the chlorine can be recycled. The relevant reactions are summarized below:

$$2C_2H_4 + 2Cl_2 \rightarrow 2C_2H_4Cl_2 \tag{2.7}$$

$$2C_2H_4Cl_2 \rightarrow 2C_2H_3Cl + 2HCl \tag{2.8}$$

$$2HCl + 1/2O_2 \rightarrow Cl_2 + H_2O \tag{2.9}$$

The overall reaction may therefore be written as:

$$2C_2H_4 + Cl_2 + 1/2O_2 \rightarrow 2C_2H_3Cl + H_2O \tag{2.10}$$

The value of the reagents and products are summarized in Table 2.1 and Figures 2.4 and 2.6 summarize the flow diagram for the more realistic case with reagent recycle.

However, a preliminary PUB analysis may ignore credits for reagent recycle as is shown in Figure 2.5. If the PUB is positive for this simple case when recycle values are ignored, then it may be increased further when the recycle effects on the PUB are included. Thus, the PUB for the simple case is a lower bound for the more realistic flowsheet which is suggested in Figures 2.4 and 2.6.

In general, for the reaction:

$$aA + bB \rightleftharpoons cC + dD \tag{2.11}$$

Table 2.1: Cost data from the Chemical Marketing Reporter, July 26, 1971.

Species	MW	\$/lb	\$/lb-mole
Ethylene	28	0.03	0.84
Chlorine	70	0.04	2.80
Vinyl Chloride	62	0.05	3.10
Oxygen	32	Use Air	Free

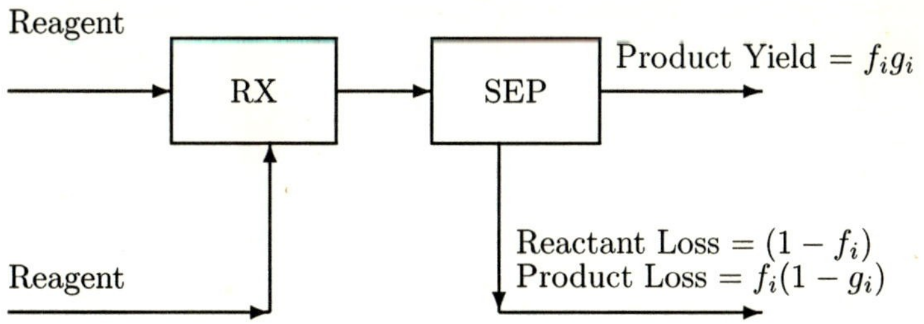

Figure 2.5: Flow diagram for partial conversion of reactants with no recycle and unit feed to reactor.

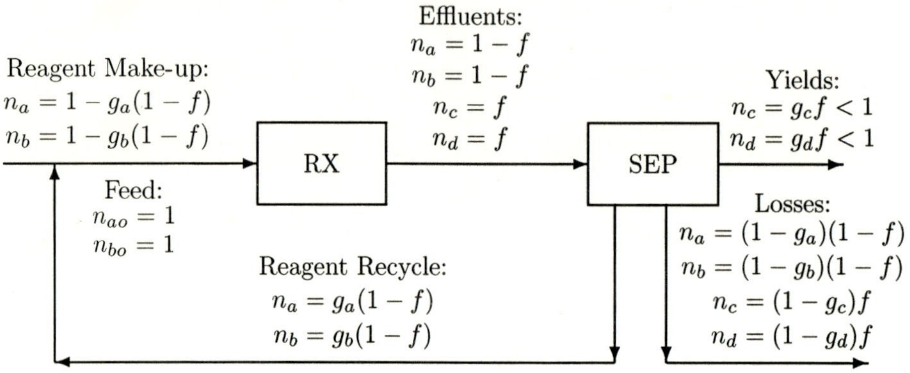

Figure 2.6: Flow diagram for steady state recycle with specified reactor feed.

the fractional conversion, f, is defined as follows for the case where species A or the moles of A entering the reactor, n_{ao}, limits the chemical reaction:

$$n_a = n_{ao}(1 - f_a) \tag{2.12}$$
$$n_b = n_{ao}(M - (b/a)f_a) \tag{2.13}$$
$$n_c = n_{ao}f_a(c/a) \tag{2.14}$$
$$n_d = n_{ao}f_a(d/a) \tag{2.15}$$

where:

$$n_{ao}, n_{bo} \rightarrow \boxed{\text{RXN}} \rightarrow n_a, n_b, n_c, n_d$$

where $M = n_{bo}/n_{ao}$

In this simple (and unrealistic) ethylene chlorination example, $f = 1$, $M = 0.5$ and none of the products are recycled. Consequently, the Profit Upper Bound is given by:

$$PUB = 3.10 - 0.84 - 2.80/2 = \$0.86/\text{lb-mole} \tag{2.16}$$

Since the estimated PUB is positive, this process should be studied further.

2.3.2 Esterification Examples (Specification Effects)

In general, chemical reagents are only partly converted to products and, therefore, it is economically attractive to separate the reaction products and recycle some

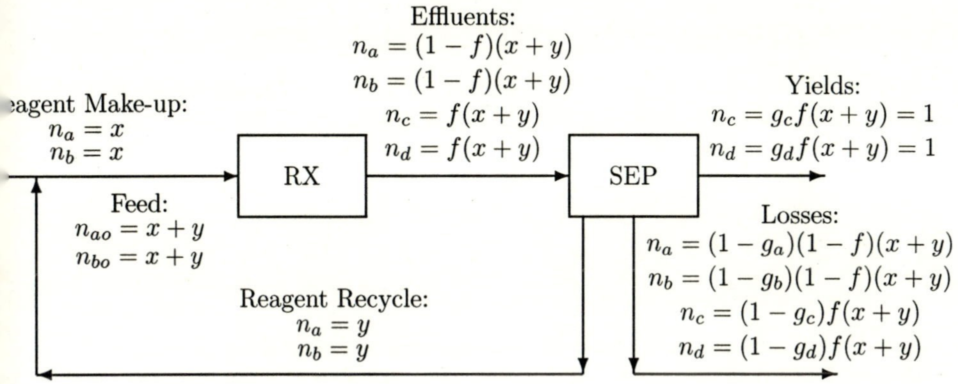

Figure 2.7: Flow diagram for steady state recycle with specified product rates.

Table 2.2: Assumed values for isopropanol, acetic acid, and isopropyl acetate.

Species	\$/lb-mole
A Isopropanol(IPOH)	4.20
B Acetic Acid (AA)	5.40
C Isopropyl Acetate (IPA)	12.25
D Water	Free

fraction of them to the reactor. Usually there are losses associated with the process and for purposes of initial screening it is convenient to define recovery factors and product yields:

where: g_i = fraction of input feed for species i that is recovered
$YIELD$ = (conversion) x (recovery) = $f_i g_i$
$LOSS$ = $f_i(1-g_i)$ for the ith species

The flow diagram for this case is similar to that for Figure 2.4, but it is slightly more complicated in that a loss term is permitted. Figure 2.5 summarizes the situation for this case. For this example, the reagents and products have the values that are summarized in Table 2.2.

The reaction is as follows:

$$\begin{aligned} C_3H_7OH + CH_3COOH &\rightleftharpoons C_3H_7OOCH_3 + H_2O \\ IPOH(A) + AA(B) &\rightleftharpoons IPA(C) + H_2O(D) \end{aligned} \tag{2.17}$$

2.3.3 Case 1: Partial Conversion of IPOH, No Recycle

In this case, the following parameters are assumed to apply: $f = 0.6$, $g_a = 0.8$, $g_b = g_a$, $g_c = 1.0$, $M = 1.0$. The flow diagram corresponds to the situation depicted in Figure 2.5. Under these conditions, the estimated PUB is calculated from:

$$PUB = [(fg_c)(12.25) - 4.20 - 5.40]/0.6 = -\$3.75/\text{lb-mole IPA} \tag{2.18}$$

Therefore, this process does not appear economical unless the reagents can be at least partly recovered and recycled. Note that the PUB basis is 0.6 lb-mole of IPA produced, so it is necessary to divide by 0.6 lb-mole to convert to a PUB/lb-mole of IPA.

2.3.4 Case 2: Partial Conversion, Partial Recycle, Fixed Reactor Feed

In this case, the same values for recovery and conversion apply as in Case 1. However, the PUB is calculated differently since some of the reagents are recycled. The analysis is based on a steady state analysis where the reactor feed is fixed at unity and used to define the other flow variables in the system. Figure 2.6 summarizes the situation.

Under these conditions:

$$\begin{aligned} PUB &= \{(fg_c)(12.25) - [1 - (1 - f)g_a](4.20 + 5.40)\}/0.6 \\ &= \$0.82/0.6 = \$1.37/\text{lb-mole} \end{aligned} \tag{2.19}$$

where $[1 - (1 - f)g_a]$ is the make-up reagent, and the PUB is normalized to a per lb-mole IPA basis by dividing by 0.6 lb-moles, the amount produced if one lb-mole of IPOH and AA are initially reacted.

This analysis suggests that the manufacturing process may be economically attractive if reagents are recovered and recycled, but remember that the PUB is only an upper bound on the true profitability because only the raw material costs have been considered. The true costs include other terms besides the raw material costs, and the true profitability is therefore less than the calculated PUB.

In this case, it is easy to evaluate the PUB because the reactor feedrate has been assigned to unity. Other formulations of the same problem, such as fixing the product rate, complicate the analysis, but can still be solved.

2.3.5 Case 3: Partial Conversion, Partial Recycle, Steady State, Specified Product Rate

This flowsheet is summarized as shown in Figure 2.7. It is similar to Figure 2.6 except for the fact that the product yield has now been specified equal to unity. Consequently, the reactor feed rate, $(x + y)$, must now be calculated by solving two simultaneous equations. Since the fractional conversions and fractional recoveries of the various species have been specified, the equations which must be solved are linear. Although the reaction path screening is still relatively simple, it has been complicated by the choice of design variables. This type of situation commonly occurs, and it illustrates how the choice of specifications can either simplify or complicate the subsequent calculations. This factor is not particularly important in this case, but it is important on large scale systems that contain large numbers of internal streams.

In this case, the original assumptions of constant fractional conversion and recoveries for the various species become useful. That is, the flowsheet can be analyzed by solving simultaneous sets of linear equations. Even in this third case, the value of these simplifying assumptions are not totally apparent, but they become more obvious as the flowsheet becomes more complex. Suppose, for example, that several reactors and separation systems are involved. In the extreme cases, it is necessary to utilize computerized linear equations solvers in order to arrive at answers.

Some key flowrates in this example are given by:

$$\text{Product Rate} = 1.0 = (x+y)fg_c \tag{2.20}$$

$$\text{Reagent Recycle Rate} = y_a = y_b = (x+y)(1-f)g_b \tag{2.21}$$

since the recovery factors for both reagents are the same. In general, y_a will not equal y_b and the subscripts must be maintained in the analysis. However, in this case these two equations can be solved simultaneously to yield $x = 1.133$ and $y = 0.533$. The yield in this case is taken as unity (the basis for the analysis) and the annual product sales term is:

$$S = (1)(12.25) = \$12.25/\text{lb-mole of IPA} \tag{2.22}$$

The variable x thus represents the reagent make-up rate to produce S:

$$MAT = x(4.20 + 5.40) = \$10.88/\text{lb-mole of IPA} \tag{2.23}$$

Hence the PUB in this case is:

$$PUB = S - MAT = \$1.37/\text{lb-mole of IPA produced} \tag{2.24}$$

Comparison of this case with Case 2 shows that the estimated PUB is the same in both cases, but it is still not clear that the process will realize an actual profit because the OTH term has not been examined. This example also illustrates how problem analysis may be simplified by the strategic specification of stream properties.

The economic loss in this last case is given by:

$$\begin{aligned} LOSS = {} & (x+y)[(1-g_a)(1-f)C_a + (1-g_b)(1-f)C_b \\ & +(1-g_c)fC_c + (1-g_d)fC_d] \end{aligned} \tag{2.25}$$

Since $g_a = g_b = 0.8$, $f = 0.6$, $g_c = C_d = 0, x = 1.133$, and $y = 0.533$, we obtain:

$$LOSS = 0.133(4.20 + 5.40) = \$1.28/\text{lb-mole} \tag{2.26}$$

Because this economic loss is about equal to the PUB, it suggests that additional work is needed to help reduce the reagent losses. A small economic loss would suggest that process improvement might be more significant in other areas.

Example: 2.3 *Use the following chemical reactions and costs information to estimate Profit Upper Bounds for alternative reaction paths and products. Cost data are taken from CMAI Market Reports in 1999*[1].

$$C_6H_6 + C_2H_4 \rightarrow C_8H_8 + H_2 \quad (2.27)$$
$$nC_8H_8 \rightarrow (C_8H_8)_n \quad (2.28)$$
$$C_2H_4 + Cl_2 \rightarrow C_2H_3Cl + HCl \quad (2.29)$$
$$2C_2H_4 + Cl_2 \rightarrow 2C_2H_3Cl + H_2 \quad (2.30)$$
$$C_2H_4 + Cl_2 \rightarrow C_2H_2Cl_2 + H_2 \quad (2.31)$$
$$nC_2H_4 \rightarrow (C_2H_4)_n \quad (2.32)$$
$$nC_2H_3Cl \rightarrow (C_2H_3Cl)_n \quad (2.33)$$

Assume that stoichiometric amounts of reagents are added to the reactor in each case, and that complete conversion is always achieved. Use the reaction equations and cost data provided to determine the Profit Upper Bound (PUB) for the following intermediates and products: (1) styrene, (2) ethylene dichloride, (3 & 4) vinyl chloride monomer (by two different routes, reactions 3 & 4 above), (5) polyvinyl chloride, (6) low density polyethylene, and (7) polystyrene. In each case, calculate the PUB per lb of ethylene added to the product or intermediate, so the results are comparable.

Species	*Formula*	*MW*	*$/lb*
Ethylene	C_2H_4	*28*	*0.228*
Benzene	C_6H_6	*78*	*0.105*
Styrene	C_8H_8	*104*	*0.268*
Vinyl Chloride Monomer (VCM)	C_2H_3Cl	*62.5*	*0.155*
Ethylene dichloride (EDC)	$C_2H_2Cl_2$	*97*	*0.072*
Low density polyethylene (LDPE)	$(C_2H_4)_n$	*28n*	*0.402*
Polyvinyl chloride (PVC)	$(C_2H_3Cl)_n$	*62.5n*	*0.32*
Polystyrene	$(C_8H_8)_n$	*104n*	*0.32*
Hydrochloric acid	HCl	*36.5*	*0.029*
Chlorine	Cl_2	*71*	*0.041*
Hydrogen	H_2	*2*	*0.123*

Solution: 2.3 *Styrene:* $PUB = [0.268(104)+0.123(2)-0.228(28)-0.105(78)]/28 = \$0.4837/lb$
EDC: $PUB = [0.072(97)+0.123(2)-0.228(28)-0.041(71)]/28 = \$-0.0738/lb$
VCM (Eqn 2.29): $PUB = [0.155(62.5)+0.029(36.5)-0.041(71)-0.228(28)]/28 = \$0.05182/lb$
VCM (Eqn 2.30): $PUB = [2(0.155)(62.5)+0.123(2)-2(0.228)(28)-0.041(71)]/28(2) = \$0.07039/lb$
PVC: $PUB = [0.32(62.5)n + n(36.5)(0.029) - n(28)(0.228) - n(71)(0.041)]/28n = \$0.42012/lb$
Polystyrene: $PUB = [104n(0.32)+n(2)(0.123)-n(28)(0.228)-n(78)(0.105)]/28n =$

[1] See http://www.cmaiglobal.com/

\$0.6769/lb
LDPE: $PUB = [28n(0.402) - 28n(0.228)]/28n = \$0.174/lb$
Polystyrene, produced from etheylene and benzene, has the greatest PUB per pound of ethylene, so it is the preferred product. Production of the sytrene monomer from ethylene and benzene is the second best choice, and the production of polyvinyl chloride is the third best. The preferred route for vinyl chloride monomer is via Reaction 2.30 where hydrogen is the byproduct. Ethylene chloride cannot be produced profitably from ethylene and chlorine.

2.4 Raw Material and Product Costs

Preferably, raw material costs should be estimated from current prices (e.g., *Chemical Marketing Reporter*[2] or *CMAI Market Reports*[3]) or from a vendor quote using the Thomas Register[4], but this information is not always available, especially for specialty chemicals or new products. As an alternative, or to cross-check price estimates from other sources, Hart and Sommerfeld [39] offer a method for estimating commercial chemicals produced in developmental or specialty quantities (e.g., up to a few tons per year), based upon price quotations from laboratory supply catalogs. They compared bulk selling prices with prices in laboratory catalogs for 24 chemicals and developed the correlation:

$$P_b = P_1 \left(\frac{Q_b}{Q_1}\right)^{-0.75} \tag{2.34}$$

where Q_1 is the laboratory quantity purchased in grams, P_1 is the laboratory selling price in \$/kg, (Q_b = 60 lb = 27,210 g) is a representative bulk amount (i.e., purchase lot), and P_b is the estimated bulk price in \$/kg for purchases of the bulk amount in 60 lb lots. Equation 2.34 is a refinement of Heuristic 6 on page 41.

Heuristic: 6 *A chemical's probable bulk selling price can be estimated by reducing the lowest unit price found in laboratory catalogs by a factor of 10.*

Example: 2.4 *A laboratory chemical sells for \$28/gm. Estimate its selling price as a bulk commodity in 60 lb lots.*
Solution: 2.4 *Using Equation 2.34,* $P_b = 28(1000)(27,216/1)^{-0.75} =$ *\$13.21/kg. Thus a reduction by a factor of over 2000 is predicted. Using Heuristic 6, the estimated cost is \$2,800/kg, a much higher number.*

The bulk amount for each commodity chemical which comprised a basis for the correlation was within the range of 20 to 120 lb. Thus, the estimate presupposes that the commodity chemical for which the estimate applies can be purchased either in the form of 50-lb sacks in the case of solids, 5- or 10-gallon carboys (glass or plastic jugs in which acids are shipped, about 10 gallons or 60 lb) in the case of liquids, or conventional gas cylinders.

---

[2]See http://www.chemexpo.com/schnell/
[3]See http://www.cmaiglobal.com
[4]See http://www.trsite.com/

Example: 2.5 *Estimate the commodity price of 99% pure benzoic acid in \$/lb from the lab price data tabulated below. Using different lab prices gives different*

Vendor	Lab Price, \$	Lab Quantity, g
A	22.55	500
B	6.10	25
C	21.80	1000
	50.40	2500
D	11.50	250

commodity price estimates. What is your estimate if an average price is used? What is your estimate if the minimum price calculated is used? If your commodity price estimate is being used to estimate a Profit Upper Bound, and the price estimate is for a product, then which estimate is preferable (i.e., the average or the minimum)? What if your price estimate is being used for a raw material? Which is then more appropriate (i.e., the average or minimum price)? Explain why in either case.

Solution: 2.5 *Calculate the \$/kg for the lab quantities, the factor $(Q_b/Q_l)^{0.75}$, the \$/kg for the bulk chemical and the \$/lb for the bulk chemical. Use the relationship:*

$$P_b = P_l \left(\frac{Q_b}{Q_l} \right)^{-0.75} \tag{2.35}$$

where $Q_b = 27,216$ g for all cases and Q_l is the lab quantity in each case. From

	Lab Data			Factor	Bulk	
Vendor	\$	g	\$/kg	$(Q_b/Q_l)^{0.75}$	\$/kg	\$/lb
A	22.55	500	45.1	20.04	2.251	1.023
B	6.10	25	244	189.52	1.287	0.585
C	21.80	1000	21.8	11.92	1.830	0.832
	50.40	2500	20.16	5.99	3.364	1.529
D	11.50	250	46	33.70	1.365	0.620

the last column, the average is \$0.918/lb. The minimum is \$0.585/lb. If the price estimate is for a product, then it makes more sense to use the minimum value calculated as that minimizes the estimated PUB and makes the calculation less favorable. If the price is being used to estimate a raw material, then it makes more sense to use the average value as that result makes the calculated PUB smaller.

Thermodynamic considerations can also be used to help predict a chemical selling price for high volume chemical commodities (i.e., volumes greater than 100 million lb/yr). Sommerfeld and Lenk [94] hypothesize that the ease or difficulty of manufacturing a chemical product is related to to its free energy of formation, and that the selling price per unit of free energy consumed is a better parameter to correlate with the commodity volume, rather than simply correlating the selling price directly with the annual product volume. They define a free energy consumption, C, that is related to the standard free energy of formation, ΔG_f^o:

$$C = \Delta G_f^o + 16 \tag{2.36}$$

where C and ΔG_f^o are in units of thousands of Btu/lb of product. They define a unit price per unit of free energy consumed, P_E (cents/thousand Btu), from the selling price in cents/lb, P_S:

$$P_E = \frac{P_S}{C} \tag{2.37}$$

and by correlating P_E with the annual volume in millions lb/yr:

$$P_E = 2.583V^{-0.3285} \tag{2.38}$$

rough estimates of the commodity selling price in 1969 can be obtained. Such estimates need to be adjusted to current prices using crude oil prices (see Tables 6.4 and 6.5.

Example: 2.6 *Estimate the 1998 selling price of styrene (*C_7H_8 *) with an annual U.S. sales volume of 5 billion lb/yr. Estimate the 2001 selling price of styrene.*
Solution: 2.6 $P_E = 2.583(5000)^{-0.3285} = 0.1574$ *cents/kBtu*
$\Delta G_f^o = 51.10$ *kcal/gmol* $= 91,980$ *Btu/lb-mole* $= 883$ *Btu/lb* $= 0.883$ *kBtu/lb*
$C = \Delta G_f^o + 16 = 0.883 + 16 = 16.883$ *kBtu/lb*
$P_S = P_E C = (0.1574$ *cents/lb*$)(16.883$*kBtu/lb*$) = 2.66$ *cents/lb styrene*
Using a crude oil price of \$3.30/barrel in 1969 (see Table 6.5) and \$10.88/barrel in 1998 (see Table 6.4), then the estimated price of styrene in 1998 was $2.66(10.88/3.30) = 8.77$ *cents/lb styrene.*
In 2001, the price of crude in Illinois was \$25.63/barrel. Using this to estimate spot prices:
(25.63/3.30)2.66 = 20.66 cents/lb styrene. Actual costs were 18-30 cents/lb that year.

2.5 Predesign Total Installed Costs

Predesign methods give very approximate estimates of the total capital investment and some of the major economic components. These methods are useful in a pinch, but should be avoided whenever energy and material balances can be completed. They can be used for preliminary screening in some instances, but they may yield large errors.

2.5.1 Lang Factors

A very simplified method for estimating total installed costs can be obtained by using *Lang* [54, 55] factors. They are also called *predesign* factors because they can be often be used without completing even material or energy balances. The basis of the predesign methods is empirical correlations and the fact that total installed costs are generally a factor of 3-5 times greater than the *delivered* purchased costs of all major equipment. A typical value for the assumed ratio is 4.25. Thus, if the delivered cost of a pump is about \$500, one would expect the installed costs to be somewhere between \$1500-2500 and probably about \$2125 or \$2100.

Typically, delivery costs are about 5% of the fob purchased cost. The abbreviation *fob* stands for *free on board* and indicates a vendor's cost to produce the item and to load it onto a common carrier at the vendor's site. The buyer must still pay the freight, taxes, and insurance costs to have the item delivered to the plant site.

Similarly, the Lang factor estimates that the ratio of the total capital investment to delivered cost of purchased equipment will be about five (i.e., $TCI/\sum_i C^a_{pi} = 5$). Hence, the working capital to delivered, purchased equipment cost ratio is typically about 0.75.

Example: 2.7 *The fob purchase cost of the major equipment items in a plant is \$10 million. Estimate the fixed investment and the total capital investment for this plant if the delivery costs are about 5% of the fob purchased costs.*

Solution: 2.7

$I_F = FCI = 4.25(10M)(1.05) \approx \$45\ M$

$TCI = I_F + I_W = 5(10M)(1.05) \approx \$53\ M$

2.5.2 Functional Predesign Methods

A number of predesign methods have been developed that that require little more than a block diagram of the process or a preliminary process flowsheet, an estimate of the plant throughput, and some measure of the process complexity. In applying these methods, one typically estimates the number of *functional units* that the process contains. A functional unit is somewhat subjective, but means a subsystem in the process that is completing some major task (*e.g.* a distillation tower with attached heat exchangers, pumps, surge vessels, etc.).

Zevnik and Buchanan [122] offer a predesign method which was further adapted by Ward [115]. Viola [113] has also developed a method which gives similar results. Allen and Page [2] review some of these methods and present their own approach.

A second key factor is the annual capacity rate, C, measured in lb/yr of product. Thirdly, when using the Zevnik-Buchanan method as modified by Ward, one also calculates a complexity factor.

Given an estimate of the significant units N_F in a flowsheet and its throughput or capacity C in lb/yr, then the next step using this predesign technique is to estimate a complexity factor, C_F, based on the process extremes in temperature, pressure, and materials of construction. After an extreme temperature has been estimated, then a temperature factor f_T is computed:

$$f_T = \begin{cases} 0.017(T/100) - 0.05, & T > 300\ K \\ -0.2(T/100) + 0.6, & T < 300\ K \end{cases} \tag{2.39}$$

Similarly, a pressure ratio P^* is defined for the extreme pressure or vacuum required by the process:

$$P^* = \begin{cases} P/100, & P > 100\ kPa \\ 100/P, & P < 100\ kPa \end{cases} \tag{2.40}$$

A pressure factor f_P is then computed from P^* as:

$$f_P = 0.10 \log P^* \tag{2.41}$$

The materials of construction also have a major impact on costs, so a materials factor f_M is computed using the following rules. For cast iron, carbon steel, and wood, $f_M = 0$. For aluminum, copper, brass, and 400-series stainless steels, $f_M = 0.1$. For monel, nickel, inconel, and 300-series stainless steels, $f_M = 0.2$. For Hastelloy, $f_M = 0.3$. For precious metals, $f_M = 0.4$.

The complexity factor for the process is then estimated using the temperature, pressure, and materials factors:

$$C_F = 2[10^{(f_T+f_P+f_M)}] \tag{2.42}$$

Investment Cost

From the complexity factor, one then estimates an investment cost per functional unit as:

$$I_F = \begin{cases} 7.5C^{0.5}C_F(I_{CE}/102), & 10^6 < C < 10^7,\ lb/yr \\ 1.5C^{0.6}C_F(I_{CE}/102), & 10^7 < C < 10^8,\ lb/yr \end{cases} \tag{2.43}$$

where I_{CE} is the CE Plant Cost Index for the year of interest. The Total Installed Cost for the process battery limits is then given by $TIC_{BL} = N_F I_F$. The Total Installed Cost for the chemical plant can be estimated using the battery limits Total Installed Cost estimate and Heuristic 9.

Example: 2.8 *An ethylene-to-ethanol conversion process utilizes eight functional units: vaporizer, reactor, hot-oil furnace, quench tower, compressor, dryer, ethylene column, and an energy recovery system. Thus, $N_F = 8$. The highest process temperature is 350° C in the reactor, and the highest pressure is 17 atm in the compressor, dryer, and ethylene column. The throughput is 22 million lb/yr. Estimate the battery limits capital investment, and the total installed cost of the plant, assuming grass-roots construction, if the equipment is made of 400-Series stainless steel. What is the estimated TIC if the new battery limits is an add-on to an existing plant? Use a CE index of 238.7.*

Solution: 2.8 $C = 22 \times 10^6\ lb/yr\ and f_M = 0.1.$
$f_T = 0.017(350 + 273)/100) - 0.05 = 0.056$
$P^* = 17(101.325)/100 = 17.23$
$f_P = 0.10 \log(17.23) = 0.10(1.236) = 0.124$
$C_F = 2[10^{(0.056+0.124+0.1)}] = 2[10^{0.28}] = 3.81$
$I_F = 1.5(22 \times 10^6)^{0.6}(3.81)(238.7/102) = \$0.34\ \times 10^6/unit$
The Total Installed Cost estimate for the battery limits is $TIC_{BL} = I_F N_F = 8(0.34 \times 10^6) = \$2.72 \times 10^6$
The Total Installed Cost for a grass-roots facility is:

$TIC = 2 \times \$2.72 \times 10^6 = \$5.44 \times 10^6$ *total*
As an add on to an existing plant, the $TIC = 1.33 TIC_{BL}$.

2.6 Adjusted PUB (APUB) Estimation

An adjusted PUB may be calculated that uses a predesign method to estimate Total Installed Cost and an approximate relationship for the Venture Cost. In this way an economic penalty for process complexity can be approximately estimated early during conceptual design.

Heuristic: 7 *Assuming that the annual Raw Material Costs (MAT) are half of the Total Product Cost and that the Total Installed Cost of the chemical plant can be estimated, then the Adjusted Profit Upper Bound (APUB) in \$/yr is given by:*

$$APUB = \frac{1}{2}[S - 2(MAT)] - \frac{1}{4}[TIC] \tag{2.44}$$

where TIC is the total installed cost for the plant and S and MAT are average sales and raw material costs in \$/year when the plant is fully operational. Dividing APUB expressed as \$/yr by the annual production rate, say in lb/yr, gives an estimated APUB in \$/lb of product.

2.7 Problems

2.1 Use Production/Consumption Analysis to screen two alternative paths for producing soda ash (sodium carbonate). Using the economics shown, what are the PUBs for the two cases? Which, if either, of the two systems is preferred?

Solvay Process	
1	$CaCO_3 \rightarrow CaO + CO_2$
2	$CaO + H_2O \rightarrow Ca(OH)_2$
3	$Ca(OH)_2 + 2NH_4Cl \rightarrow 2NH_3 + CaCl_2 + 2H_2O$
4	$NH_3 + H_2O \rightarrow NH_4OH$
5	$NH_4OH + CO_2 + NaCl \rightarrow NH_4Cl + NaHCO_3$
6	$2NaHCO_3 \rightarrow Na_2CO_3 + CO_2 + H_2O$
7	$NH_4OH + CO_2 + NaCl \rightarrow NH_4Cl + NaHCO_3$
LeBlanc Process	
1	$2NaCl + H_2SO_4 \rightarrow Na_2SO_4 + 2HCl$
2	$Na_2SO_4 + 2C \rightarrow Na_2S + 2CO_2$
3	$Na_2S + CaCO_3 \rightarrow Na_2CO_3 + CaS$

Cost basis for producing soda ash.	
Species	$/Ton
Calcium Sulfide	1
Sodium Chloride	14
Sulfuric Acid	30
Sodium Sulfate	24
Hydrogen Chloride	110
Carbon	20
Carbon Dioxide	0
Calcium Carbonate	12
Sodium Carbonate	16
Calcium Chloride	3
Ammonium Hydroxide	95
Ammonium Chloride	215
Sodium Sulfide	3

2.2 Draw a flow diagram depicting a system that consists of two chemical reactor and separation subsystems in series. In this system the first reactor partially converts species A and B to species C and D. The unconverted reagents are recycled to the first reactor after being separated from the species C and D. The species C and D are then mixed with species E and fed to the second reactor where they are partially converted to species F. The second reactor is followed by a second separation system where F is isolated and eventually sold while the unused reactants are recycled back to the second reactor. The relevant equations are:

$$\begin{aligned} A + B &\rightarrow C + D \\ D + C + E &\rightarrow F \end{aligned}$$

Derive relationships describing this system. Assume that the recycle rates and internal feeds can be described in terms of linear parameters such as fraction conversions from the reactors and recovery factors. What effect does it have on the system evaluation if the second reaction is:

$$D + C + E \rightarrow F + 0.5A + 0.5B$$

and the A and B species are separated from the other reactor effluents in the second separation system and recycled to the first reactor? How many equations must be solved simultaneously in these two cases if final product rate is specified? How can this problem be simplified by specifying internal streams?

2.3 Derive material balance equations similar to those in Figures 2.5 and 2.6 for the complete system which is shown in Figure 2.4.

2.4 Derive material balance equations similar to those in Figure 2.7 for the complete system which is shown in Figure 2.4.

2.5 A process flowsheet is under consideration for the manufacture of species C. In this process, raw materials A and B are reacted to yield C and D by the pathway:

$$A + B \rightleftharpoons C + D$$

However, only partial conversion is achieved in the reactor and species B is added in excess. Species D is a waste, but a disposal cost must be paid for its management.

The conceptual flowsheet under evaluation consists of the following steps. A mixer is used to combine recycled A and B with makeup A and B. The composite effluent from this mixer is sent to the reactor. The reactor effluent is treated using a two-step separation scheme. First species C is recovered, then species A and B are recovered for recycle with species D discarded as a waste for disposal.

The process will operate with the following characteristics (expressed on a molar basis per pass through the reactor):

- 1.1 moles of B/mole of A are fed to the reactor
- 6% of species A is converted
- 93% of species C is recovered
- 98% of species B is recovered
- 89% of species A is recovered
- 100% of species D is sent to waste

The economic assumptions for this process are tabulated below in terms of costs or sales values.

Species	Cost or Value ($/kg-mole)
A	5.59
B	7.55
C	53.99
D	8.95

Using this information, estimate an upper bound for profits (PUB).

2.6 A process flowsheet is under consideration for the manufacture of valuable species C. It will use the chemistry:

$$A + B \rightarrow C + D$$

Species D is a waste byproduct. For every mole of species A added to the reactor, 1.3 moles of B are added. Unfortunately, only partial conversion is achieved and unreacted A and B must be recycled to the reactor after C is recovered and D is removed and sent to waste disposal.

Based on experimental data, the reactor and separator train are expected to perform as tabulated below.

19	% of A is converted in the reactor
88	% recovery of C is achieved
72	% recovery of B is achieved
69	% recovery of A is achieved
100	% of D is sent to the waste

Relevant raw material and waste management costs and the sales value of C are tabulated below.

Species	$/kg-mole
A	17.85
B	9.40
C	86.50
D	5.80

Based on this information, estimate the Profit Upper Bound (PUB) for this process.

Part I

Engineering Economics

Chapter 3

Project Economics

The definitions in this chapter focus on two general concepts — the Accounting Entity and Project Economics. The former is described in order to explain the general environment in which the latter exists. In fact, the attractiveness of a particular project usually depends strongly upon the characteristics of the accounting entity. Small projects, for example, may be of little or no interest to large corporations while small firms may be unable to raise the necessary capital to undertake larger projects. Of course, the terms *small* and *large* are in themselves relative and depend upon the nature of the accounting entity.

In this chapter, no consideration is given to the time value of money which is discussed in Chapter 7 along with various measures of project economic performance. Here terms are defined on an annual or quarterly basis.

3.1 The Accounting Entity

All economic analyses begin with a definition of the basic accounting entity. This entity could be, for example, an individual, a partnership, or a corporation. It could consist of either a single manufacturing plant or a large firm with separate headquarters and many manufacturing sites. In any case, the economic definitions are the same and based on the general accounting principles as described here.

3.1.1 The Balance Statement

The accounting entity uses double entry accounting procedures that enable the generation of several financial statements. With double entry accounting, each transaction requires that one account in the books be credited while another account is debited. At all time on the books:

$$\text{Debits} = \text{Credits} \tag{3.1}$$

otherwise the books are not in balance. If the financial transactions are ever recorded in such a way that Equation 3.1 is not satisfied, then a bookkeeping error has occurred.

In order for Equation 3.1 to work, the rules for liability and equity accounts on the entity books must be the opposite of the rules for their asset accounts. The accepted accounting procedure is that all asset accounts are increased by debits and decreased by credits. On the other hand, to increase a liability or equity account, it receives a credit. To decrease a liability or equity account, it receives a debit.

The *balance statement* results from accounting procedures using double entry accounting, and it is one of the most important measures of the financial position of a firm. This statement is based upon one central accounting relationship:

$$\text{Assets} = \text{Liabilities} + \text{Equity} \tag{3.2}$$

By examining the meaning of the terms in Equation 3.2, we will answer basic questions about how firms raise money or *capital* and how that capital is used to make more money.

All U.S. corporations are required to provide a balance statement on an annual basis as part of their corporate income tax return, and most corporations provide balance statements as part of their financial summary at the annual stockholders meeting. As such, the annual balance statement is a partial picture of the financial situation for the firm at that time.

The Assets are usually summarized on the left hand side of the balance statement while the Liabilities and Equity are organized on the right hand side. Alternatively, that part of the balance statement summarizing the Assets may be presented as a table preceding the Liabilities and Equities table as shown in this chapter. Normally, the balance statement will include entries for the previous and current years of operation. Tables 3.1 and 3.2 are provided to illustrate the features of a balance statement.

Assets

Assets are either *Current* or *Fixed.* When a balance statement is prepared, all assets are listed as line items by categories with actual dollar amounts shown to the right of each entry in the assets table. Assets within each of these categories are described below.

Current Assets Current assets consist of items such as cash, marketable securities, accounts receivable, inventories, and prepaid expenses. This order of current assets also roughly corresponds to the relative ease with which they can be *liquidated* or converted into cash. The term *cash* generally refers to monies that are either held in banks or on the company premises. Cash is considered to be a completely liquid asset because it can be used immediately to either pay off debts or to make purchases as required by the firm.

Table 3.1: Typical statement of assets.

Balance Statement
E. Z. Chemicals, Inc.

ASSETS		
December 31	Last Year	This Year
Current Assets		
Cash	484,714	1,308,017
Marketable securities	20,000	15,000
Accounts receivable	276,000	355,000
Inventories	185,000	104,000
Prepaid expenses	5,000	14,000
Total current assets	970,714	1,796,017
Fixed assets		
Land and buildings	154,000	388,000
Machinery and equipment	303,000	660,000
Startup expenses	100,000	200,000
Total cost of fixed assets	557,000	1,248,000
(Amortization, depreciation)	(83,550)	(421,140)
Net fixed assets	473,450	826,860
TOTAL ASSETS	1,444,164	2,622,877

Table 3.2: Typical liabilities and equity statement.

Balance Statement
E. Z. Chemicals, Inc.
LIABILITIES AND STOCKHOLDERS' EQUITY

December 31	Last Year	This Year
Current liabilities		
Accounts payable	209,000	144,000
Salaries payable	35,000	42,000
Taxes payable	5,000	15,000
Dividends payable	24,688	112,500
Short-term notes payable	4,500	11,000
Mortgage notes payable	43,000	55,000
Interest payable	1,000	2,000
Total current liabilities	322,188	381,500
Long-term liabilities		
Industrial revenue bonds	371,333	646,333
Total liabilities	693,521	1,027,833
Stockholders' equity		
Preferred stock ($2/share)	250,000	760,000
Common stock ($1/share)	20,000	300,000
Retained earnings	480,643	535,043
Total stockholders equity	750,643	1,595,043
TOTAL L & E	1,444,164	2,622,877

Marketable securities include such items as stocks, bonds, and Treasury bills which are purchased to earn interest, but which may be sold at any time on either the stock or bond markets and easily converted to cash assets.

Accounts receivable consist of those amounts that are owed to the firm by clients or customers, but for which payment has not yet been received. A company that operates on a cash basis may exclude accounts receivable from its balance statement.

Inventory consists of material goods that are purchased by the firm for use in the sale and manufacture of final products. In chemical companies, *inventory* usually refers both to those raw materials that have been purchased to support various manufacturing operations as well as the firm's finished products that have not been sold at the time the balance statement is prepared.

Prepaid expenses include all expenses, such as rent and insurance, that have been prepaid for time periods beyond the time of the balance statement. Such payments may be made several years in advance, for example, in order to obtain more favorable agreements with vendors and lessors.

Fixed Assets Fixed assets include items like land, buildings, machinery, and equipment. They are referred to as *fixed* because they are not easily liquidated and converted into cash. Usually, when fixed assets are described, their original costs are shown as a line item. Immediately below them an *amortization, depreciation* term is shown and subtracted from their original costs. The amortization and depreciation are the *accumulative* values. The difference between the original purchase cost and the depreciation is then shown as the *net fixed assets* or the net value (NV) of such assets. (More information about depreciation and amortization is provided in Sections 3.2.5 and 3.2.5.) However, when reference is made to *capital investment*, it usually refers primarily to the acquisition of fixed assets as opposed to the acquisition of current assets.

Liabilities and Equity

Liabilities are usually considered as being either *current* or *long-term.* Current liabilities are often divided into categories such as Accounts payable, Salaries payable, Taxes payable, Dividends payable, Short-term notes payable, Mortgage notes payable, and Interest payable. These accounts include items that are payable within 30 working days or less.

Long-term liabilities include debts on notes and industrial revenue bonds. In many instances, such debts may have resulted from the purchase of *fixed assets* or other *fixed capital investments.* Long-term liabilities are usually not payable within 90 days, and the period of repayment may extend over years.

The *Stockholders' Equity* includes the actual purchase prices for all stocks issued by the firm plus the retained earnings. Several stock classes may have been issued such as preferred or common stocks. Preferred stocks often pay higher dividends than common stock, and some or all of the dividends may be guaranteed by the firm. The terms *proprietorship* and *net worth* are synonyms for the stockholders' equity.

One feature of the balance statement is that the *working capital* is easily computed from it. It is simply the difference between the current assets and the current liabilities or current debts. The *current ratio* is the ratio of the total current assets divided by the current liabilities.

Example: 3.1 *What was the working capital for E. Z. Chemicals, Inc. at the end of their fiscal year (FY) 1987?*
Solution: 3.1 *From Table 3.1, the current assets at the end of FY-87 were \$970,714. From Table 3.2, their current liabilities at the end of FY-1987 were \$322,188.*
Working Capital $= \$970,714 - \$322,188 = \$648,526$

Example: 3.2 *What is the net worth of a corporation with assets of \$10 million and \$4 million in liabilities? What will be debt to equity ratio if they borrow an additional \$2 million?*
Solution: 3.2
Net worth = Assets - Liabilities = \$10 - \$4 = \$6 million dollars
Equity = Net Worth
Current Debt/Equity = \$4/\$6 = 0.67
If they borrow \$2 million more, then:
Future Debt/Equity = \$(4+2)/\$6 = 1.0

Companies with excessively high Debt/Equity ratios are said to be *leveraged.*

3.1.2 The Income Statement

The annual income statement provides information that complements the balance sheet and interfaces with it. While the balance sheet provides information about the net worth of a corporation, the income statement gives information about its annual earnings. The income statement excludes, however, information about nontaxable sources of capital which are not reported as ordinary income. Nontaxable sources of capital appear on the balance sheet as assets and as equity.

Table 3.3 is an example annual income statement for E. Z. Chemicals, Inc. This statement is based on one fundamental accounting relationship:

$$\text{Total income} = \text{Costs} + \text{Profits} \tag{3.3}$$

The profits may be further categorized as either being *gross* or *net.* The gross profits refer to those profits before income taxes while net profits are those computed after income taxes and capital recovery have been subtracted from the gross profits. The net profits may be further divided into *retained earnings* or *dividends.*

The first major heading of the income statement summarizes the *annual revenues* for the firm. These revenues are *ordinary taxable* income categories only, and they include such items as total sales, earned interest, rental receipts, and dividends and royalties paid to the firm.

Table 3.3: Typical income and expense statement.

Income & Expense Statement E. Z. Chemicals, Inc. December 31	Last Year	This Year
REVENUES		
Total sales	1,800,000	2,500,000
Interest earned	5,400	76,000
Dividends earned	12,000	14,000
Royalties earned	85,000	123,000
Gross income	1,902,400	2,713,000
COSTS AND EXPENSES		
Production and operating	760,000	1,100,000
General and administrative	105,000	111,000
Engineering	45,000	86,000
Total	910,000	1,297,000
GROSS PROFITS	992,400	1,416,000
(Taxes on income)	(347,340)	(495,600)
CASH FLOW	645,060	920,400
(Capital recovery)	(185,667)	(416,000)
NET PROFITS	459,393	504,400
CASH DIVIDENDS		
Preferred stock	93,750	356,250
Common stock	5,000	93,750
Total dividends	98,750	450,000
RETAINED EARNINGS		
Beginning of year	120,000	480,643
Current year	360,643	54,400
End of year	480,643	535,043
Dividends paid/share		
Preferred stock	\$0.75	\$0.94
Common stock	\$0.25	\$0.31

There are, however, several sources of capital for firms that are not included as taxable income. These sources include such capital generating items as stock sales, capital contributions to the firm by an owner or stockholder, and monies obtained from loans and bonds. Other sources of capital that are not ordinary taxable income include leasehold improvements, the appreciation of property, and materials received by the firm on consignment. If property appreciates (or increases in value) over time, then the net appreciation may be categorized as *capital gains* during the fiscal year in which that property is liquidated or sold. Otherwise, the net gains are ordinary income in that fiscal year.

The second major heading in the income statement summarizes the *annual costs and expenses.* Costs and expenses are monies spent for production and operating costs, administrative and engineering payrolls, raw materials, and the like. For manufacturers, production costs include the costs of all raw materials or parts purchased for manufacture into a finished product. In addition, the direct labor costs to manufacture the product is an operating cost. Other operating costs include the costs of containers and packages, freight-in, express-in, and cartage-in on raw materials and merchandise. Overhead expenses such as rent, heat, light, power, insurance, taxes, maintenance, labor, and supervision are also included. Operating expenses often involve consumable items.

In general, operating costs are not associated with items that have a fixed life. So the purchase costs of large machinery with a life expectancy greater than one year or land purchases are not operating costs. Instead, such expenses are classified as fixed capital investment and are treated as part of the firm's fixed assets as described in the balance statement and below.

The annual *gross profits* are calculated as the gross revenues less all costs and expenses in the income statement:

$$\text{Gross Profits(GP)} = \text{Revenues} - \text{Costs and Expenses} \tag{3.4}$$

The change in Gross Profits from year-to-year may be calculated as $\Delta GP = GP_j - GP_{j-1}$ for the year end in the jth and $j-1$th years. In order to compute the firm's *net profits*, two additional items must be deducted from the gross profits:

$$\text{Net Profit(NP)} = GP - \text{Income Taxes(IT)} - \text{Capital Recovery(CR)} \tag{3.5}$$

Similarly, $\Delta NP = NP_j - NP_{j-1}$ when figuring the change in net profits from year to year.

Corporate income taxes are computed from the net taxable income as described below. The goal is generally to minimize taxes in a given year. On the other hand, capital recovery expenditures are based on the firm's internal policy. Such expenditures are also often referred to as *depreciation* expenses. Here, however, a distinction is made since depreciation is used only to compute *net taxable income,* while capital recovery expenses may be either greater or less than depreciation for taxation purposes.

Typically, for example, companies like to recover their capital investment in 3-5 years although long-range investment horizons may extend to ten years. Generally, however, such long-range capital recovery periods are limited to institutional investors such as government research organizations. On the other hand, *depreciation* for fixed assets like chemical plants cannot occur in less than five years by law. Therefore, the capital recovery time period may be three years or less while the depreciation time period is five years for the same fixed assets.

In any event, the *annual cash flow*[1] is given by:

$$\text{Cash Flow} = CF = NP + \text{Capital Recovery(CR)} \tag{3.6}$$

and the change in year-end cash flow from year-to-year is $\Delta CF = CF_j - CF_{j-1}$. The annual depreciation may be substituted for the capital recovery term in Equation 3.6 if they are equal. However, from Table 3.3 it should be clear that the cash flow funds may be distributed either to capital recovery, dividends, or retained earnings based upon corporate policy. The income taxes always have to be paid from the gross profits, but the capital recovery may be somewhat adjustable.

After estimating the net profits, management makes a decision about the annual dividends that are paid to the stockholders. The retained earnings for that year are then:

$$\text{Retained Earnings(RE)} = NP - \text{Cash Dividends(CD)} \tag{3.7}$$

The annual retained earnings are then added to those of the previous year to give the End-of-year retained earnings as shown in Table 3.3 and Table 3.1. Also, the retained earnings at the beginning of the next year equal those at the end of the previous year. So retained earnings are simply the sum of all net profits, less dividends paid by the corporation, that have been retained or held by the firm since its beginning.

The ability of a firm to raise capital either as taxable income or else nontaxable income in the form of loans and stock sales ultimately determines the economic environment in which projects are evaluated. For example, banks will examine the financial statements before loaning funds and may refuse to loan if they feel the corporate debt-to-equity ratio from the balance statement is too high.

On the other hand, prospective stockholders will look closely at the annual revenues of the firm and its historical policy of declaring dividends versus retained earnings. Low gross profits and cash dividend payments will make stock purchases appear less attractive and will ultimately limit the cash assets of the corporation. On the other hand, excessive cash dividends can lead to longer capital recovery periods and excessively high debt-to-equity ratios.

3.2 Project Constraints

As described in Section 9.3 , most engineering projects evolve through many levels of detail. Upon completion of one level, a more accurate and detailed cost estimate is usually generated. Regardless of the project scope, however, the project future usually depends on three central economic estimates:

[1]This definition is the same as that given by Peters and Timmerhaus[76] if the Capital Recovery equals the Depreciation.

1. the estimated total capital investment to complete the project,

2. the projected gross revenues from the project during each year of its life, and

3. the projected costs to operate the project or process.

From these estimates, one may compute a variety of economic measures of the overall project merit. Many of these measures lead to the same conclusions, and the actual measures that are used are often determined by the preferences of management and those individuals completing the study.

Details of how such measures are computed are provided in Chapter 8. Here, the underlying concepts leading to such measures are summarized.

3.2.1 Project Funding

In order to understand *why* economic terms are defined as they are, it is helpful to consider the total capital investment and cash flow associated with a hypothetical grass roots project. Before a chemical plant can be operated, for example, it must first be designed and built. Also, operators must be trained, safety manuals must be prepared, and raw materials must be purchased. All of these initial costs comprise parts of the *total capital investment* as is explained below.

After the plant has been constructed and started-up, it begins to generate product. After some additional time, gross revenues are returned to the plant owners through product sales. Of course, additional expenses continue during the plant operation. For example, the owners must buy additional raw materials and pay laborers and supervisors to run the plant.

Capital Investment generally refers to a one-time expenditure that occurs before the plant is in operation. *Operating costs* usually refer to recurring expenses while the plant is in operation. If one considers the total expenditures of the owners, then a large capital investment is often required several years in advance of the plant start-up. This initial expenditure is often followed by several lean years where the cash flows are small until the plant is in full operation, and all marketing resources are being fully utilized to sell product. Net profits may be lean for several additional years even after the maximum annual sales are realized, especially until all of the initial capital investment has been recovered.

There are at least three primary reasons for drawing a distinction between operating costs and capital investment. One reason is the fact that operating costs can be totally subtracted from sales revenues during the year in which they occur when estimating net taxable income, but fixed capital investment can only be depreciated at a schedule that is fixed by law over several years. Since it is generally advantageous to maximize the deductions totally in the year that they occur and thereby minimize income taxes, the tax laws are one reason for distinguishing between the two. A project with higher operating costs and lower initial capital investment may be preferable because of taxes alone.

A second reason is the fact that capital investment may require outside financing by the owner, especially for the construction of large plants. Also, most contractors will not initiate construction until the buyer's ability to pay has been clearly demonstrated. Often the owner is required to place a significant fraction of the initial capital investment into escrow. *Escrow* refers to monies that are placed into the hands of third parties, such as banks and lending institutions, under legal seal and for delivery to the grantee (*i.e.* the creditor or contractor) upon delivery of contractual terms (*e.g.* a completely installed chemical manufacturing plant). Escrow requirements usually mean that considerable amounts of up-front monies must be provided by the eventual owners. In part, such monies may have to be borrowed from a financial institution, but the primary owner (*i.e.* the eventual plant operator) usually must provide significant collateral.

When the project owner must borrow to finance the initial investment, then the capital funds will have financing and interest costs associated with them. If additional stock issues are sold by a corporation, then additional dividends will be payable to the stockholders in subsequent years. As more shares are sold to investors (become outstanding), the ability of the corporation to pay high dividends/share of outstanding stock is diminished. Also, the influence of the initial stockholders is diluted by the sale of the additional shares.

A third reason for distinguishing between capital investment and operating costs are the levels of financial risk that are associated with these two categories. Because operating funds are much more liquid, substantially less risk is associated with them than with capital assets. Fixed or capital assets are often difficult to convert back into cash, and the liquidation of capital assets may involve a three-party transaction (*i.e.* the owner, the buyer, and the financier).

In order to take these diverse factors into account, one must resort to the evaluation of suitable economic criteria that incorporate both of these financial aspects. However, it is necessary to begin by first describing the net profits and the net costs on an annual basis in more detail. Subsequently, in Chapter 7 the effects of the time-value of money over the project life are discussed.

From the above comments, one might conclude that the initial fixed capital investment for a project and the subsequent annual operating costs result in engineering tradeoffs that must be evaluated. This conclusion is, in fact, correct. On the one hand, it is often possible to reduce operating costs in some categories (*e.g.* utility costs) at the expense of increased capital investment. For example, a heat exchanger can be made to transfer heat more reversibly by increasing the transfer area. That is, if the transfer area is increased for any given process load, then the utility approach temperature can be reduced. Consequently, some reduction in utility costs will result.

For another example, a distillation tower [47] can in principle be made to operate reversibly by adding an infinite number of trays, adding heat reversibly at each point below the feed tray, and removing heat reversibly at each point above the feed tray. Theoretically, this approach will reduce utility costs, but it also results in substantially increased tower dimensions (*i.e.*an infinitely tall column) and infinitely higher capital investment.

Therefore, it is not clear in either of these two cases that an overall reduction in production costs will result or that a company working toward these end goals will increase its net profits because of the competition between subsequent plant operating costs and the initial capital investment. However, the measure of performance described in Chapter 7 will provide a basis for making rational decisions in such cases. At this point, each of these terms is simply described in somewhat more detail on an annualized basis.

3.2.2 Capital Investment

Loosely speaking, capital investment refers to the acquisition of fixed assets. More specifically, however, capital investment may be broken down into at least five major categories:

1. I_P, process battery limits,
2. I_W, working capital,
3. I_L, land investment,
4. I_A, auxiliary investment, and
5. I_S, start-up costs.

The total capital investment (TCI) for a project is generally given by:

$$TCI = I = I_P + I_A + I_W + I_S + I_L \tag{3.8}$$

where:

I_P = The process battery limit consists of all specialty items that have been purchased to construct the actual process. These items generally require the most design work, are often fabricated specially for the project, and are of the least value to others for resale if they do not work. Battery limits equipment can usually be depreciated.

In most chemical processes, the fixed investment is largely confined to such things as the actual processing equipment, but in nuclear chemical plants, for example, this is not the case. Once the process has been contaminated with radioactive materials, both the equipment and the building containing the equipment may be contaminated. Since the nuclear facilities housing the chemical processing equipment are much more expensive than the equipment itself, the battery limits investment in that case is largely determined by the building costs rather than the process.

The other extreme is represented by process units that are constructed and operated completely out of doors.

I_A = The auxiliary investment generally consists of items that are of common use. For example, a cooling tower or steam generation plant is usually not peculiar to any one facility. Therefore, this type of auxiliary equipment may be more easily liquidated than the process battery limits investment. Also, auxiliary equipment generally has a higher salvage value than process battery limits equipment, and less risk is associated with its purchase. Trade magazines, such as *Chemical Engineering*, commonly advertise used auxiliary equipment that is for sale. Auxiliary equipment can usually be depreciated.

I_W = Working capital is defined as those net current assets that are required to finance day to day expenses, primarily in the form of inventories. That is, a certain amount of raw materials and supplies must usually be stockpiled at the plant site in order to maintain operations. Working capital is also needed to meet payroll expenses, and it may neither be depreciated nor deducted as an operating cost during the year in which it occurs. Working capital is part of the owners' basis for the business, and it may be divided into cash and inventory. Changes in the amounts of cash on hand and raw material and product inventories therefore affect annual cash flow.

Coal fired steam generation plants, for example, stockpile coal at the plant in the form of large conspicuous piles. The size of these coal piles is usually a function of the plant size, the distances from coal mines, the time of year, and labor conditions in the mining industry.

I_L = Investments in land, minerals, or standing timber which may in some cases be depleted. Although land is not depreciated, its mineral assets may be depleted.

I_S = Costs incurred during the start-up of a plant. These one-time expenses, such as research and development costs, operator training programs, initial testing, and debugging operations. These expenses can usually be amortized.

In general, process battery limits investment is riskier than investment in auxiliary equipment, and auxiliary equipment investment is riskier than working capital investment. Consequently, it is frequently justifiable to spend more money and effort in designing a specialty item in order to ensure that it works properly than the effort to design an auxiliary item which may more easily be sold and replaced if it does not work properly.

In determining the best level of design effort, it is important to consider both the cost of replacing the item if it does not work properly as well as the initial installed cost. Rudd and Watson [86] give some guidelines involving probabilistic arguments.

The total capital investment, I, may be related to the balance sheet by identifying the fixed investment, I_F, and I_C, the current investment. The fixed assets, I_F, are given by:

$$I_F = I_P + I_A + I_S + I_L \tag{3.9}$$

Hence the current assets, I_C, consist only of the working capital:

$$I_C = I_W \tag{3.10}$$

The total installed cost, TIC, of a facility is computed to estimate capital investment as discussed below. It excludes both the working capital and the start-up costs:

$$TIC = I_P + I_A + I_L \tag{3.11}$$

and the total capital investment is, therefore:

$$TCI = I = TIC + I_S + I_W \tag{3.12}$$

The depreciable assets, I_D, are discussed in more detail below, but they are usually equal to the installed costs of the auxiliary and process equipment:

$$I_D = I_P + I_A \tag{3.13}$$

3.2.3 Manufacturing and Total Product Costs

Manufacturing or operating costs may be divided into at least three distinct categories. These terms have units of dollars per unit of time, and it is frequently convenient to choose a per annum time basis. They are generally either proportional to the fixed investment, C_I, or to the production rate, C_Q, or to the labor requirements, C_L. All operating costs may be deducted from the gross profits in the year which they are incurred.

The total annual operating or manufacturing costs are estimated as:

$$C = C_I + C_Q + C_L \tag{3.14}$$

where:

C_I = aI. The coefficient, a, is the sum of several terms which are roughly proportional to the capital investment that is required for the venture. These operating costs include:

1. Maintenance costs, such as labor and materials (typically 5-9% of I_D).
2. Property taxes (typically 1-4% of TIC).
3. Safety expenses, such as charges for fire protection, first aid, and security requirements (typically 10-20% of I_D). In Table 3.6, this expense is listed as part of the Plant Overhead Cost.
4. General services, including things like laboratory, cafeteria, road, transportation, and central computing services (typically 30-40% of I_D). In Table 3.6, this expense is listed as part of the Plant Overhead Cost.
5. Administrative support services, *e.g.* payroll, travel, and benefits departments and other management offices such as purchasing and legal (typically 10-20% of I_D). This expense appears in Table 3.7 as part of the Total Product Cost, but is not a direct manufacturing cost.
6. Interest payments for capital loans, but excluding the capital recovery payments.

C_Q = bQ_P. This term includes those operating costs which are roughly proportional to the annual production rate, Q_P. The coefficient, b, is the sum of several terms in units of \$ per unit of Q_P which include the following:

1. Raw material costs.
2. Utility costs such as steam, cooling water, refrigeration, electricity, fuel oil, *etc.*

3. Chemicals, such as catalysts, lubricants, or other materials which are proportional to the production rate, but separate from the raw materials costs.
4. Other charges such as royalty fees, shipping and storage costs, analytical analyses, and other quality control costs.
5. Product distribution, promotion, and other sales costs.

$C_L = cL$. This term includes those operating costs which are roughly proportional to the annual labor requirements, L, in employee-hours per year. The coefficient, c, is the sum of several terms including:

1. Direct operating labor.
2. Supervision and management.
3. Overhead, such as vacation, retirement, health benefits, life insurance, and social security.

In some cases, if the labor requirements are essentially identical among several alternatives that are being considered, then it is not necessary to include labor costs explicitly in the analysis. In other cases, the various alternatives do involve significantly different labor requirements, and these costs can affect the project conclusions. In these latter instances, it may be convenient to estimate the annual person-hour requirements from a suitable correlation (see Chapter 6, for example) or to estimate an average $/person-hour rate for all personnel on the plant site.

3.2.4 Project Revenues

Project gross income gross revenues are usually estimated from projected annual sales:

$$S = sQ_P \tag{3.15}$$

where: S = annual revenues from sales/year.
s = sales price/unit of product (*e.g.* $/lb).
Q_P = annual production rate of the product by the process if constructed as envisioned by the project planners.

As is indicated in Table 3.3, the accounting entity for the project basis may realize revenues other than those from sales. It is only necessary that such terms be included if they can affect the project conclusions. In some instances, detailed marketing surveys are justified. In other cases, the estimated revenues may be based on sales to a single customer.

3.2.5 Net Taxable Income

The net taxable income[2] also must be computed in order to estimate the annual income taxes that are paid by the accounting entity. In simple terms, the net

[2]Net taxable income is defined following the U.S. Internal Revenue Tax Code. It is the main reason a distinction is drawn between depreciation and capital recovery.

taxable income is simply the gross profits, or gross income, after certain deductions have been made to it. These deductions are established by the tax codes, and the definitions below are derived from those sources. A general equation defining the net taxable income is given by:

$$NTI = GP - eI_S - dI_D - lI_L \tag{3.16}$$

where: e = amortization factor, usually 0.2
d = depreciation factor
l = depletion factor

Details about how the e, d, and l factors may be calculated are provided below.

Example: 3.3 *A company achieves gross sales of \$95 million in one year, on a \$26 million investment with annual operating expenses of \$72 million during the same year. It is the first year in which it is allowed 15% ACRS depreciation on investment. Its investors require capital recovery in four years (straight line) and they expect 25% rate of return on investment and it pays an income tax of 42% of Net Taxable Income (NTI). Calculate: Gross Profits, Net Taxable Income, Income Taxes, Cash Flow, Net Profits, Venture Profits, and Rate of Return for the year.*

Solution: 3.3 *(All in millions of dollars.)*
$GP = S - C = 95 - 72 = 23$
Net Taxable Income $(NTI) = GP - dI = 23 - 0.15(26) = 19.1$
Taxes $= t(NTI) = 0.42(19.1) = 8.022$
Cash Flow: $CF = GP - t(NTI) = 23 - 8.022 = 14.978$
Net Profit: $NP = CF - gI = 14.978 - 0.25(26) = 8.478$
Venture Profit: $VP = NP - i_{min}I = 8.478 - 0.25(26) = 1.978$
Rate of Return: $ROR = NP/I = 8.478/26 = 0.3261$ *or 32.6%*

Depreciation

The term "depreciation" is commonly used in several ways. As described here, the term "depreciation" is used to mean a deduction from Gross Profits as defined by the Internal Revenue Service for purposes of computing Net Taxable Income and Income Taxes. So depreciation is used to compute a tax deduction.

Depreciation is also used to mean Capital Recovery. In this instance, depreciation is a fictitious expense that is added to a Manufacturing Cost estimate to account for the decline in value of fixed assets. Capital Recovery more generally refers to a payment schedule in which the Total Installed Cost of a plant or the Total Capital Investment is recovered by periodic payments over a set period of time.

Depreciation is sometimes loosely applied to mean Maintenance & Repair costs. These expenses occur routinely during plant operation and are necessary expenses to maintain the equipment in working order. In spite of Maintenance & Repair expenditures, however, it is still often necessary to replace equipment periodically (e.g., from long-term corrosion). While Maintenance & Repair can reduce the frequence of equipment replacement, Replacement Costs cannot always be avoided.

Depreciation is also sometimes used to mean a fixed charge to represent periodic Replacement Costs for equipment.

A fixed asset or property is depreciable for computing Income Taxes if, and only if, it meets certain requirements. Specifically, the asset must be:

1. used in business or held for the production of income,
2. have a determinable life greater than one year,
3. decay, become obsolete, or lose its value due to natural causes, and
4. meet all of the above requirements.

Both tangible and intangible items can be depreciated. For example, equipment, machinery, buildings, vehicles, patents, copyrights, and furniture can all be depreciated. Moreover, depreciable equipment usually includes two important categories already defined:

$$I_D = I_P + I_A \tag{3.17}$$

where I_D is the depreciable capital investment associated with the project.

This definition automatically excludes rented property, land, inventory, and goodwill. Also, an item which is depreciated cannot be listed as an operating or manufacturing expense.

Depreciation may be calculated from one of several methods. The ACRS method, set up by the U.S. tax code, defines different categories of depreciation (5, 10, 15, or 20 years). Each category has a fixed depreciation schedule. A summary of the ACRS factors for three, five, and ten year depreciation periods is provided in Table 3.4. Generally, it is advantageous for companies to depreciate items as quickly as possible, but the rate is determined by law. For example, 3-year property has a midpoint class life of 4 years or less, or it is used in research and experimentation, or it is a race horse more than two years old when first placed in service. Examples of 3-year property include automobiles, light duty trucks, and machinery and equipment used in experiments.

Five-year property has a longer life. Some examples are computers and peripheral equipment, office furniture, and single purpose agricultural and horticultural structures (other than a building and its structural components).

Ten-year property includes items with a midpoint class life of more than 18 years and no more than 25 years. It includes items such as manufactured homes, railroad tank cars, and qualified coal utilization property.

Depletion

A depletion allowance is given for land, minerals, oil or standing timber which will be *depleted* or *used up* over a period of years. An item must have a life greater than 2 years to qualify for a depletion allowance. Depletion allowances are calculated by one of two methods — cost depletion or percentage depletion. Each item is treated on an individual basis.

1. Cost Depletion = (\$/unit)(units/year)/(total number of units).

Table 3.4: ACRS annual depreciation factors.

Period	Percentages									
Year =	1	2	3	4	5	6	7	8	9	10
3-Year	25	38	37							
5-Year	15	22	21	21	21					
10-Year	8	14	12	10	10	10	9	9	9	9

2. Percentage depletion is calculated as a certain fraction of the gross income of the property for each mineral.

   Depletion Percentages
   18% for oil and gas (old)
   16% for oil and gas (1983)
   15% for oil and gas (1984+)

It should be noted that small producers will have separate depletion tables and will be exempt from the above specifications.

Amortization

Amortization is calculated as a straight line depreciation with a service life of 5 years. It is a deduction in lieu of depreciation, and anything that is amortized cannot be depreciated.

In order to be amortized, an item must *not* have a fixed life. Examples of expenses that may be amortized include such things as start-up costs, pilot testing, research and experimentation costs, market surveys, the training of employees, and advertisements for business opening.

It should be noted that the costs of going into a business are viewed as capital expenditures that comprise part of the owner's basis in the business. These expenses cannot be recovered unless the business fails or the owner goes out of business. At that time, they may be declared as business losses or capital gains.

3.2.6 Annual Cash Flow, Net Profits & Costs

The gross profits are simply the difference between annual sales and costs (excluding taxes), such that:

$$GP_j = S_j - C_j \tag{3.18}$$

where $S_j = sQ_{Pj}$ = (\$/unit)(number of units sold/year)
The costs are the operating or manufacturing costs as described in Section 3.2.3. The annual cash flow in the jth year is the annual gross profits less the corporate income taxes.

$$CF_j = GP_j - (GP_j - e_jI_{Sj} - d_jI_{Dj} - l_jI_{Lj})t_j \tag{3.19}$$

where t_j is the tax rate in the jth year of operation. Equivalently, $CF_j = GP_j - IT_j$. Since cash flow is also affected by changes in capital investment, working capital, and inventories from year to year, we can write:

$$CF_j = GP_j - IT_j - \Delta I \tag{3.20}$$

where $\Delta I = I_j - I_{j-1}$ is the incremental investment made during the jth year. If additional funds are invested in equipment, the cash flow decreases. If the inventory is reduced during the year, then the cash flow is increased. (See Example 3.5.)

The net profits are then defined as the annual cash flow less the sinking funds or capital recovery expenditures in the jth year:

$$NP_j = CF_j - g_j I_j \tag{3.21}$$

where: g_j = the capital recovery, annuity or sinking fund factor which may be based on the actual plant life. In that case: $g_j = 1/N$

N = the expected service life in years.

and

$$I = \sum_{j=1}^{N} I_j \tag{3.22}$$

is the total capital investment for the project. Also:

$$I = \sum_{j=1}^{N} CR_j = I \sum_{j=1}^{N} g_j = \sum_{j=1}^{N} I_j \tag{3.23}$$

and I_j is the capital investment during the jth year.

Today capital recovery policy is more frequently set by corporate goals rather than by actual asset life. For example, investors (*i.e.* stockholders) often demand that capital be recovered in three to five years. This approach reduces the ability to pay dividends, but it quickly reduces the debt/equity ratio and the total interest payments. Since a project must also realize net profits, a rapid payback policy penalizes projects which require large capital investment, but it tends to protect a corporation from hostile takeovers.

In some cases, capital recovery payments may equal the depreciation in a given year. In such cases:

$$CF_j = NP_j + CR_j = NP_j + DEP_j \tag{3.24}$$

It is more common, however, for the capital recovery to be larger than the depreciation allowed by law in a particular year.

For a given project, all significant cost terms must be identified and correctly used to accurately describe the project environment. It will usually be necessary to distinguish between capital investment and operating costs, and to correctly estimate finance charges, depreciation schedules, and capital recovery charges when they are appropriate.

If the gross revenues term is deleted from the analysis, and the remaining terms in the net profit definition have the signs reversed, then the result is an expression known as the net cost:

$$NC_j = (1 - t_j)C_j - (eI_{Sj} + dI_{Dj} + lI_{Lj})t_j + g_jI \tag{3.25}$$

In some cases, it is more convenient to analyze a project in terms of its net cost rather than its net profit. For example, if the gross revenues term is always constant in the project analysis, then there is no reason to include it, and the analysis is simplified by using net cost.

Example: 3.4 *E. Z. Chemicals, Inc. began a new project. During the first year they spent $500,000 on research and development. In the second year they spent $250,000 on land and $1.5 million on construction. During the third year they spent $4.5 million to complete the plant installation. They also spent another $800,000 on start-up costs in the third year including operator training and the development of emergency shutdown procedures. In the fourth year, they went into production and began depreciating and amortizing expenses. Their operating costs that first year in production were $6 million and they realized $20 million in gross revenues.*

The firm management wants to recover all capital expenditures within three years of operation. They are using 5-year ACRS depreciation on all fixed assets and amortization where allowed. Ignoring the time value of money, estimate the fourth year gross and net profits, cash flow, net taxable income, income tax (at 38% of the net taxable income) and capital recovery during that year.

Solution: 3.4 *(in millions of dollars)*
GP = S-C 20 - 6 = $14 M
I_S = *0.5 + 0.8 = $1.3 M*
I_L = *$0.25 M*
$I_D = 1.5 + 4.5 =$ *$6 M*
$I = I_S + I_L + I_D$ = *1.3 + 0.25 + 6 = $7.55 M*
Capital Recovery = CR = 7.55/3 = $2.52 M
$NTI = GP - dI_D - eI_S$ = *14 - 1.3/5 - 0.15(6) = $12.84 M*
Taxes = 0.38(NTI) = 0.38(12.84) = $4.88 M
CF = GP - Taxes = 14 - 4.88 = $9.12 M
NP = CF - CR = 9.12 - 2.52 = $6.6 M
Note: This analysis ignores any Equipment Replacement Costs during the first year of operation.

Example: 3.5 *A company raised $1550k in stock sales to invest in a new venture that will have a full-depreciable installed cost of $1200k. Startup will require an initial inventory of raw materials at $250k. An income tax credit amounting to $25k will be realized at the end of the second year of operation. The venture will be operated four years and then shut down with no salvage value. Capital recovery is uniform over the four years of operation. Receipts, year end inventories, and expenses, excluding depreciation are:*

Year	*Receipts (k$)*	*Expenses (k$)*	*Inventories (k$)*
1	0	0	250
2	1500	1000	300
3	1700	900	200
4	1400	850	250
5	780	400	0

Working and fixed capital are invested in the first year. ACRS depreciation is used for tax purposes with a useful life of three years. The income tax rate is 46%. Develop a Balance Statement for this project, and an Income & Expense Statement. Generate a Cash Flow Analysis table. Then calculate the discounted cash flow rate of return for this project. Capital recovery is uniform over the four years of operation.

Solution: 3.5 *The next year beginning working capital equals the year end working capital. The beginning The change in working capital equals the year end working capital less the beginning working capital in the same year:*

$$\Delta\,Working\ Capital = \Delta Fixed\ Capital + \Delta Inventory + Gross\ Profit - Income\ Tax \tag{3.26}$$

Where all terms are evaluated over the same year and $\Delta WC = WC_j - WC_{j-1}$ *is the change in working capital from the end of the* $j-1$*th year to the end of the* j*th year.*

$$End\ Cash\ Balance = Beginning\ Cash\ Balance + \Delta\,Working\ Capital \tag{3.27}$$

The requisite tables are listed below. The cash flow internal rate of return is determined from:
$\sum_{j=1}^{5} CF_j/(1+i)^j = 0$, *and by trial-and-error,* $i = 0.1493$.
Using these equations to evaluate the cash flows for this problem results in the following table.

Cash Flow Analysis ($1000)

Year End	0	1	2	3	4	5
1. Sales, S			1500	1700	1400	780
2. Costs, C			1000	900	850	400
3. Gross Profits			500	800	550	380
4. ACRS Factors			0.25	0.38	0.37	0
ACRS Depreciation			300	456	444	0
5. Net Taxable Income			200	344	106	380
Income Tax Credit				25		
6. Net Income Tax			92	133	49	175
7. Fixed Capital, I_d		1200	1200	1200	1200	1200
8. Inventories		250	300	200	250	0
9. Cash Flow		-1450	358	767	451	455
10. Year End Working Capital	1550	100	458	1225	1676	2131
11. Beginning Cash Balance		1550	100	458	1225	1676
12. Ending Cash Balance		100	458	1225	1676	2131
13. Capital Recovery			300	300	300	300
14. Net Profit			58	467	151	155

Balance Statement ($1000)

Year End	1	2	3	4	5
Current Assets					
Cash	100	70	925	1376	1831
Inventories	250	300	200	250	0
Total Current Assets	350	370	1125	1626	1831
Fixed Assets					
Startup Expenses	0	0	0	0	0
Total Fixed Assets	1200	1200	900	444	0
Annual Depreciation		300	456	444	0
Net Fixed Assets	1200	900	444	0	0
Total Assets	1550	1270	1569	1626	1831
Liabilities & Stockholders Equity					
Current Liabilities	0	0	0	0	0
Long-Term Liabilities	0	0	0	0	0
Total Liabilities	0	0	0	0	0
Stockholders Equity					
Common Stock	1550	1550	1550	1550	1550
Retained Earnings	0	-280	19	76	281
Total Stockholders' Equity	1550	1270	1569	1626	1831
Stock Value ($/share)	1.00	0.82	1.01	1.05	1.18
Total Liability & Equity	1550	1270	1569	1626	1831

Income & Expense Statement ($1000)

Year End	*1*	*2*	*3*	*4*	*5*
Revenues					
Sales	*0*	*1500*	*1700*	*1400*	*780*
Gross Income	*0*	*1500*	*1700*	*1400*	*780*
Costs and Expenses					
Total Costs	*0*	*1000*	*900*	*850*	*400*
Gross Profits	*0*	*500*	*800*	*550*	*380*
Taxes on Income	*0*	*92*	*133*	*49*	*175*
Cash Flow (CF_j)	*-1450*	*358*	*767*	*451*	*455*
Capital Recovery	*0*	*300*	*300*	*300*	*300*
Net Profits (NP_j)	*0*	*58*	*467*	*151*	*155*
Total Dividends	*0*	*0*	*0*	*0*	*0*
Retained Earnings					
Beginning of year	*0*	*0*	*-280*	*-261*	*-185*
Current year	*0*	*-280*	*19*	*76*	*281*
End of year	*0*	*-280*	*-261*	*-185*	*96*
Venture Profits[a] (VP_j) ($i_{min} = 0.06$)	*-150*	*-92*	*317*	*1.2*	*5.2*

[a] $VP_j = NP_j - i_{min} I$

3.3 Total Product Cost

The capital investment required to build a chemical plant is a major contributor to the total product cost, but other factors (e.g, chemical plant operating costs and product selling costs) are also important and should be considered as part of the total product cost. Peters and Timmerhaus [76] suggest breaking down the total product costs into two general categories: (1) manufacturing costs and (2) general expenses. The former costs are sometimes referred to as operating or production costs, and it is important to consider all factors that may be important in order to generate an unbiased estimate of the total product cost. Typically, total product costs are figured on an annual basis to smooth out seasonal effects.

The best information source for accurate estimation of total product cost are actual experimental data, say based on previous corporate experiences with similar products. In the absence of such data, and for purposes of conceptual design, total product costs can be estimated using the approximation methods described below and adapted from Peters and Timmerhaus [76].

Table 3.5 summarizes the approximate contributions to an estimate of the total capital investment (TCI) required to construct a manufacturing facility to produce chemical products. The basis for the TCI estimate is usually the nominal annual production rate for the facility, say operating 320 days per annum. (Some days must be allowed for maintenance & repair of equipment.) The nominal operating days also depend upon how the plant is to be operated. Small volume plants, producing high value added products, may be operated only five days per week, 40 hours per week, or less. High volume plants, producing commodity products with relatively low value added, are more commonly operated 24 hours per day and with as little downtime per annum as possible. For a conceptual design, it is not uncommon to assume the plant operates 320 days per annum, 24 hours per day, to obtain a nominal design basis of 7680 operating hours per annum which is then used to compute a nominal production rate per hour, based on the desired annual production rate.

Example: 3.6 *A company wishes to produce 100 million lbs per year of a chemical commodity. Estimate the nominal throughput on an hourly basis.*
Solution: 3.6 *Assume the plant operates continuously for 320 days per year. The hourly rate is then*
$Rate = 100 \times 10^6/(320 \times 24) = 13,021$ *lbs/hour.*
This production rate is then used to develop the process material and energy balances, size equipment, and estimate the total installed cost of the facility.

Some of the terms in Table 3.5 are described in more detail in Chapter 4. In particular, Chapter 4 describes how the Total Direct and Indirect Costs in Table 3.5 are computed for each process module, using factors that are appropriate for that piece of equipment. In addition to the battery limits direct and indirect costs, the bare module costs for process and auxiliary buildings, and service facilities and yard improvements should be added to the estimated costs for any land purchases to obtain the Bare Project Cost.

Siting a chemical plant can also be a significant cost item. Unfortunately, chemical plants have something in common with nuclear power plants. They both suffer from the "not in my back yard" (NIMBY) syndrome. Public perceptions toward chemical facilities usually are not favorable and, as a consequence, significant hurdles must often be overcome before a new chemical plant can be located. For this reason, it is often preferable to co-site a new facility with an existing plant if space is available, or else to seek space adjacent to other chemical manufacturing operations.

Expenses associated with locating a new plant site include searches for available space, meetings with local officials and prospective neighbors, presentations of plant manufacturing operations in town meetings, payments to public relations firms, licensing and permitting costs, and environmental impact studies. At least in North America, chemical plants are required to meet strict emissions and waste management standards and management must be able to clearly explain hazards and emissions issues to prospective neighbors. Such activities typically costs \$10,000–\$500,000 (1998\$) and may be as high as 1–2% of the Total Installed Cost of a new chemical plant, in addition to the actual cost of land acquisitions. Acceptance for a new chemical plant requires convincing evidence that management and operating staff are competent and experienced, that management personnel at the proposed facility are cooperative, honest, and communicative with the host community, that

Table 3.5: Approximate contributions to capital investment costs based on individual components. This information should only be used verify estimates for general accuracy. The percentages listed are not recommended as the basis of a cost estimate. The table basis is for the construction of facilities to produce product at a specified annual rate. Adapted from Peters and Timmerhaus [76].

TOTAL DIRECTS[a] **(TD)** = Battery Limits Directs+ Buildings + Service Facilities & Yard Improvements + Land, often ≈ 70–85% of the TOTAL INSTALLED COST (TIC)

Battery Limits Directs[a] = Purchased Equipment (f.o.b.)[b] + Installation + Instrumentation & Controls + Piping + Electrical, often ≈ 50–60% of TIC.)

Purchased Equipment Cost (PEC), (f.o.b.)[b], often ≈ 15–40% of TIC

Installation[c], including insulation and painting, often ≈ 25–55% of the PEC[c]

Instrumentation & controls, installed, often ≈ 6–30% of the PEC[c]

Piping, installed ≈ 10–80% of the PEC[d]

Electrical, installed, often ≈ 10–40% of the PEC[c]

Buildings, process and auxiliary, often ≈ 10–70% of the PEC[c]

Service facilities & yard improvements, often ≈ 40–100% of the PEC[c]

Land, often ≈ 1–2% of TIC, or ≈ 4–8% of the PEC[c]

TOTAL INDIRECTS (TI) = expenses which are not directly involved with material and labor cost of installing the complete facility, often ≈ 15–30% of the TIC.

Engineering and supervision, often ≈ 5–30% of the TOTAL DIRECTS

Construction expense, often ≈ 6–30% of the TOTAL DIRECTS

BARE PROJECT COST[e] = TOTAL DIRECTS + TOTAL INDIRECTS

CONTINGENCY & FEES[f], often ≈ 5–18% of the BARE PROJECT COST

TOTAL INSTALLED COST[g] = BARE PROJECT COST + CONTINGENCY & FEES

WORKING CAPITAL, often ≈ 11–25% of the TOTAL INSTALLED COST

TOTAL CAPITAL INVESTMENT = TOTAL INSTALLED COST + WORKING CAPITAL + START UP COSTS

---

[a] Estimate using Guthrie's method. See Section 4.3 for details.

[b] Free On Board (f.o.b.) The cost after packaging and loading on a common carrier at the vendor's site.

[c] low-pressure, carbon-steel basis. Estimate using methods in Section 4.3.

[d] Seventy percent of the piping costs are for process piping which may require special materials of construction.

[e] Includes both the battery limits and the offsites.

[f] Use higher percentages for conceptual estimates, typically ≈ 20–40% of TIC

[g] Also known as the FIXED CAPITAL INVESTMENT (FCI)

emissions and waste management will not be significant issues for neighbors, and that the new facility will offer new high-paying jobs and contribute significantly to the tax digest at the new location.

The Total Installed Cost for the facility is obtained by adding Contingency and Fees to the Bare Project Cost estimate. The Contingencies and Fees listed in Table 3.5 refer to those incurred during the construction phase of the project. Because of the complexity in designing a chemical plant, additional funds are put aside to cover unexpected costs during plant construction. Contingency funds are added to adjust for unlisted pieces of equipment or insufficient scope definition, typically 10–20%, but for an early conceptual design should perhaps be increased to 20–40%, depending upon the degree of detail that is available when the cost estimate is performed. Contractors fees (e.g., for an experienced architect/engineering firm) are typically in the range of 3–5% of the Bare Project Module Cost estimate. Hence, Contingency and Fees for a very preliminary design should be around 35%, but may be reduced to around 18% when sufficient experimental data are available to support the process design and firm estimates of the most expensive equipment items in the major equipment list are available.

3.3.1 Working Capital

Working Capital in Table 3.5 is required to initially startup and operate the process. After production begins, some product is accumulated before shipments to buyers actually begin. Actual payments from buyers are not received until after product shipment, the buyer has been invoiced, and some additional time period until payment is received (typically up to 30 days). As a consequence, working capital is needed to meet actual expenses incurred prior to receiving payments from buyers.

Working capital includes funds invested in raw materials and supplies (e.g, a 30 day inventory), finished products and inventories of chemical intermediates, accounts receivable, cash requirements to meet actual operating expenses (e.g., payroll, taxes, and raw material and packaging purchases), and any other funds required for accounts payable (usually short-term liabilities, payable within 30 days). The working capital typically is 11–25% of the Bare Project Cost. For conceptual design purposes, it is sufficient to use a percentage of the Bare Project Cost to estimate working capital. In subsequent design studies, working capital should be estimated based on projected inventory requirements for raw materials and products, as well as more careful estimates of actual payroll and projected expenses. Many of these factors cannot be known precisely until the actual plant location is established.

3.3.2 Manufacturing Costs

Manufacturing costs may be divided into Direct Production Costs, Fixed Charges, and Plant Overhead Costs [76]. Table 3.6 summarizes the major contributing factors to these three elements in manufacturing costs. Direct production costs typically include raw materials, operating labor, direct supervisory and clerical labor, utilities, maintence and repairs, operating supplies, laboratory charges, and patent and royalty expenses. Each is discussed below. Typically, Directe Production Costs are about 50–70% of the Total Product Cost (TPC), but individual contributions should be estimated using the guidelines described below and in Table 3.6.

Direct Production Costs

These costs are a function of the production rate. Typically, they are 50–70% of the Total Product Cost.

Raw Materials Raw material costs vary with the product, but typically range from 10–50% of the Total Product Cost. Usually, they are a major portion of the cost and can be used to determine a Profit Upper Bound (PUB) (see Section 12). Even in a preliminary design, raw material costs should be based upon the best available material balance estimates in hand at the time, preferably upon experimental data, but at least realistically estimated with a process flow diagram. Raw material costs are typically a higher percentage of the Total Product Cost when producing high-volume, low-value-added products. They are often a smaller percentage when producing high-value added, specialty products.

Costs for raw materials should be based upon commodity prices (e.g., *Chemical Marketing Reporter*) or from supplier quotes. Substantial, high-volume purchases of raw materials, particularly for an extended time periods, can often be obtained at prices below those quoted in the commodities market, but firm quotes typically require negotiations and contractual arrangements. When estimating raw material costs, one should also consider possible seasonal effects (e.g., as with gasoline or fuel oil products), and it may be worthwhile to base estimates upon annual averages in the commodities markets rather than from a single weekly quote. The *Wall Street Journal* can be helpful in obtaining quotes of some raw materials (e.g., fuels and materials such as metal bar stocks. If commodity price estimates cannot be found, preliminary estimates of a specialty chemical price in bulk may be estimated using Equations 2.34 or 2.37.

Freight and transportation charges should be added to raw material cost estimates, based upon the plant and supplier locations and the form in which the raw materials are to be shipped and stored at the plant site. Bulk shipments are less expensive than for smaller quantities, but the former case requires more on-site storage capacity which increases the total installed cost of the plant. Thus, a balance is required. In some cases, storage costs can be reduced by renting storage equipment for on-site use (e.g., ships, freight cars or trucks) and paying demurrage. In this way capital investment for storage may be reduced, but at the expense of higher operating costs. This approach may also be preferred for safety reasons if it can reduce on-site inventories of hazardous materials and avoid the purchase of higher-priced containment vessels.

Several considerations are important when deciding how much on-site storage to provide. On the one hand, larger less-frequent shipments have the advantages of reducing transportation costs, the hazards associated with shipping, and the visibility of the plant to its neighbors. On the other hand, smaller more-frequent shipments have the advantages of reducing on-site inventories and storage costs, and may reduce on-site operational risks (e.g., from uncontrolled releases of toxic

Table 3.6: Approximate contributions to the annual manufacturing cost for chemical products. This information should only be used verify estimates for general accuracy. The percentages listed are approximate. The basis is one year of production. Adapted from Peters and Timmerhaus [76].

MANUFACTURING COST = Direct Production Costs + Fixed Charges + Plant Overhead Costs

---

DIRECT PRODUCTION COSTS = Raw Materials + Operating Labor + Direct Supervisory & Clerical Labor + Utilities + Maintenance & Repairs + Operating Supplies + Laboratory Charges + Patents & Royalties + Catalysts & Solvents + Waste Management, often $\approx$ 55–75% of the TOTAL PRODUCT COST (TPC).

- **Raw Materials,** estimate from material balances and projected costs (e.g., from the *Chemical Marketing Reporter*), often $\approx$ 10–50% the TPC
- **Operating Labor (OL),** estimate from Stein correlation or from plant complexity, often $\approx$ 10–20% of the TPC
- **Direct Supervisory & Clerical Labor (DS&CL),** estimate from Operating Labor, often $\approx$ 10–20% of the Operating Labor
- **Utilities,** estimate from energy balances and internal prices for utilities, often $\approx$ 10–20% of the TPC
- **Maintenance & Repairs (M&R),** estimate from TIC (see Table 3.5) & plant complexity, $\approx$ 2–10% of the TIC
- **Operating Supplies,** estimate from M&R or from TIC, usually $\approx$ 2–10% of the M&R or $\approx$ 0.5–1% of the TIC
- **Laboratory Charges,** estimate as $\approx$ 10–20% of the Operating Labor
- **Patents & Royalties,** estimate as $\approx$ 0–6% of the TPC
- **Catalysts & Solvents,** estimate from approximate makeup requirements and costs. Typically, 1–5% of inventory per year must be replaced, but varies with specific application.
- **Waste Management,** estimate from losses, toxicities, and cost of appropriate disposal. Typically, 5–10% of the Raw Material Costs.

FIXED CHARGES = Replacement Costs + Local Taxes + Insurance + Rent + Financing, often $\approx$ 10–25% of the TPC (see Table 3.7)

- **Replacement Costs,** estimate from the life period, salvage value, and method of calculation, often $\approx$ 5–15% of the TIC for machinery & equipment and $\approx$ 2-3% of the TIC for buildings
- **Local Taxes,** estimate from the local tax rate, often $\approx$ 1–4% of the TIC
- **Insurance,** estimate from vendors, often $\approx$ 0.4–1% of the TIC
- **Rent,** estimate from actual use, often $\approx$ 8–12% of the value of rented land & buildings
- **Financing,** estimate interest paid from assumed financing schedule, often $\approx$ 0–10% of the TOTAL CAPITAL INVESTMENT (TCI)

PLANT OVERHEAD COSTS include general plant upkeep, payroll and other overhead, packaging, medical services, safety & protection, salvage operations, laboratories, and storage facilities. Typically, $\approx$ 50–70% of the OL+ DS&CL + M&R, or $\approx$ 5–15% of the TPC.

materials), but smaller inventories may also result in unplanned downtime if raw material shipments to the plant site are late and neighbors will observe more traffic into and out of the plant site. Since all of these considerations are important, some compromise is required. Often it is adequate to base a preliminary design around the size of bulk carriers and to assume shipments every one to two weeks. Storage of a 30 day inventory on-site may be desirable, but often uneconomical. Preliminary calculations based on material balance considerations can be used to identify the major concerns.

Operating Labor Both skilled and unskilled labor are required at most chemical plants. Estimates for hourly rates can be obtained from the U.S. Bureau of Labor publication, *Monthly Labor Review*. Typically, operating labor is about 15% of the Total Product Cost. The Stein correlation (see Section 6) can be helpful in some instances. If labor requirements for an existing plant are known, a power rule of $\approx$ 0.2–0.25 may be helpful to adjust for changes in plant capacities [76]. When a process flow diagram is available, operating labor may be estimated by counting unit operations and analyzing the operating labor for each step. Generally, more modern plants are designed to minimize operating labor by utilizing centralized process control schemes and a high degree of automation. In this way, a single operator can monitor multiple complex operations that are controlled from a single area and, in many instances, tasks are completed through a programmed sequence of steps (e.g., for batch processes). Additional information on labor cost estimation are provided by Johnnie and Aggarwal [48], Kiley [51], Wessel [117], and Peters and Timmerhaus [76]. The *Engineering News-Record* labor indices provide indices to permit estimation of prevailing wage rates by location.

Direct Supervisory & Clerical Labor Direct Supervisory & Clerical Labor is usually required in manufacturing operations. The exact amount depends upon the complexity of the operation, product quality standards, and corporate philosophy. Typically, the cost is about 15% of the operating labor cost.

Utilities Utility costs should be estimated from energy balances and the conceptual process flow diagram. They can be figured from estimates of hourly usage rates, hours per year in operation, and a price for the utility. For conceptual studies where capital investment estimates are based upon the battery limits, it is helpful to use internal price estimates for utilities (e.g., in Btu transferred either to or from the process). Typical values are given in Section 6. Internal prices for utilities are based on estimated Total Product Costs for producing the utility on the plant site. In addition to the costs for fuels and other energy sources, they all of the manufacturing, fixed charges, and plant overhead costs associated with the production of the utility.

Maintenance & Repairs Maintenance and repair expenses include costs for labor, supervision and labor. Annual costs typically vary from 2–10% of the total installed cost of the chemical plant. Costs vary with the unit operations and process complexity. The service environment is also a factor that determines fouling and corrosion rates, and the frequency of maintenance and repair. Some down time is usually provided for routine and preventative maintenance. Unplanned outages may also occur whenever equipment breaks down. As a general rule, unplanned outages tend to increase with the amount of mechanical equipment (i.e., with moving parts) in the process, but process chemistry is also an important factor. Peters and Timmerhaus [76] provide additional guidance. Even with a good M&R program, it is still necessary to replace equipment because it wears out. Replacement costs are in addition to M&R costs.

Operating Supplies Miscellaneous supplies are also required to operate a manufacturing process. These supplies are not raw materials, or maintenance or repair supplies, but they are consumables required to operate the process. Examples include lubricants, test chemicals, charts, custodial supplies, batteries, etc. On an annual basis, such costs are typically 15% of the total costs for Maintenance & Repairs [76].

Patents & Royalties Patent & Royalty fees are payable whenever portions of a process use patented techniques. Actual prices are determined by negotiations with the patent owners. For conceptual designs, such fees may be estimated as 0–6% of the Total Product Cost [76].

Catalysts & Solvents Catalysts and solvents are normally recycled in processes and, therefore, are not raw materials. They may be excluded from initial Profit Upper Bound (PUB) analyses (see Chapter 2). Conceptual cost estimates should consider the costs of replacing catalysts lost from attrition or irreversible poisoning. Solvent losses can easily make extraction processes economically unattractive. Consequently, conceptual designs utilizing solvent extraction frequently include steam stripping of raffinates to recover as much solvent as possible, but losses still occur. As a first approximation, one may assume that 1–5% of a catalyst or solvent inventory must be replaced annually. However, it is preferable to base makeup costs on actual experimental data.

Fixed Charges

Fixed charges are those which are a function of the Total Installed Cost. They do not vary with the amount of production and, therefore, are described as fixed charges. Examples of fixed charges include depreciation, insurance, rental costs, local property taxes, and financing charges for borrowed capital. As a rough approximation, they are about 10–25% of the Total Product Cost.

Replacement Costs Equipment, buildings and other material objects that are used in a manufacturing operation are part of the Total Capital Investment for a manufacturing operation. Equipment items often wear out and must be periodically replaced over the life of the project. Replacement costs depend upon the effective working life of the item, its salvage value (if any), and the actual cost of replacing the equipment. Typically, replacements are scheduled during normal down time. Otherwise, additional expenses result from plant outage and reduced production rates.

Replacement Cost (RC) estimates should be based on the actual rates at which equipment, or other fixed assets, must be replaced due to wear and tear. Eventually, these costs should be based on operating experience, but during the Preliminary Design Phase, experimental rates of equipment failure are not always available, and failure rates vary, depending upon the unit operation and the service. The annual replacement rate for machinery and equipment is often about 10% of the Fixed Capital Investment. Replacement costs for buildings are often assumed to be about 3% per annum.

During Preliminary Design, Replacement Costs (RC) may be estimated using the straight line formula and engineering judgement. In this case, a stated effective lifetime of n years is used to decrease the asset value by equal increments over its estimated lifetime. The RC in the jth year is simply:

$$RC = \frac{j}{n} I_o, \text{with } j < n \tag{3.28}$$

However, the property may have a final salvage value at the end of its life. In this case, the RC in the jth year is:

$$RC = \frac{j}{n} (I_o - I_{sv}) \tag{3.29}$$

where: I_o = the original purchase price for the property
I_{sv} = the sales price or *salvage value* at the end of its service life
n = the expected service life

Using the straight line method, equal amounts are charged for RC each year. The present or net asset or book value (P_v) in the ith year may then be calculated as:

$$P_v = I_o - RC \tag{3.30}$$

The sum-of-the-years-digits method is another estimation technique for RC that may be used. It estimates larger costs per year than the straight line method, and the RC more realistically parallels actual losses in many cases (e.g., the blue book value of a car). For the jth year, the RC by this method is:

$$RC = (I_o - I_{sv}) \left[1 - \frac{(n-j+1)}{\sum_{j=1}^{n} j}\right] = (I_o - I_{sv}) \left[1 - \frac{2(n-j+1)}{n(n+1)}\right] \tag{3.31}$$

where $j \leq n$.

Example: 3.7 *A heat exchanger costs $100,000 to install in 1987 and it has an estimated life of 5 years with a final salvage value at the end of that time of $5,000. Compute the first year depreciation that can be deducted from Gross Profits using ACRS. Estimate the annual Maintenance & Repairs for this exchanger. Estimate* *the replacement cost for this exchanger in the 3rd year of service using both straight line and sum-of-the-years-digits.*

Solution: 3.7

Using 5-year ACRS, $d = 0.15$ *from Table 3.4. Therefore, in the first year:*
Depreciation $= d_1(I_o) = 0.15(100,000) = \$15,000$

From Figure 3.6, Maintenance & Repairs $= 0.06(100,000) = \$6,000$ *per annum.*

Using straight-line to estimate Replacement Costs in the 3rd year:
$RC = (3/5)(100,000 - 5,000) = \$57,000$
Using sum-of-the-years-digits:
$h = (1 - 2(5 - 3 + 1)/(5(6))) = 0.80$
$RC = h(I_o - I_{sv}) = 0.80(95{,}000) = \$76{,}000$

Financing Financing expenses consists primarily of interest paid on borrowed funds. They are typically 0–10% of the Total Capital Investment, but are best computed using a per annum interest rate, an estimate of the amount of capital that must be borrowed, and a time period for paying off the loan. From Section 7, the present worth of a sum P is related to a future annual, uniform cash flow A by:

$$P = \frac{A}{i}\frac{(1+i)^n - 1}{(1+i)^n} \tag{3.32}$$

where n is the number of years to pay off the loan, and i is the interest rate. The total amount paid over n years is nA, if one payment is made at the end of each of n years. So the Total Finance Charge or interest paid over the life of the loan is $(nA - P)$ and

$$nA - P = P\left[\frac{(ni-1)(1+i)^n + 1}{(1+i)^n - 1}\right] \tag{3.33}$$

Example: 3.8 *A chemical company must borrow \$10 million for 20 years at 5% p.a. to pay for a new chemical plant. The loan will be repaid in 20 equal payments. For purposes of analysis, depreciation is taken as straight line over 20 years with zero salvage value at the end of that time. Determine the annual depreciation, average annual interest payment, and total finance charge for the loan.*

Solution: 3.8 *Annual depreciation is* $10 \times 10^6/20 =$ *\$500,000 p.a.*
Total cost of the loan $= nA = nPi(1+i)^n/[(1+i)^n - 1]$
$nA = 20\left[0.05P(1.05^{20})\right]/(1.05^{20} - 1) = 1.6048P$
Total finance charge $= nA - P = 1.6048P - P = 0.6048P$
Average annual interest payment $= A - P/n = 0.6048P/20 = 0.03024P$
So the total paid back is \$16.048 $\times 10^6$
The total interest paid is \$6,048,000
The average annual interest payment is \$302,400.
The depreciation and average interest payment over the life of the loan is $(500,000 + 302,400)/10 \times 10^6 x 100\% = 8.029\%$ *p.a. of the loan amount.*

Property Taxes Property taxes can only be accurately estimated after the exact plant location has been determined. Rates vary with locality, but are typically 1–4% per annum of Fixed Capital Investment.

Property Insurance Insurance rates depend upon the type of process that will be operated. Risks are greater whenever toxic or hazardous chemicals are used, and if significant excursions in temperature or pressure are required. Fire hazards are also a significant consideration when insurance rates are determined. Typically, insurance rates are 1% per annum or less of the Total Installed Cost.

Rental Costs Annual rental costs are often about 8–12% of the value of the rented property. Demurrage for on-site use of transport vehicles is one of the more common rental charges, but such costs should only be included if they are part of the conceptual design. Often preliminary estimates for demurrage can be obtained by contacting the manufacturer who will provide the raw material in question. As a product manufacturer, you may consider purchasing a fleet of transport containers for rental by your end-users too, if that approach seems reasonable as part of your overall marketing strategy.

Plant Overhead Costs

Manufacturing operations incur costs that are not directly related to actual production. These expenses fall into the category of Plant Overhead Costs and are summarized in Table 3.6. Typical plant overhead expenses include: cafeteria facilities, control laboratories, hospital and medical services, general engineering, general plant maintenance, payroll overhead (i.e., the employer portions of social security, medicare, and employee benefits), product packaging, plant security, fire and safety equipment, janitorial services, employment offices and personel recruiting, utility distribution, plant shops, electrical services, communications, warehouses, shipping and receiving facilities.

Some of these activities, particularly product packaging, warehouses, shipping and receiving facilities, may be detailed as line items in a conceptual design even though they are not part of the battery limits. In other cases they are factored. Typically, plant overhead is 50–70% of the total expense for Operating Labor (OL), Direct Superviory & Clerical Labor (DS&CL), plus Maintenance & Repairs (M&R).

General Expenses

The Total Product Cost also includes General Expenses (GE) in addition to the Manufacturing Costs from Tables 3.5 and 3.6. These GE costs are summarized in Table 3.7. Major contributors are Administrative Costs, Distribution & Selling Costs, Research & Development Costs, and GE Contingency.

Administrative Costs Expenses associated with the upper levels of management at a plant site are not usually considered a manufacturing cost, but rather

Table 3.7: Approximate contributions to the annual Total Product Cost (TPC) for chemical products. This information should only be used verify estimates for general accuracy. The percentages listed are approximate. The basis is one year of production. Adapted from Peters and Timmerhaus [76].

TOTAL PRODUCT COST (TPC) = Manufacturing Cost (see Table 3.6) + General Expenses

---

GENERAL EXPENSES (GE) = Administrative Costs + Distribution & Selling Costs + Research & Development Costs + GE Contingency

Administrative Costs include executive salaries, clerical wages, legal fees, office supplies & communications, often ≈ 2–6% of the TPC or ≈ 10–20% of the OL + DS&CL + M&R (see Table 3.6)

Distribution & Selling Costs, includes costs for sales offices, salespersons, shipping & advertising, often ≈ 2–20% of the TPC

Research & Development Costs, ≈ 2–5% of the TPC

GE Contingency, added to account for unforeseen events (e.g., strikes, floods, & market uncertainties), ≈ 1–5% of the TPC. Note: This GE Contingency is in addition to that Contingency added to the Fixed Capital Investment to account for uncertainties in capital expenditures.

GROSS PROFITS (GP) = Total Income (Sales Revenues) - Total Product Cost (TPC)

CASH FLOW (CF) = Gross Profits - Income Taxes (IT)

NET PROFITS (NP) = Cash Flow - Capital Recovery (CR)

RETAINED EARNINGS (RE) = Net Profits - Dividends Paid to Investors

Income Taxes (IT), estimate from Net Taxable Income and Income Tax Rate, typically ≈ 20–40% of Gross Profits (see Section 3)

fall into the category of an administrative cost. They should be included if the overall economic analysis is to be complete. Administrative expenses include charges for salaries and wages of adminstrators, secretaries, accountants, and other administrative workers, costs for office supplies and equipment, communications, and administrative buildings which may or may not be located at the actual plant site. For a quick estimate, such costs should be estimated as 20–30% of the Operating Labor (OL).

Distribution & Selling Costs Expenses that should be included in this category include salaries, commissions, and travel expenses for sales personnel, sales office salares, wages and supplies, shipping expenses, container costs, advertising, and technical sales support services. Distribution costs vary widely. For a conceptual design, they may be assumed as 2–20% of the Total Product Cost. These costs decrease on a percentage basis as the product volume increases.

Research & Development Costs These costs include salaries and wages for research personnel, laboratory space and equipment, supplies and other similar costs. Typically, research and development costs are about 2–5% of the annual sales.

Development costs may include pilot testing at vendor sites to obtain process guarantees. In many cases, equipment manufacturers will offer process guarantees if they complete the final design of a piece of equipment, based upon their pilot studies. This procedure is recommended to improve certainty in the final design and to facilitate startup, but it requires production of sufficient quantities of process streams so that samples can be shipped to the equipment manufacturer's site and tested. The amounts required vary with the unit operation, but frequently are less than 55 gallons for liquids. Testing for drying operations (e.g., with fluidized bed or rotating drum dryers) may require larger quantities of solids.

GE Contingency This Contingency is different from that listed in Table 3.5. The Contingency listed in Table 3.5 is added to compensate for uncertainty in actual construction costs when building the manufacturing facilities. Here the GE Contingency is added to compensate for uncertainties in the actual Total Product Cost. Unforseen events that may affect actual manufacturing costs include strikes, hurricanes, raw material price fluctuations, variations in fuel prices, and the like. These uncertainties are accounted for my adding 1–5% contingency to the Total Product Cost.

Capital Recovery The total cost of plant construction is usually recovered over time (i.e., written off as a manufacturing expense). Manufacturing costs are therefore higher during the time over which capital expenditures are recovered, but are lower when the manufacturing facilities are paid for. Partly for this reason, commodity prices for chemicals tend to decrease over time.

The Internal Revenue Service defines depreciation rates for purposes of calculating Net Taxable Income (see Section 3.2.5). It is usually economically desirable to recovery capital expenditures as quickly as possible, and for this reason it is not unusual to write off manufacturing investment costs in three to five years, even if the effective plant life is 10 years or more, or may be realistically considered to be a perpetuity. The feasibility of a short write-off depends upon the Total Installed Cost of the facilities as well as the anticipated annual Cash Flow.

Example: 3.9 *Estimate the Total Product Cost per pound of product based on the following information. The Total Capital Investment for the manufacturing plant was $10 million when the CE Index was 402, and it will produce 10 million lbs/year of product. Raw materials cost $1.23/lb and utilities are $0.52/lb. Average labor costs for Operating Labor and Direct Supervision & Clerical Labor is $30/hour. There are no rental charges, nor are there any solvent or catalyst costs. All other costs may be estimated using Tables 3.5 through 3.7 assuming average percentages and the Stein correlation.*

Solution: 3.9 *Use the Stein correlation to estimate OL + DS&CL for the plant. Assume that the Working Capital is 18% of the Total Installed Cost.*

$TIC = 10/1.18 = 8.4745$

$I = 8.4745(228.8/402) = 4.82334$

$P = 7I^{0.88} = 28$ *employees*

*Annual OL +DS&CL Payroll = ($30/h)(28 emp)(52 weeks/yr)(40 h/week) /*$(10 \times 10^6 lb/year)$ *= $ 0.174/lb*

$DS\&CL = 0.15OL$, *OL = 0.17443/1.15 = $0.152/lb*

DS&CL = $0.022/lb

TCI = TIC + WC = TIC + 0.18TIC = $ 10 $\times 10^6/(10 \times 10^6)$ *lbs = $1.00/lb*

TIC = 1.00/1.18 = $0.8475 /lb

WC = 0.18TIC = 0.18(0.8475) = $0.1525/lb

Maintenance & Repairs (M&R) $= 0.06(TIC) = 0.06(0.8475) =$ $0.0509/*lb*

Operating Supplies (OS)= $0.06(M\&R) =$ $0.0031 */lb*

Laboratory Charges (LC)= $0.15OL = 0.15(0.152) =$ $0.02275 */lb*

Solvents & Catalyst = 0.0

Patents & Royalties (P&R) = 0.03TPC

Waste Management = 0.075(RMC) = 0.075(1.23) = $0.0922

Direct Production Costs (DPC) less Patents & Royalties = 1.23 + 0.52 +0.1744 +0.0509 + 0.0031 + 0.02275 + 0.0922

DPC - P&R = $2.09335/lb

Replacement Costs = 0.10 TIC = $0.085/lb

Local Taxes = 0.025 TIC = $0.021/lb

Insurance = 0.007 TIC = $0.006/lb

Rent = 0

Financing = 0.05 TCI = $0.05/lb

Fixed Charges = 0.085 + 0.021 + 0.006 + 0.05 = $0.162/lb

Plant Overhead = 0.6(OL+DS&CL + M&R) = 0.6(0.15168 + 0.02275+ 0.051) = $0.13517/lb

Manufacturing Cost(MC) - P&R = 2.09335 + 0.162 + 0.13517 = $2.54177/lb

MC = $2.54177 + 0.03 TPC

Administrative Costs = 0.15(OL + DS&CL + M&R) = 0.15(0.17443+0.051) =

$0.034/lb
Distribution & Selling Costs (D&S) = 0.11 TPC
Research & Development (R&D) = 0.035 TPC
GE Contingency = 0.03 TPC
TPC = MC + GE = 2.39052 + 0.03 TPC + 0.034 + 0.11 TPC + 0.035 TPC + 0.03 TPC
0.795 TPC = $2.42452/lb
TPC = $3.0497/lb

3.4 Summary

The accounting entity determines the environment in which a project must be evaluated. Small firms may not be able to fund large projects while large ones may not be interested in small projects. The balance and income statements for a corporation provide useful information that describes the economic project environment. Such statements determine the project tax rate, for example, and they define the availability of working capital for the project entity.

Gross profits, income taxes, and capital recovery expenses appear as line items in the income statement, and they must be clearly defined in order to complete project studies on either a net profit or cost basis. Net taxable income must be computed using the gross profits and depreciation, depletion, and amortization deductions as allowed by law. Capital recovery may be approximately equal to depreciation, but it is set by corporate policy rather than the income tax laws.

3.5 Problems

3.1 What differences exist between operating costs and capital investment?

3.2 What differences exist between process battery, auxiliary, working capital, start-up, and land investment?

3.3 Prepare a balance sheet for E. Z. Chemicals, Inc. using the following line-item entries:

Cash	$80,000	Stock sold	$180,000
ST payables	15,000	Equipment (TV)	90,000
Receivables	35,000	Buildings (TV)	350,000
Inventories	7,500	Treasury notes	100,000
LT payables	350,000	Land	32,000
Retained earnings	???	Depreciation	40,000

What are current and fixed assets, the liabilities, the retained earnings, and the stockholders' equity?

3.4 During the same year as in Problem 3, E. Z. Chemicals paid $250,000 in salaries, $178,000 for raw materials and utilities, $7,500 for shipping and freight. In that year their gross revenues were \$865,000, while their taxes were 42% of their net taxable income. Depreciation was their only tax deduction. Prepare an income statement for that year. During that year, their capital recovery charges were \$15,000 and retained earnings were \$20,000.

3.5 A chemical company is considering a project. It is estimated to cost \$1 M in start-up and require \$0.25 M working capital. The installed cost of the plant is \$8 M, and the land is already owned. The estimated annual revenues are \$15 M after full production is achieved with annual operating costs of \$8 M. What will be the first year gross and net profits, cash flow, net taxable income, income taxes (52% of NTI), and capital recovery expenses? The corporation wishes to recover capital in three years using straight line with 100% recovery of all costs.

3.6 A plant produces 10 million lb/yr of product X which sells for \$0.90/lb with operating costs of \$0.15/lb. The capital recovery and depreciation are equal at \$1 million/yr. The income tax rate is 52% of the net taxable income. Compute the gross profits, net taxable income, income taxes, cash flow and net profits. (Answer in millions of \$: $GP = 7.5$, $NTI = 6.5$, $IT = 3.38$, $CF = 4.12$, $NP = 3.12$)

3.7 A large corporation is operating with an income tax rate of 38%. It requires a minimum annual rate of return equal to 8% of the investment. A new \$44 million investment is under consideration. It can be fully depreciated over five years with an ACRS factor of 15% in the first year, and the investment must be recovered in 5 years using straight line. The projected gross sales associated with this project are \$74 million annually with an estimated total annual operating cost of \$23 million. Insurance and maintenance costs are 5% of the investment. In your analysis, ignore the time value of money. What are the estimated gross profits for this investment?

3.8 In Problem 7 what is the estimated net taxable income for the first year?

3.9 In Problem 7 what is the estimated cash flow during the first year?

Chapter 4

Capital Investment

Estimation of the total capital investment that is required to construct a chemical process or plant is an important engineering task which should be initiated as early as possible in any design study. Initially, the uncertainty in such estimates is often quite high, but it decreases as the project plans become finalized and more details about the project are precisely known. Typical levels of estimate uncertainty, developed by the American Association of Cost Engineers, are summarized in Table 9.6 as a function of the estimate type.

The definitions of the major items in the total capital investment (TCI) and the total installed costs (TIC) are summarized from Chapter 3:

$$\begin{aligned} I &= TCI = I_P + I_A + I_L + I_S + I_W \\ I &= TCI = I_F + I_W \\ TIC &= I_P + I_A + I_L = TCI - I_W - I_S \end{aligned}$$

and the total installed cost is simply the total capital investment less the working capital and the start up costs.

The total installed cost (TIC) of a facility is simply the sum of all costs that are required to completely construct a new chemical facility so that it is ready to start-up and to operate. In addition to purchasing the equipment components, it is also necessary to install these components into the system, and this installation entails additional costs such as the field materials and labor that are required to put the installation together.

4.1 Cost Components

For purposes of estimation, it is useful to break down the total installed costs into those costs which are *direct* and those which are *indirect*. Following [35, 36]:

$$TIC = \begin{bmatrix}\text{Total} & \text{Direct}\\ \text{Cost} & \end{bmatrix} + [\text{Indirect Costs}] \qquad (4.1)$$

A better appreciation of these costs and their significance can be obtained by examining each in more detail.

4.1.1 Direct Costs

These costs are divided into several categories as shown below.

1. Land — these costs include actual property costs, surveys and fees.

2. Yard improvements — these are costs for site clearing, grading, roads, walkways, railroads, fences, parking areas, wharves and piers, recreational facilities, landscaping, etc.

3. Buildings — these costs are subdivided as shown below.

   (a) Process — substructures, superstructures, platforms, supports, stairways, ladders, access ways, cranes, monorails, hoists, elevators as required in the finished process.

   (b) Auxiliary — administration and office, medical or dispensary, cafeteria, garage, product warehouse, parts warehouse, guard and safety, fire station, change house, personnel building, shipping office and platform, research laboratory and control laboratory.

   (c) Maintenance — shops to provide services for electrical, plumbing, sheet metal, machine, welding, carpentry, and instrument support are typical.

   (d) Building services — facilities for plumbing, heating, ventilation, dust collection, air conditioning, building lighting, elevators, escalators, telephones, computer network, sprinkler systems, fire alarms, and the like.

4. Equipment & Materials

   (a) Purchased Equipment — this consists of all major equipment items as they are listed on the equipment list for the project. Both process and auxiliary equipment are included.

   (b) Piping — this includes all process, electrical, and safety piping for service supply and waste disposal.

   (c) Concrete — all foundations and supports for the equipment and piping are included.

   (d) Steel — this includes any steel supports or superstructure components that are required to orient and position the equipment within the process. It may also include ladders, elevators, and observation platforms.

   (e) Instruments — all process control instrumentation that is needed.

   (f) Electrical — any electrical wiring for operating machinery with electric motors and the process control circuits.

   (g) Insulation — as required for the process and, in some cases, to protect the operators.

   (h) Paint — as required by the operating service and designed life of the equipment.

5. Installation Costs

(a) Material erection — the labor and equipment costs to position the equipment and piping for installation.

(b) Equipment setting — the labor and equipment costs to fix the equipment and piping in place for eventual use (*e.g.* welding, pipe hangers, etc.).

4.1.2 Indirect Costs

These costs are conveniently categorized as shown below.

1. Freight, insurance, & taxes — these costs result from shipping the equipment and field materials to the plant site location.

2. Construction overhead — the costs of on-site supervision and equipment (*e.g.* cranes and lifts) to erect the equipment and set it in place. These costs also include temporary facilities, offices, roads, construction tools and equipment, construction timekeeping, purchasing, expediting, environmental and building permits, safety, medical, and construction fringe benefits.

3. Engineering — these costs result from the design activities that occur primarily during the detailed engineering step. They include administrative, process design, general engineering, drafting, computer-aided design, cost engineering, procuring, expediting, reproduction, communications, scale models, consultant fees and travel expenses for the project.

4. Contingency — these are funds that are held in reserve to cover any unexpected costs during construction.

5. Fees — these costs are paid to the contractor upon the successful completion of the project construction.

While the primary design emphasis is usually placed on the process itself and its major equipment components, the other items in the enumeration above are also significant factors in the total installed costs. Typically, for example, the TIC can be 3-5 times greater than the actual purchased equipment costs, but the TIC is normally evaluated from the major equipment list and the purchased equipment costs.

4.2 Time Factors

Capital estimates are often factored from one time frame to another. This adjustment is needed because the correlations and vendor quotes that are used for estimating purchased equipment costs are only valid for a *particular* time period and because the costs of most items are continually changing. In general, a cost is factored from one time period to another using an equation of the form:

$$\frac{C(t_2)}{C(t_1)} = \left[\frac{Index(t_2)}{Index(t_1)}\right] \tag{4.2}$$

where: $C(t_i)$ = the estimated cost at time t_i
$Index(t_i)$ = index value at time t_i

This adjustment is easily made once an appropriate index has been found. For example, *Chemical Engineering* magazine is published biweekly, and every issue has a one-page summary of two of the indices which are of particular interest to chemical engineers whenever they are adjusting capital costs to different time frames. These indices are the CE plant cost index and the M&S equipment cost index.

Table 4.1 and 4.2 summarize data as it is presented in *Chemical Engineering.* The data in Table 4.2 are plotted in Figure 4.1. Table 4.1 indicates the breakdown in the factors that are used to generate either index. As can be seen, the CE Plant Cost Index is based on unit functionality while the M&S Index is results from averages for various process industries. Both indices give about the same results, but differences exist in terms of how they are computed and used. Specifically, the CE index has four major components:

1. equipment, machinery and supports,
2. construction labor,
3. building materials and labor, and
4. engineering supervision and manpower.

Details of how these various components are weighted can be found elsewhere [3, 20]. However, values for the index components are summarized in Table 4.1 for mid-1968 and mid-1981 from Matley [62, 63].

For example, in mid-1968 the CE plant index had a value of 113.6, and the components of the index were as shown in Table 4.1. The components of this index correspond to the components which are involved in estimating a total installed cost from a purchased fob[1] cost. The components in the CE plant index can be used to factor the various cost components which are involved in the installed cost. Alternatively, the installed cost for a piece of equipment may be factored using the overall CE plant cost index with the assumption that the relative increases in the various cost components (*i.e.* equipment, other materials, construction labor, indirects, *etc.*) are about the same.

Example: 4.1 *You are the cost engineer for the Filthy Rich independent oil conglomerate. Your boss, the famous oilman Jet Rink II has decided to repeat an old family tradition. He wants to give his son a distillation column for his birthday. He has no cost information except that his father gave him a one million dollar column in 1955. Estimate the cost in mid-1981 dollars.*

Solution: 4.1

$$C(t_2) = C(t_1)\frac{I(t_2)}{I(t_1)}$$

[1]free on board. The cost of equipment to the buyer to have it fabricated by a vendor and placed on a common carrier at the vendor's site. The fob cost does not include transportation to the buyer's plant site or other subsequent costs.

Table 4.1: Component factors used to generate the CE Plant and the M&S Indices

Component	Mid-1968	Mid-1981
CE Plant Cost Index	**113.6**	**297.0**
Equipment, machinery, supports	111.5	323.0
Construction labor	120.9	242.4
Buildings	115.7	274.9
Engineering & supervision	108.6	268.5
Fabricated equipment	109.9	321.8
Process machinery	112.1	301.5
Pipes, valves & fittings	117.4	360.1
Process instruments & controls	120.9	287.9
Pumps & compressors	115.2	388.5
Electrical equipment & materials	91.4	222.4
Structural supports, insulation & paint	105.7	322.0
M&S Equipment cost index	**273.1**	**721.3**
Averages for process industries		
Cement	268.3	745.3
Chemical	271.6	733.8
Clay products	260.7	721.7
Glass	256.3	691.3
Paint	267.4	738.2
Paper	261.6	699.4
Petroleum products	275.2	790.4
Rubber	276.5	772.9
Averages for related industries		
Electrical power	264.9	707.6
Mining & milling	273.2	740.2
Refrigeration	312.5	850.2
Steam power	261.8	749.6

Using the M&S index:

$$C(1955) = \$1,000,000 \quad I_{M\&S}(1981) = 721.3 \quad I_{M\&S}(1955) = 190.6$$
$$C(1981) = \$1,000,000\left(\tfrac{721.3}{190.6}\right) = \$3,784,000$$

Using the CE Plant index:

$$C(1955) = \$1,000,000 \quad I_{CE}(1981) = 299.0 \quad I_{CE}(1955) = 88.3$$
$$C(1981) = \$1,000,000\left(\tfrac{299.0}{88.3}\right) = \$3,386,000$$

Hence the two indices agree with each other within about 12%.

Table 4.2: Mid-year M&S equipment and plant cost indices. Taken from *Chemical Engineering* magazine.

Year	M&S	CE	Year	M&S	CE	Year	M&S	CE
1950	167.9	73.9	1973	344.1	144.1	1996	1039.2	386.5
1951	180.3	80.4	1974	398.4	165.4	1997	1056.8	386.5
1952	180.5	81.3	1975	444.3	182.4	1998	1061.9	389.5
1953	182.5	84.7	1976	472.1	192.1	1999	1068.3	390.6
1954	184.6	86.1	1977	505.4	204.1	2000	1009.0	394.1
1955	190.6	88.3	1978	545.0	218.8	2001	1093.9	394.3
1956	208.8	93.9	1979	599.4	238.7	2002		
1957	225.1	98.5	1980	659.6	261.2	2003		
1958	229.2	99.7	1981	721.3	299.0	2004		
1959	234.5	101.8	1982	745.6	314.0	2005		
1960	237.7	102.0	1983	760.8	316.9	2006		
1961	237.2	101.5	1984	780.4	322.7	2007		
1962	238.5	102.0	1985	789.6	325.3	2008		
1963	239.2	104.2	1986	797.6	318.4	2009		
1964	241.8	103.3	1987	813.6	323.8	2010		
1965	244.9	104.2	1988	852.0	342.5	2011		
1966	252.5	107.2	1989	895.1	355.4	2012		
1967	262.9	109.7	1990	915.1	357.6	2013		
1968	273.1	113.6	1991	930.6	361.3	2014		
1969	285.0	119.0	1992	943.1	358.2	2015		
1970	303.3	125.7	1993	964.2	359.2	2016		
1971	321.3	132.2	1994	993.4	368.1	2017		
1972	332.0	137.2	1995	1027.5	381.1	2018		

4.3 Factored Estimates for Battery Limits

More accurate cost estimates result from completing detailed energy and material balances, an equipment list, and sizing the items in the list. This procedure enables one to compute a factored estimate in which the installed cost of each major equipment item is factored separately rather than factoring the entire plant as a single entity.

Guthrie [35, 36] suggests a modular approach to capital cost estimation where the installation costs are factored from the *base* and the fob *purchased* equipment costs. For each equipment item in the equipment list, both base and purchased costs are estimated. These costs are then used to estimate a *bare module* cost for each line item. One then computes a total installed cost from the bare module cost estimate for the plant battery limits:

$$TIC = \begin{bmatrix}\text{Bare Module}\\ \text{Cost}\end{bmatrix} + \begin{bmatrix}\text{Contingency}\\ \text{and Fees}\end{bmatrix} \tag{4.3}$$

Contingency and Fees are usually expressed as a percentage of the Bare Module Cost

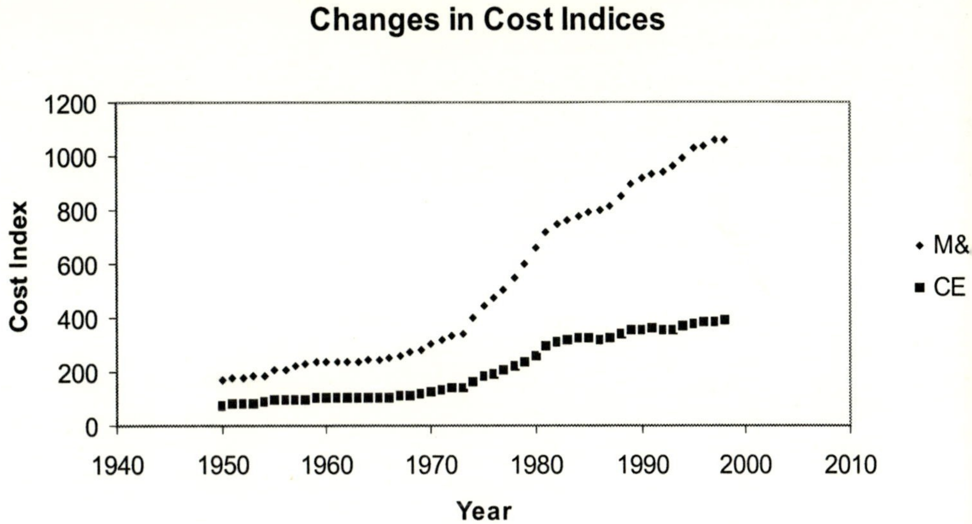

Figure 4.1: Change in the M&S and CE Indices over time.

which is estimated from the purchased equipment costs. The purchased equipment costs are estimated either from vendor quotes or base equipment cost correlations. Therefore, the entire cost estimate rests primarily on the battery limits equipment list. The steps leading to a TIC estimate are then as follows:

1. Complete a process flow diagram (PFD)
2. Complete energy and material balances
3. Size all major equipment items
4. Develop a major equipment cost list
5. Estimate the Battery Limits Bare Module Cost:
   (a) Estimate fob purchased equipment costs (E_i')
   (b) Estimate Bare Module Costs (BMC_i) for each item
6. Estimate Other Plant Site Costs:
   (a) Off-site facilities (cooling towers, steam plants, *etc.*
   (b) Site development
   (c) Industrial buildings
7. Estimate the Contingency and Fees

8. Estimate the Total Installed Cost for the Battery Limits

9. Estimate the Total Installed Cost for the Plant

Thus, the sum of the bare module costs from the major equipment list is used to generate factored estimates for the total installed costs. Each step is described in greater detail below.

4.3.1 Base Equipment Costs

In each category, the *base* equipment cost for the *ith* line item in the equipment list usually refers to equipment items with carbon steel construction and standard features.

The base equipment costs (E or C_p^{cs}) may be obtained from cost correlations for specific pieces of equipment or from vendor cost quotes. Typically, the base fob equipment costs are correlated using power rules of the form:

$$E = C_p^{cs} = E_0 \left(\frac{S}{S_0} \right)^a \tag{4.4}$$

where E_0 = a reference cost associated with the unit of size, S_0.
S = a size parameter such as volume, transfer area, or horsepower requirements.
a = a power dependency for the equipment unit.

Typical values for E_0, S_0, and a are summarized in Tables 4.4 through 4.12 for a number of equipment items based on mid-1968 costs and carbon steel construction. In addition to the parameters needed to estimate C_p^{cs}, these tables also include the bare module factors, F_{bm}^{cs}, and the tubing or piping factors, f_t, that are needed to estimate the actual bare module cost, BMC_i, or C_{bm}^a, for each type of equipment that is purchased and installed.

Equation 4.4 is an oversimplification of the cost picture and is usually only valid over a limited range of conditions. Base costs for equipment (i.e., fob purchased costs) do not go to zero with the size factor as predicted by Equation 4.4. Also, the power dependence on size is not constant over all possible sizes. In fact, upper limits on size exist, and eventually, multiple units are required in parallel if the process demand is sufficiently large. So a more general expression for fob purchased costs is given by:

$$E = C_p^{cs} = b_0 + b_1 S + b_2 S^{b_3} \tag{4.5}$$

where the parameters, $b_0 - b_3$, are empirical coefficients which are a function of the nature of the equipment, the fabrication techniques used by the manufacturer, and the purchasing organization's overhead. The first parameter, b_0, for example, is primarily due to the fact that most chemical companies employ a purchasing agent, and this agent must follow certain procedures in making any purchase. For a large equipment item, this first term on the right hand side of Equation 4.5 is negligible, but for smaller items it results in significant deviations from the behavior predicted by Equation 4.4. When base equipment costs are extrapolated over a wide range

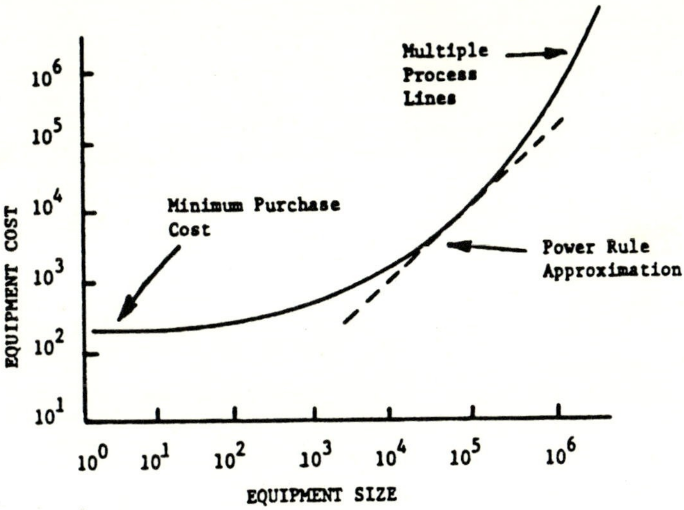

Figure 4.2: Typical power rule deviations experienced for base equipment costs.

of conditions, the cost behavior indicated by Figure 4.2 is more representative of the actual situation.

When large equipment items are required, parallel processing lines eventually must be implemented. For example, a rail car or flat bed truck has size and weight limitations. These limitations may be overcome through on-site fabrication, with increased costs, or through the purchase of duplicate equipment for parallel lines. Either way, the costs are increased, Therefore, the change in cost with respect to the size increment also becomes larger.

The base equipment cost correlations presented in Table 4.4 were derived from Guthrie's [35] data and converted to equations by statistical analysis. *All base costs are given in mid-1968 dollars and carbon steel construction.* They use slightly different mathematical forms from either Equation 4.4 or 4.5.

Example: 4.2 *You are the assistant to the assistant cost engineer for the Windfall Profits Oil Company (WIPOCO). Your boss has asked you to estimate the cost of a cabon steel, floating head, shell and tube heat exchanger that has 2100 ft² of transfer area and will operate at just above 1 atm pressure.*

Solution: 4.2 *From Equation 4.4:* $E = E_0(S/S_0)^a$

From Table 4.9:

$E_0 = \$2000, a = 0.65, S_0 = 100$ *ft²*

$$E = \$2000\left(\frac{2100}{100}\right)^{0.65} = \$14,470$$

where the estimated cost is in mid-1968 dollars.

Table 4.3: Reference equations for estimating capital costs.

$$E = E_o\left(\frac{S}{S_o}\right)^a = C_p^{cs}$$
$$E' = F_m F_p E = C_p^a = F_m F_p C_p^{cs}$$
$$C_t^a = 0.7 C_p^{cs} F_p F_m f_t = 0.7 E' f_t$$
$$F_{bm}^a = F_{bm}^{cs} + (1.0 + 0.7 f_t)(F_m F_p - 1)$$
$$C_{bm}^a = C_p^{cs} F_{bm}^a = C_p^{cs}\left[F_{bm}^{cs} + (1 + 0.7 f_t)(F_m F_p - 1)\right]$$

a, power rule for unit size
C_t^a, actual cost for process piping (or tubing)
C_p^a, actual fob purchased equipment cost
C_p^{cs}, fob purchased equipment cost, low pressure, carbon steel basis ($C_p^{cs} = E$)
E_o, fop purchased cost for reference size, S_o
E, fob purchased equipment cost, low pressure carbon steel basis
E', actual fob purchased equipment cost
F_{bm}^a, actual bare module factor, adjusted for pressure or materials of construction
F_{bm}^{cs}, bare module factor, carbon steel, low pressure basis
F_m, materials factor
F_p, pressure factor
f_t, tubing or piping factor
S, actual equipment size
S_o, reference equipment size at which the fob purchased cost is E_o

Example: 4.3 *As the cost estimation engineer for the Super Colossal-Whizbang Motion Picture Company, you get some relatively strange assignments. None were stranger than the time your boss, Mr. Mogul, bought the movie rights to the hit record entitled,* We all live in a vertical pressure vessel, *by the Small Insects Band. The movie called for a vertical pressure vessel with a height of 100 ft, and a diameter of 30 ft. You were told to estimate the base equipment cost for a carbon steel vessel in mid-1980 dollars.*

Solution: 4.3 *From Table 4.4:*
$E = (E_0)(D/C3)^{1.047}$
D = *30 ft; Height* = *100 ft*
$C_3 = (33.134/HT) + 0.127 = 0.45834$
$E_0 = 2000(659.6/273.1) = 4830$
$E = 4830(30/0.45834)^{1.047}$
E = *\$384,795*

Table 4.4: Correlations for estimating base equipment costs for process vessels. Referenced to mid-1968 dollars with (M&S = 273.1). Correlated using Guthrie's figures. $E = C_p^{cs}$ (Low pressure, carbon steel basis)

Item	Correlation
Vertical Process Vessels[a]	
	$E = 2000\left(\frac{D}{C_3}\right)^{1.047}\left(\frac{M\&S}{273.1}\right)$
	$C_3 = \begin{cases} 20.867/HT + 0.778 & HT < 15ft \\ 33.143/HT + 0.127 & HT \geq 15ft \end{cases}$
	D =nominal column diameter, ft
	HT =nominal column height, ft, actual trays plus 10 ft
Horizontal Process Vessels[b]	
	$E = 1375\left(\frac{D}{C_3}\right)^{0.9747}\left(\frac{M\&S}{273.1}\right)$
	$C_3 = \begin{cases} 20.867/HT + 0.778 & HT < 15ft \\ 33.143/HT + 0.127 & HT \geq 15ft \end{cases}$
	HT =nominal horizontal length of vessel, ft
	D =nominal diameter of horizontal vessel, ft

[a]Distillation columns, absorbers, scrubber, strippers, flash drums, etc.
[b]Accumulator drums, surge vessels, decanters, etc.

Table 4.5: Correlations for estimating fob purchased costs for internals (e.g., distillation trays, absorber packings, etc. Referenced to mid-1968 dollars with (M&S = 273.1). Correlated using Guthrie's figures. $E_i = C_{pi}^{cs}$ See Table 4.6 for tray adjustment factors.

Item	Correlation
Sieve Trays, 24 inch spacing, shop installed[a,b]	
	$E_i = C_4(HT-10)\left(\frac{M\&S}{273.1}\right)$
	$C_4 = \begin{cases} 8.665D - 1.218 & D < 4ft \\ 16.67D - 33.35 & 4 \le D \le 6ft \\ 23.33D - 73.31 & 6 < D \le 7ft \\ 25D - 75 & 7 < D \le 10ft \end{cases}$
	D = nominal column diameter, ft
	HT = nominal column height, ft, actual trays plus 10 ft
Sieve Trays, 24 inch spacing, field installed[c]	
	$E_i = C_4(HT-10)\left(\frac{M\&S}{273.1}\right)$
	$C_4 = 20D - 60 \quad D > 10ft$
	D = nominal column diameter, ft
	HT = nominal column height, ft, actual trays plus 10 ft

[a] Add $0.08E_i$ as indirects for freight, taxes and insurance.
[b] Add $0.05E_i$ to estimate shop labor for installation.
[c] Add $0.58E_i$ for field labor and $0.75E_i$ for indirects.

Table 4.6: Adjustments for fob purchased costs for tray internals (e.g., distillation trays, absorbers, stripper columns, etc.). Correlated from Guthrie. $E_i = C_{pi}^{cs}$ (Also see Table 4.5.)

$$E_i' = \textbf{Tray Cost} = E_i(F_s + F_t + F_m) = E_i \sum F_i$$

Adjustment Factors

Tray[a] Spacing, in	F_s	Tray Type	F_t^b	Tray Material	F_m^b
24	1.0	Grid[c]	0.0	Carbon steel	0.0
18	1.4	Plate	0.0	Stainless steel	1.7
12	2.2	Sieve	0.0	Monel	8.9
		Trough or valve	0.4		
		Bubble cap	1.8		
		Koch Kascade	3.9		

[a] More generally, Tray spacing, TS, and F_s are related by: $F_s = 13.53/(TS)^{0.8}$.
[b] Add 1.0 to these values if these factors are used individually (i.e., $\sum F_i \geq 1.0$).
[c] No downcomer is provided with Grid trays.

Table 4.7: Cost factors for column packings and linings. Use for distillation, stripping, absorber and adsorption columns. Referenced to mid-1968 dollars with (M&S = 273.1). $E_i = C_{pi}^{cs}$ Taken Guthrie. $E_i = C_{pi}^{cs}$

Packing costs[a], $/ft^3

Raschig Rings				
Size, in	1.0	1.5	2.0	3.0
Stoneware	5.2	4.3	3.5	2.9
Procelain	7.0	5.8	4.7	3.9
Stainless	70.2	45.8	32.5	22.8

Berl saddles			
Size, in	3/4	1.0	1.5
Stoneware	18.8	14.5	7.8
Porcelain	20.7	15.9	8.7

Activated Carbon	14.2
Alumina	12.6
Coke	3.5
Crushed limestone	5.8
Silica gel	27.2

Lining Costs[a], $/ft^2

Lining Material	Thickness inches	$/ft^2
Acid brick	3.0	3.80
	4.0	5.50
	6.0	8.25
Firebrick	4.5	7.16
	9.0	10.79
Rubber	3/16	4.37
	1/4	4.75
Refractory	2.0	7.50
	4.0	10.52
Gunite	2.0	3.20
	4.0	4.55
Chemical lead[b]	5 lb	6.25
	10 lb	7.13
	15 lb	8.86

[a]Field Material and Labor (M&L)

[b]Measured in pounds of lead per square foot covered.

Table 4.8: Correlations for estimating fob purchased costs for liquid pumps. Referenced to mid-1968 dollars with (M&S = 273.1). Correlated using Guthrie's figures. $E = C_p^{cs}$.

Item	Correlation
Centrifugal pumps and drivers	
	$E = \begin{cases} 440\,[q(TDH)/100]^{0.1047} & q(TDH) \leq 10^3 \\ 560\,[q(TDH)/1000]^{0.2952} & 10^3 < q(TDH) \leq 10^4 \\ 1100\,[q(TDH)/10000]^{0.5607} & 10^4 \leq q(TDH) \leq 10^5 \end{cases}$
Motor driven reciprocating pumps	
	$E = \begin{cases} 10\,[q(TDH)/100]^{1.30} & 100 \leq q(TDH) \leq 10^4 \\ 4000\,[q(TDH)/10,000]^{0.743} & 10^4 \leq q(TDH) \leq 3 \times 10^5 \end{cases}$
Steam driven reciprocating pumps[a]	
	$E = 200\,[q(TDH)/1000]^{0.596} \quad 10^3 \leq q(TDH) \leq 3 \times 10^5$
	q =volumetric flowrate in gal/min
	TDH =total developed head[b] delivered to fluid in psi
	C/H =$q \times TDH$ in units of gpm × psi

[a]Direct acting

[b]Also see Equation 11.15

4.3.2 Purchased Equipment Costs

The data in Tables 4.4 through 4.12 correlate *base* equipment costs (i.e., E or C_p^{cs}, fob purchase costs for carbon steel equipment with wall thicknesses that are appropriate for atomospheric operation). If the equipment actually purchased corresponds to these requirements, then the actual *purchased* cost, C_p^a, will be the same as the base cost, C_p^{cs}. However, the purchased cost will be greater than the base cost whenever special materials of construction or high pressure operation are required. The costs are also affected by the purchase of multiple units (a slight reduction occurs).

In general, the materials of construction must be accounted for as well as the operating conditions. Higher operating pressures require thicker equipment walls which increase the fob purchased equipment costs.

To compute a *purchased* cost, the base cost, E_i, must be adjusted by the appropriate factors, F_m and F_p using the equation:

$$E' = C_p^a = F_m F_p C_p^{cs} = E F_m F_p \tag{4.6}$$

where:		
E	=	the low-pressure, carbon steel cost (base cost)
E'	=	the adjusted unit cost (purchased cost)
F_m	=	the materials adjustment factor
F_p	=	the pressure adjustment factor

Table 4.9: Factors for estimating process equipment costs in mid-1968 dollar (M&S = 273.1). Taken from Guthrie. $E = C_p^{cs} = E_0(S/S_0)^a$

Item	Size		S_o	E_o	a	F_{bm}^{cs}	f_t
	Range	Units		1968 $			
Crystallizers							
Growth	$1–10^2$	tpd[a]	1.0	5,500	0.65	2.24	0.18
Forced Circulation	$1–10^3$	tpd[a]	1.0	7,900	0.55	2.24	0.18
Batch	$1–10^3$	gal	1.0	170	0.70	2.05	0.18
Dryers							
Drum	$10–10^3$	ft^2	1.0	3,000	0.45	2.23	0.15
Pan	$1–10^2$	ft^2	1.0	1,900	0.38	2.23	0.15
Rotary vacuum	10–5000	ft^2	1.0	3,100	0.45	2.23	0.15
Gas compressor[b]	$30–10^4$	Bhp[c]	600	10^5	0.82	2.160	0.141
Evaporators							
Forced Circulation	10–5,000	ft^2	1.0	6,000	0.70	2.55	0.20
Vertical Tube	10–5,000	ft^2	1.0	1,200	0.53	2.55	0.20
Horizontal Tube	10–5,000	ft^2	1.0	800	0.53	2.55	0.20
Jacketed vessel[d]	$10–10^3$	gal	1.0	1000	0.50	2.33	0.20
Filters							
Plate & press	1–5,000	ft^2	1.0	330	0.58	2.40	0.28
Pressure leaf, wet	1–5,000	ft^2	1.0	410	0.58	2.40	0.28
Pressure leaf, dry	1–5,000	ft^2	1.0	1,500	0.53	2.40	0.28
Rotary drum	10–5,000	ft^2	1.0	1,400	0.63	2.144	0.28
Rotary disk	$10^2–10^4$	ft^2	1.0	10^3	0.78	2.144	0.28
Flakers, drum	10–5,000	ft^2	1.0	1,300	0.64	2.13	0.15
Heaters							
Direct Fired	1–40	MBH[e]	1.0	5,200	0.85	2.177	0.148
Process	10–400	MBH[e]	10	40,000	0.85	2.130	0.177
Heat Exchangers[f]							
Floating Head	$10^2–10^4$	ft^2	10^2	2,000	0.65	3.178	0.447
Double pipe	$2–10^2$	ft^2	6	300	0.08	3.291	0.456
Fixed tube sheet	$10^2–10^4$	ft^2	10^2	1,600	0.65	3.178	0.447
Kettle, reboiler	$10^2–10^4$	ft^2	10^2	2,700	0.65	3.178	0.447
U tube	$10^2–10^4$	ft^2	10^2	1,700	0.65	3.178	0.447

[a] tons per day
[b] Includes driver
[c] Brake horsepower
[d] Glass lined
[e] Millions Btu/hour
[f] Shell and Tube

Table 4.10: Factors for estimating offsite equipment costs in mid-1968 dollar (M&S = 273.1). Taken from Guthrie. $E = C_p^{cs} = E_0(S/S_0)^a$

Item	Size Range	Size Units	S_o	E_o 1968 $	a	F_{bm}^{cs}	f_t
Air Coolers[a]	10^2–10^4	ft^2	2,000	20,000	0.80	2.185	0.139
Boilers, large industrial							
Packaged	10–10^4	klb[b]/h	300	200,000	0.70	1.80	0.07
Field erected	20–2000	klb[b]/h	10^2	5×10^5	0.80	1.82	0.08
Boilers, small industrial							
15 psig	10–10^4	lb/h	1.0	400	0.50	1.80	0.07
150 psig	10–10^4	lb/h	1.0	440	0.50	1.80	0.07
300 psig	10–10^4	lb/h	1.0	500	0.50	1.80	0.07
600 psig	10–10^4	lb/h	1.0	560	0.50	1.80	0.07
Cooling towers	10^3–10^5	gpm[c]	3,000	85,000	0.60	1.70	0.08
Ductwork (shop fabricated and field errected)							
Aluminum	10^2–10^4	ft	1.0	5.42	0.55	1.0	na
Galvanized	10^2–10^4	ft	1.0	8.00	0.55	1.0	na
Stainless	10^2–10^4	ft	1.0	15.12	0.55	1.0	na
Dust collectors							
Cyclones	10–10^4	ft^3/min	1.0	3.0	0.80	2.26	0.15
Cloth filter	10–10^4	ft^3/min	1.0	25	0.68	2.26	0.15
Precipitators	10–10^4	ft^3/min	1.0	390	0.75	2.26	0.15
Ejectors							
Power generators	500–3×10^4	Kva[d]	6,000	2×10^6	0.75	1.46	0.0
Refrigeration[e]	50–3,000	tons	200	60,000	0.70	1.80	0.06
Storage tanks, atmospheric							
Field erected	10^4–5×10^5	gal	4×10^5	40,000	0.63	1.937	0.128
Shop fabricated							
light gage	20–2000	gal	50	824	0.28	1.63	0.04
API conical	800–7×10^4	gal	10,000	7,000	0.30	1.63	0.045
Storage tanks, pressurized							
Horizontal	10^3-10^5	gal	9,000	6,000	0.65	2.084	0.153
Spherical	10^4–5×10^5	gal	40,000	35,000	0.70	1.957	0.108

[a]Based on aircooled area = estimated area/15.50 ft^2

[b]Steam capacity, 10^3 lb/h

[c]System capacity, gal/min of water recirculated

[d]Generating capacity, Kva

[e]Systems, with evaporator temperature at +40 °F

Table 4.11: Factors for estimating special equipment costs in mid-1968 dollar (M&S = 273.1). Taken from Guthrie. $E = C_p^{cs} = E_0(S/S_0)^a$

Item	Size		S_o	E_o	a	F_{bm}^{cs}	f_t
	Range	Units		1968 $			
Agitators							
Propellers	$10–10^3$	hp	1.0	350	0.50	2.09	na
Turbines	$10^2–10^4$	hp	1.0	750	0.30	2.09	na
Air Compressors[a]	$10–10^3$	cfm	1.0	2,900	0.28	2.06	na
Air Conditioners							
Window vent	na	Ea	na	300	na	1.44	na
Floor-mounted	na	Ea	na	200	na	1.44	na
Rooftop							
10 tons	na	Ea	na	3,800	na	1.54	na
20 tons	na	Ea	na	6,500	na	1.54	na
30 tons	na	Ea	na	8,100	na	1.54	na
Air Dryers	10–10,000	cfm	1.0	200	0.56	2.23	0.25
Bagging machines							
By weight	1–20	bags/min	1.0	3,300	0.80	1.88	0.15
By volume	1–20	bags/min	1.0	1,000	0.80	1.88	0.15
Blenders	1–1000	cfm	1.0	850	0.52	2.06	0.18
Blowers & fans	1–1000	cfm	1.0	7	0.68	2.04	0.20
Centrifuges							
Horizontal basket	6–48	in[b]	1.0	140	1.25	2.01	0.12
Vertical basket	6–48	in[b]	1.0	310	1.00	2.01	0.12
Solid bowl (SS)	6–72	Hp	1.0	1,900	0.73	2.05	0.12
Sharples (SS)	6–72	Hp	1.0	5,200	0.68	2.05	0.12

[a] 125 psig output

[b] basket diameter

Table 4.12: Factors for estimating solids conveyor and crane costs in mid-1968 dollar (M&S = 273.1). Taken from Guthrie. $E = C_p^{cs} = E_0(S/S_0)^a$

Item	Size		S_o	E_o	a	F_{bm}^{cs}	f_t
	Range	Units		1968 $			
Conveyors							
Belt (by width)							
18 in	1–1,000	ft[a]	1.0	450	0.65	2.16	na
24 in	1–1,000	ft[a]	1.0	540	0.65	2.16	na
36 in	1–1,000	ft[a]	1.0	620	0.65	2.10	na
42 in	1–1,000	ft[a]	1.0	700	0.65	2.10	na
48 in	1–1,000	ft[a]	1.0	750	0.65	2.10	na
Bucket							
30 tons/h	1–250	ft[b]	1.0	220	0.65	2.36	na
75 tons/h	1–250	ft[b]	1.0	400	0.83	2.36	na
120 tons/h	1–250	ft[b]	1.0	500	0.83	2.36	na
Roller (by width)							
12 in	1–500	ft[a]	1.0	7	0.90	2.16	na
15 in	1–500	ft[a]	1.0	8	0.90	2.16	na
18 in	1–500	ft[a]	1.0	9	0.90	2.11	na
20 in	1–500	ft[a]	1.0	10	0.90	2.16	na
Screw (by diameter)							
6 in	1–500	ft[a]	1.0	230	0.90	2.04	na
12 in	1–500	ft[a]	1.0	270	0.80	2.04	na
14 in	1–500	ft[a]	1.0	290	0.75	2.04	na
16 in	1–500	ft[a]	1.0	300	0.60	2.04	na
Vibrating (by width)							
12 inch	1–500	ft[a]	1.0	80	0.80	2.10	na
18 inch	1–500	ft[a]	1.0	110	0.80	2.10	na
24 inch	1–500	ft[a]	1.0	120	0.90	2.05	na
36 inch	1–500	ft[a]	1.0	150	0.90	2.05	na
Cranes, by span							
10 ft	1–100	tons	1.0	1,800	0.60	1.0	na
20 ft	1–100	tons	1.0	2,400	0.60	1.0	na
30 ft	1–100	tons	1.0	3,800	0.60	1.0	na
40 ft	1–100	tons	1.0	4,800	0.60	1.0	na
50 ft	1–100	tons	1.0	6,300	0.60	1.0	na
100 ft	1–100	tons	1.0	8,500	0.60	1.0	na
Crushers							
Cone	1–100	ton/h	1.0	750	0.85	2.18	na
Gyratory	1–100	ton/h	1.0	55	1.20	2.18	na
Jaw	1–100	ton/h	1.0	85	1.20	2.18	na
Pulverizers	10–1000	lb/h	1.0	520	0.35	2.04	na

[a] conveyor length
[b] lift height

Table 4.13: Material factors. Taken from Guthrie.

	Construction Material					
Item	CS[a]	CM[b]	316[c]	Mon[d]	Ti[e]	Al[f]
Air Coolers	1.00		2.85	3.20		1.50
Fired Heaters	1.00	1.45	1.50		4.23	
Offsite Piping	1.00	2.70	3.80	3.20		
Pressure Vessels						
Clad	1.00		2.25	3.89	4.23	
Solid	1.00		3.67	6.34	7.89	
Process Furnaces	1.00	1.35	1.75			

[a]Carbon steel
[b]Chrome/Moly
[c]316 Stainless Steel
[d]Monel
[e]Titanium
[f]Aluminum

Typical values for F_m are summarized in Table 4.13 through 4.15 for several types of equipment. Table 4.15 summarizes factors for shell and tube heat exchangers (all design types) as a function of the material on each side of the exchanger and the transfer area.

The F_p factor are correlated using the equation:

$$F_p = \begin{cases} 1 & P \leq P^* \\ 1 + d(P - P^*) & P > P^* \end{cases} \tag{4.7}$$

where: P^* = a reference pressure at which costs begin to increase significantly
d = an empirical parameter that measures the incremental effects of pressure on cost

Typical values for d and P^* are summarized in Table 4.16.

Example: 4.4 *Estimate the fob purchased equipment cost of a monel air cooler requiring 6000 ft² of transfer area in mid-1968. From Table 4.10,* $S_0 = 2000\ ft^2$, $E_0 = \$20,000$, *and* $a = 0.80$.

Solution: 4.4 *From Table 4.13,* $F_m = 3.20$.
Aircooled Area = AA = 6000/15.50 = 387 ft². (see footnote a in Table 4.10)
Carbon steel cooler (E) $= 20,000(387/2000)^{0.8} = 5,375$
Monel cooler (E′) $= 3.2(5375) = \$17,200$

As can be seen from this example, the installation of a monel unit rather than a carbon steel unit roughly doubles the total installed cost. The cost of process piping (i.e., piping in direct contact with process fluids) and the fob purchased cost of the air cooler are increased. However, other direct costs, the field labor charges. and

Table 4.14: Material factors, F_m, for pumps and storage tanks. Taken from Gruthrie.

Material	Pump Type		Storage Tank	
	Centrifugal	Reciprocating	Small[a]	Large[b]
Cast iron	1.00	1.00		
Bronze	1.28	1.25		
Cast steel	1.32	1.55		
Stainless	1.93	2.10	3.20	2.20
Carpenter 20	2.10			
Worthite	2.44			
Hastelloy C	2.89			
Monel	3.23			
Nickel	3.48			
Titanium	8.98			
Carbon steel			1.0	1.0
Aluminum			1.40	
Rubber lined			1.48	1.48
Lead lined			1.55	1.55
Glass lined			4.25[c]	

[a]Up to 40,000 gal
[b]Above 40,000 gal
[c]Up to 10,000 gal (max)

Table 4.15: Material factors for Shell and Tube Heat Exchangers, all designs. Taken from Guthrie.

Surface Area ft^2	Shell/Tube Materials, F_m								
	CS/ CS	CS/ Brass	CS/ Mo	CS/ SS	SS/ SS	CS/ Monel	Monel/ Monel	CS/ Ti	Ti/ Ti
1–99	1.00	1.05	1.60	1.54	2.50	2.00	3.20	4.10	10.28
100–499	1.00	1.10	1.75	1.78	3.10	2.30	3.50	5.20	10.60
500–999	1.00	1.15	1.82	2.25	3.26	2.50	3.65	6.15	10.75
1,000–4,999	1.00	1.30	2.15	2.81	3.75	3.10	4.25	8.95	13.05
5,000–10,000	1.00	1.52	2.50	3.52	4.50	3.75	4.95	11.10	16.60

CS, Carbon Steel
SS, Stainless Steel

Table 4.16: Pressure factors. Taken from Guthrie. For $P > P^*, F_p = 1+d(P-P^*)$.

Item	Pressure Range	Pressure Units	d	P^*
Air Coolers	150–1,000	psia	0.0005	150
Boilers				
Field erected		psia	0.0002	400
Packaged		psia	0.001	250
Fired heaters	500–1,500	psia	0.0003	500
Pumps (liquid)				
Centrifugal	150–1,000	psia[a]	0.001	150[a]
Reciprocating	150–1,000	psia[a]	0.0005	150
Process furnace	500–3,000	psia	0.0002	500
Pressure vessels				
Horizontal storage		psia	0.003	150
Spherical storage		psia	0.004	30
Process	50–1,000	psia	0.001	50
Heat exchangers				
Shell & Tube	150–1,000	psia	0.0007	150

[a]Suction pressure

the indirect costs are not affected by the materials of construction. Contingency and fees are affected because they are based on the bare module cost which includes the process piping and the fob purchased cost of the air cooler.

4.3.3 Installed Costs

The total bare module cost of any chemical processing unit consists of three primary direct elements. These elements are:

1. the purchased fob equipment costs (E),
2. the other direct materials costs (M), and
3. the direct labor costs (L).

If contingency is assumed as 15% and the A/E fees are 3% of the Bare Module Cost (BMC or C^a_{bm}) for the entire plant site, then the Total Installed Cost (TIC) can be approximated as:

$$TIC = 1.18BMC = 1.18\sum_{i=1}^{N} BMC_i = 1.18\sum_{i=1}^{N} C^a_{bmi} \tag{4.8}$$

where: BMC_i = the bare module cost of the ith item in the purchased equipment list

N = the number of line items in the purchased equipment list

The bare module cost of each purchased equipment item is estimated from the indirect cost factor, the purchased equipment cost, and the direct field materials and labor. For carbon steel at low P:

$$BMC_i = (1 + f_i^{ind})(E_i + M_i + L_i) \tag{4.9}$$

where:
- E_i = the fob purchased equipment cost for carbon steel at low pressures
- f_i^{ind} = the indirect cost factor for the ith equipment item
- M_i = the direct field materials (except E_i)
- L_i = the field labor for installing the ith equipment item

In the case where special materials of construction or high pressures are required for the equipment, then for the ith line entry item:

$$\begin{aligned} BMC_i &= (1 + f_i^{ind})E_i \left[1 + f_i^{lab} + \sum_j^Q f_{i,j}^{mat}\right] \\ &+ (E_i' - E_i)\left[1 + 0.7 f_{i,pipe}^{mat}\right] \end{aligned} \tag{4.10}$$

where:
- E_i = the base equipment cost in carbon steel
- E_i' = the fob purchased equipment cost
- f_j^{mat} = the materials factor for the jth of Q_i entries in the direct materials list for the ith entry in the equipment list
- f_i^{lab} = the labor cost factor for the ith base equipment cost
- f_i^{ind} = the indirect cost factor for the ith base equipment cost
- Q_i = the number of line entries in the direct field materials, excluding the purchased equipment cost E_i'

Typically, the purchased equipment costs represent only about 25-35% of the total installed cost of the equipment in an operating plant. The remaining costs include direct and indirect costs. The direct costs include such things as piping, foundations, instrumentation, insulation, paint, safety devices, and installation labor.

The indirect costs include taxes, freight, insurance, overhead and engineering.

Direct Costs

Besides the purchased equipment cost, E', the remaining direct materials costs (M_i) include:

1. piping,
2. concrete,
3. steel (structural support),
4. instrumentation,
5. electrical connections,

6. insulation, and
7. paint.

These costs may generally be estimated as percentages of the base equipment costs, E, using factors summarized in Tables 4.17-4.21. However, about 70% of the piping [36] costs are directly associated with the processing equipment and, therefore, are necessarily constructed of the same material as the purchased equipment. If a high-pressure pump is required which must be fabricated of titanium, then about 70% of the piping (i.e., the process piping) associated with that pump must also be constructed of titanium and able to withstand the same pressures. The remaining 30% of the piping is associated with support services (*e.g.* steam utilities, drains, electrical conduits, *etc.*) and is not affected by the materials of construction for the equipment. Similarly, the remaining direct material costs are not considered to be affected by special material requirements.

Consequently, 70% of the piping costs (i.e., the process piping) should be adjusted to correspond to the purchased equipment cost (which has been adjusted for the special materials of construction) while the remaining 30% of the piping costs, and the balance of the direct materials costs, should be factored from the base equipment cost.

Heuristic: 8 (70/30 Rule) *Process piping costs are 70% of the total piping costs for a unit operation. Non-process piping costs are 30% of the total piping costs for the unit operation. Piping costs are estimated using the factor f_t and the fob purchased cost for that equipment. The factor f_t estimates the total piping cost for the unit operation. The factor $0.7f_t$ estimates cost for process piping. If special materials are required for a unit operation, then they are also required for process piping for that equipment. The factor $0.3f_t$ estimates costs for non process piping (e.g., electrical conduits, sumps, safety showers, etc.) which do not normally require special materials of contruction and are not affected by high operating pressures in the process.*

Typical installation cost factors for several major equipment items are summarized in Tables 4.17-4.21. Figure 4.3 graphically summarizes the method of using these factors. For units where the base and purchased costs are the same, one simply multiplies the equipment fob purchase cost by the total installed cost factors to estimate the total installed cost in mid-1968. Hence, the installed cost of an air cooler is about three times its purchased cost whereas the installed costs of vertical pressure vessels are about five times their purchased costs.

As an exercise, one may study Figure 4.3 and determine how to compute the factors which are summarized at the bottom of Tables 4.17-4.21 (particularly the $L_i/(E_i + M_i))$ ratio and the indirect costs factor f_i^{ind}).

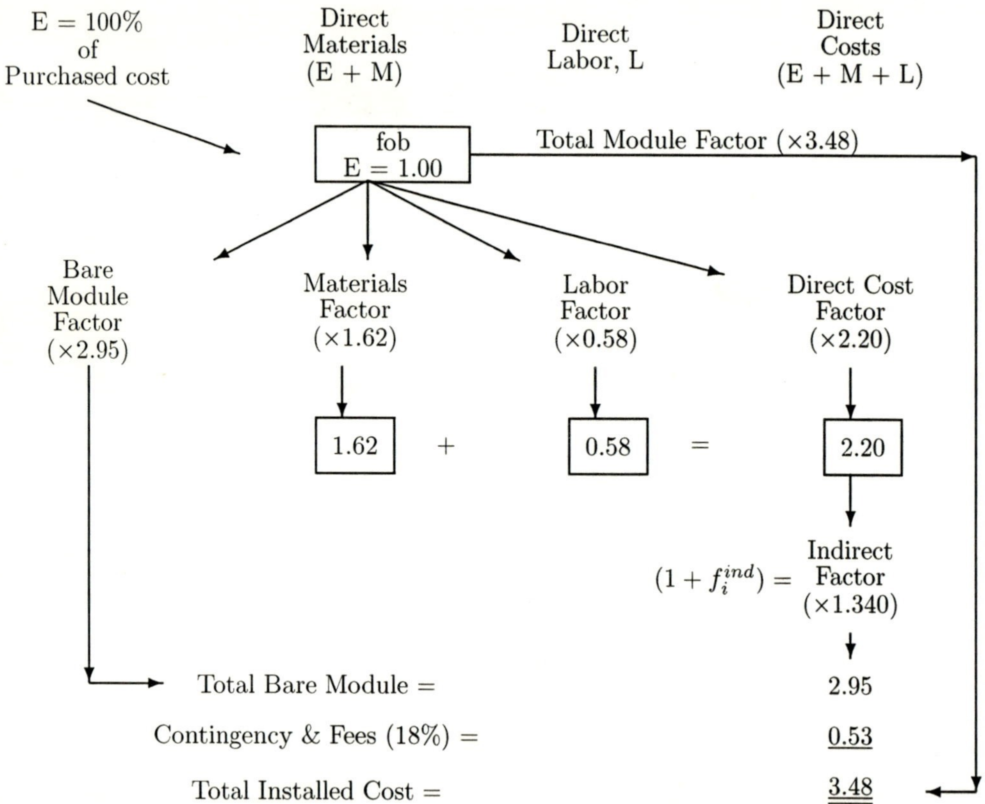

Figure 4.3: Use of Guthrie factors to compute Total Installed Costs from fob purchased costs (average or "norm" chemical process module factors). All factors are expressed as a percentage of E, the fob purchased cost for a low-pressure, carbon steel unit.

Table 4.17: Time independent installed cost factors per unit dollar of base equipment. Taken from Guthrie.

Cost Element	Process Furnaces[a]	Fired Heaters[b]	Fired Heaters[a]	Shell[b] & Tube Heat Exchangers
Base Cost (E_i)	1.000	1.000	1.000	1.000
Direct Materials (M_i)				
Piping	0.177	0.148	0.118	0.447
Concrete	0.081	0.084	0.050	0.050
Steel	0.000	0.000	0.000	0.030
Instruments	0.039	0.048	0.048	0.100
Electrical	0.020	0.020	0.032	0.020
Insulation	0.018	0.015	0.012	0.047
Paint	0.000	0.000	0.000	0.005
Total Field Mat (M_i)	0.335	0.315	0.260	0.699
Field Labor (L_i)	0.291	0.286	0.215	0.604
Total Directs[c]	1.626	1.601	1.475	2.303
Total Indirects	0.504	0.576	0.369	0.875
Bare Module Cost	2.130	2.177	1.844	3.178
Cont & Fee (18%)	0.383	0.392	0.332	0.572
Total Installed Cost	2.513	2.569	2.176	3.750
Factor Summary				
$L_i/(E_i + M_i)$	0.218	0.217	0.171	0.356
Materials ($E_i + M_i$)	1.335	1.315	1.260	1.699
Directs ($E_i + M_i + L_i$)	1.626	1.601	1.475	2.303
Indirects ($1 + f_i^{ind}$)	1.310	1.360	1.250	1.380

[a] Field fabricated
[b] Shop fabricated
[c] $(E_i + M_i + L_i)$

Table 4.18: Time independent installed cost factors per unit dollar of base equipment. Taken from Guthrie.

Cost Element	Air Coolers[a]	Vertical Process Vessels	Horizontal Process Vessels	Pumps & Drivers
Base Cost (E_i)	1.000	1.000	1.000	1.000
Direct Materials (M_i)				
Piping	0.139	0.595	0.397	0.293
Concrete	0.014	0.098	0.060	0.039
Steel	0.000	0.078	0.000	0.000
Instruments	0.037	0.114	0.060	0.029
Electrical	0.093	0.049	0.050	0.303
Insulation	0.000	0.080	0.050	0.028
Paint	0.006	0.013	0.005	0.008
Total Field Mat (M_i)	0.289	1.027	0.622	0.700
Field Labor (L_i)	0.294	0.972	0.580	0.679
Total Directs[b]	1.583	2.999	2.202	2.379
Total Indirects	0.602	1.140	0.837	0.904
Bare Module Cost	2.185	4.139	3.039	3.283
Cont & Fee (18%)	0.393	0.745	0.547	0.591
Total Installed Cost	2.578	4.884	3.586	3.874
Factor Summary				
$L_i/(E_i + M_i)$	0.228	0.480	0.358	0.399
Materials $(E_i + M_i)$	1.289	2.027	1.622	1.700
Directs $(E_i + M_i + L_i)$	1.583	2.999	2.202	2.379
Indirects $(1 + f_i^{ind})$	1.380	1.380	1.380	1.380

[a]Field assembled

[b]$(E_i + M_i + L_i)$

Table 4.19: Time independent installed cost factors per unit dollar of base equipment. Taken from Guthrie.

Cost Element	Gas Compressors	Boiler Plants[a]	Boiler Plants[b]	Turbine Electricity Generators[a]
Base Cost (E_i)	1.000	1.000	1.000	1.000
Direct Materials (M_i)				
Piping	0.141	0.082	0.068	0.028
Concrete	0.043	0.065	0.052	0.032
Steel	0.000	0.000	0.000	0.000
Instruments	0.013	0.042	0.032	0.008
Electrical	0.068	0.026	0.023	0.058
Insulation	0.005	0.008	0.004	0.002
Paint	0.000	0.001	0.001	0.000
Total Field Mat (M_i)	0.270	0.224	0.180	0.128
Field Labor (L_i)	0.295	0.195	0.306	0.156
Total Directs[c]	1.565	1.419	1.486	1.284
Total Indirects	0.595	0.397	0.312	0.193
Bare Module Cost	2.160	1.816	1.798	1.477
Cont & Fee (18%)	0.389	0.327	0.324	0.266
Total Installed Cost	2.548	2.143	2.122	1.742
Factor Summary				
$L_i/(E_i + M_i)$	0.232	0.159	0.259	0.138
Materials $(E_i + M_i)$	1.270	1.224	1.180	1.128
Directs $(E_i + M_i + L_i)$	1.565	1.419	1.486	1.284
Indirects $(1 + f_i^{ind})$	1.380	1.280	1.210	1.150

[a]Field erected

[b]Packaged units

[c]$(E_i + M_i + L_i)$

Table 4.20: Time independent installed cost factors per unit dollar of base equipment. Taken from Guthrie.

Cost Element	Cooling Towers[a]	Mechanical Refrigeration[a]	Storage Tanks[b]
Base Cost (E_i)	1.000	1.000	1.000
Direct Materials (M_i)			
Piping	0.082	0.058	0.043
Concrete	0.108	0.043	0.028
Steel	0.000	0.008	0.000
Instruments	0.006	0.023	0.026
Electrical	0.023	0.052	0.008
Insulation	0.000	0.000	0.000
Paint	0.000	0.000	0.003
Total Field Mat (M_i)	0.219	0.184	0.108
Field Labor (L_i)	0.218	0.168	0.275
Total Directs[c]	1.437	1.352	1.383
Total Indirects	0.259	0.176	0.249
Bare Module Cost	1.696	1.528	1.632
Cont & Fee (18%)	0.305	0.275	0.294
Total Installed Cost	2.001	1.803	1.926
Factor Summary			
$L_i/(E_i + M_i)$	0.179	0.142	0.248
Materials ($E_i + M_i$)	1.219	1.184	1.108
Directs ($E_i + M_i + L_i$)	1.437	1.352	1.383
Indirects ($1 + f_i^{ind}$)	1.180	1.130	1.180

[a]Field erected
[b]Shop fabricated
[c]$(E_i + M_i + L_i)$

Table 4.21: Time independent installed cost factors per unit dollar of base equipment. Taken from Guthrie.

Cost Element	Storage Tanks[a]	Horizontal Pressure Storage Tanks	Spherical Pressure Storage Tanks
Base Cost (E_i)	1.000	1.000	1.000
Direct Materials (M_i)			
Piping	0.128	0.153	0.108
Concrete	0.086	0.080	0.066
Steel	0.000	0.000	0.000
Instruments	0.038	0.118	0.086
Electrical	0.006	0.007	0.004
Insulation	0.000	0.000	0.000
Paint	0.012	0.012	0.009
Total Field Mat (M_i)	0.270	0.370	0.273
Field Labor (L_i)	0.267	0.352	0.318
Total Directs[b]	1.537	1.722	1.591
Total Indirects	0.400	0.362	0.366
Bare Module Cost	1.937	2.084	1.957
Cont & Fee (18%)	0.349	0.375	0.352
Total Installed Cost	2.285	2.459	2.309
Factor Summary			
$L_i/(E_i + M_i)$	0.210	0.257	0.250
Materials $(E_i + M_i)$	1.270	1.370	1.273
Directs $(E_i + M_i + L_i)$	1.537	1.722	1.591
Indirects $(1 + f_i^{ind})$	1.260	1.210	1.230

[a]Field erected

[b]$(E_i + M_i + L_i)$

Example: 4.5 *Estimate the installed cost of the monel air cooler in Example 4.4 in mid-1968. From Table 4.18, the piping factor is 0.139 and 70% of the piping is monel. From Example 4.4,* $E = 5,375$, $E' = 17,200$.

Solution: 4.5 *Using the factors in Table 4.18 gives the following:*

Purchased equipment cost(E')			=	*17,200*
Process Piping (monel)	=	*0.7(0.139)(17,200)*	=	*1,674*
All other directs	=	*[0.583 -0.7(0.139)](5,375)*	=	*2,611*
Total direct costs(E' + M + L)			=	*21,485*
Indirect Costs	=	*0.602(5,375)*	=	*3,236*
Bare module cost			=	*24,721*
Contingency + fee (18%)			=	*4,450*
Total Installed Cost			=	*29,171*

Check these calculations using the Factors in Tables 4.10 and 4.13, $F_{bm}^{cs} = 2.185$, $f_t = 0.139$, *and* $F_m = 3.20$. *At low P,* $F_p = 1.0$. *From Table 4.3:*
$F_{bm}^{a} = F_{bm}^{cs} + (1 + 0.7f_t)(F_m F_p - 1) = 2.185 + (1 + 0.7(0.139))(3.20 - 1) = 4.599$
$TIC = 1.18(BMC) = 1.18F_{bm}^{a} C_p^{cs} = 1.18(4.599)(5,375) = 29,169 \approx 29,171$
The Process Piping Cost is: $0.7f_t C_P^a = 0.7(0.139)(17,200) = 1,674$

Example: 4.6 *Estimate the installed cost of the carbon steel air cooler in Example 4.4 in mid-1968.*

Solution: 4.6 *From Table 4.18, the installed cost factor is 2.578. Total Installed Cost = 2.578(5,375) = $13,857*
Using the Factors in Tables 4.10, $F_{bm}^{cs} = 2.185$, $f_t = 0.139$ *For a carbon steel unit, the total installed cost is:*
$TIC = 1.18F_{bm}^{cs} C_p^{cs} = 1.18(2.185)(5,375) = 13,858$
The Process Piping Cost is: $0.7f_t C_p^{cs} = 0.7(0.139)(5,375) = 523$

Tables 4.17 through 4.21 summarize representative cost factors that can be used to provide more detailed estimates of where money is likely to be spent installing the unit operations described or, by analogy, similar unit operations which are not explicitly listed in the tables. The bare module costs can also be estimated without detailing the information in Tables 4.17 through 4.21, but simply by using the factors provided in Tables 4.9 through 4.12 and the equations in Table 4.3, along with the pressure and material factors in Tables 4.13 through 4.16, as needed.

The actual fob purchase cost for any item is:

$$C_p^a = F_m F_p C_p^{cs} = F_m F_p E_0 \left(\frac{S}{S_0}\right)^a \tag{4.11}$$

The cost for the process piping (70% of all installed piping) is

$$C_t^a = 0.7f_t C_p^a \tag{4.12}$$

where f_t is the piping (or tubing) factor taken either from Tables 4.9 through 4.12 or from Tables 4.17 through 4.21. (In either case, the tubing factors, f_t, should be the same for the same unit operations.)

None of the remaining installation costs, except for the contingencies, are affected by special materials, F_m, or high operating pressures, F_p. So the actual bare module factor, F_{bm}^a, can be computed from the low-pressure, carbon steel bare module factor, F_{bm}^{cs}, by

$$F_{bm}^a = F_{bm}^{cs} + (1.0 + 0.7f_t)(F_m F_p - 1) \tag{4.13}$$

and the actual bare module cost for the unit operation, considering pressure and special materials of construction, is

$$C_{bm}^a = C_p^{cs} F_{bm}^a = C_p^{cs}\left[F_{bm}^{cs} + (1 + 0.7f_t)(F_m F_p - 1)\right] \tag{4.14}$$

The steps in cost estimation based on the battery limits design are summarized as follows. First develop a process flow diagram and a major equipment list for everything in the battery limits of the process. For each line item in the major equipment list, estimate a low-pressure, carbon steel fob purchase cost. Then estimate the actual bare module factor, F_{bm}^a, considering materials and pressure. Use them to compute C_{bm}^a for each item in the battery limits. The sum is then the estimated bare module cost for the battery limits.

Example: 4.7 *Estimate the fob purchased costs and Total Installed Costs for distillation towers consisting of vertical process vessels and sieve tray internals in mid-year 1990 (M&S = 915.1). The towers operate at 150 psia. One is constructed of carbon steel (CS); the second is constructed of 316 stainless steel (SS). Both are 110 ft tall and 9 ft in diameter. The internals are shop installed, 2 ft spacing, sieve trays. Assume contingency & fees are 30% of the bare module costs. Assume that the cost of shop installation of internals is 5% of their fob purchase cost, carbon steel basis.*

Solution: 4.7 *From Table 4.16,* $F_p = 1.0 + 0.001(150 - 50) = 1.10$
From Table 4.13, $F_m = 1.0$ *for CS VPV, and* $F_m = 3.67$*, for SS VPV*
From Table 4.6, $\sum F_i = 1.0$ *for CS trays, and* $\sum F_i = 2.7$ *for SS trays.*
From Table 4.4, $C_3 = 33.143/110 + 0.127 = 0.4283$
From Table 4.4, $E = 2000(9/0.4283)^{1.047}(915.1/273.1) = 162,491$
$E' = F_p E = 1.1E = 178,740$, *for carbon steel*
$E'' = F_m E' = 3.67E' = 655,975$, *for a stainless VPV*
For the trays, from Table 4.5, $C_4 = 25(9) - 75 = 150$
From Table 4.5, $E_i = E_i' = 150(110 - 10)(915.1/273.1) = 50,262$
For the internals $E_i = E_i'$ *because operating the column at high (or low) pressure does not result in a pressure gradient across the internals, since they are entirely inside the column.*
$E_i'' = \sum F_i E_i' = 2.7E_i' = 135,707$
Table 4.22 was constructed using this information and the factors in Table 4.18, together with the following observations. The shop installation costs for the internals are taken as 5% of E_i*. Since the internals are shop installed, the only indirects incurred are those of freight, insurance, and shipping taxes, taken as 8% of* E_i*.*

Table 4.22: Estimated Total Installed Costs for carbon steel and stainless steel vertical process vessels and sieve tray internals, shop installed.

Item	CS Unit (1990 \$)	SS Unit (1990 \$)
Vertical Process Vessel (VPV) (E' or E'')	178,740	655,975
Process Piping [0.7(0.595)(E' or E'')	74,445	273,214
Other Directs $(1.999 - 0.4165)E$	257,142	257,142
Total Directs	510,327	1,186,331
Indirects ($1.14E$)	185,240	185,240
Bare Module Costs (BMC)	695,567	1,271,571
Contingency & Fees (30% of BMC)	208,670	411,471
VPV Total Installed Costs (TIC)	904,237	1,783,042
Internals[a] (Sieve Trays) (E_i' or E_i'')	50,262	135,707
Installation, Shop Labor $0.05E_i'$	2,513	2,513
Total Direct Costs	52,775	138,220
Freight, Insurance, Taxes[b] $0.08E_i'$	4,021	4,021
Bare Module Costs (BMC)	56,796	142,241
Contingency & Fees (30% of BMC)	17,039	42,672
Total Installed Costs for Trays	73,835	184,913
Total Installed Costs for VPV and Trays	**978,072**	**1,967,995**

[a]Shop Installed, Total Direct Costs
[b]Average Value

Example: 4.8 *Estimate the fob purchased costs and Total Installed Costs for distillation towers consisting of vertical process vessels and sieve tray internals in mid-year 1990 (M&S = 915.1). The towers operate at 150 psia. One is constructed of carbon steel (CS); the second is constructed of 316 stainless steel (SS). Both are 110 ft tall and 12 ft in diameter. Because of the column diameter, the internals are field installed, 2 ft spacing, sieve trays. Assume contingency & fees are 18% of the bare module costs.*

Solution: 4.8 *From Table 4.16,* $F_p = 1.0 + 0.001(150 - 50) = 1.10$
From Table 4.13, $F_m = 1.0$ *for CS VPV, and* $F_m = 3.67$, *for SS VPV*
From Table 4.6, $F_{mi} = 1.0$ *for CS trays, and* $F_{mi} = 1.7$ *for SS trays.*
From Table 4.4, $C_3 = 33.143/110 + 0.127 = 0.4283$
From Table 4.4, $E = 2000(12/0.4283)^{1.047}(915.1/273.1) = 219,604$
$E' = F_pE = 1.1E = 241,564$, *for carbon steel*
$E'' = F_mE' = 3.67E' = 886,540$, *for a stainless VPV*
For the trays, from Table 4.5, $C_4 = 20(12) - 60 = 180$
From Table 4.5, $E_i = E_i' = 180(110 - 10)(915.1/273.1) = 60,314$
For the internals $E_i = E_i'$ *because operating the column at high (or low) pressure does not result in a pressure gradient across the internals, since they are entirely inside the column.*
$E_i'' = F_mE_i' = 1.7E_i' = 102,534$
Table 4.23 was constructed using this information and the factors in Table 4.18, together with the following observations. Since the internals are field installed, field labor cost is incurred. This is figured using Figure 4.3 which is an average for chemical processing modules (i.e., $L = 0.58E_i$*). Similarly, the indirects are also figured using Figure 4.3 (i.e.,* $Ind = 0.75E_i$*).*

Example: 4.9 *A company is evaluating two options for an absorber. In the first option, a vertical pressure (or process) vessel (VPV) will be filled with porcelain Raschig rings (see Table 4.7). Field materials and labor* ($M_I + L_i$) *for the installed Raschig rings cost $7.0/ft*^{3} *in mid-1968 dollars. The VPV in this case will contain a Raschig ring bed that is 100 ft high and 6 ft in diameter. The total VPV height is 110 ft. The packing material in this case will be field installed. The Total Indirects for the Raschig rings are taken as* $f_i^{ind}(M_i + L_i) = 0.34(M_i + L_i)$ *(assumed to be average, based on Figure 4.3).*

In the second case, the VPV will contain a high efficiency packing that costs (E_i) *$120/ft*^{3} *in mid-1968 dollars, but the required VPV bed height is only 40 feet and the total VPV height is 50 ft. The VPV nominal diameter in this second case is 4 ft. The bed packing material will be shop installed with Other Directs for the high efficiency packing consisting of shop labor with a factor of* $0.05E_i$. *The Total Indirects for the high efficiency packing are* $0.08E_i$ *(freight, taxes, and insurance only).*

All other costs and risk factors are the same for these two options and may be ignored. In either case, the column will operate at atmospheric pressure and can be constructed of carbon steel. Total directs and indirects for the two VPVs should be figured using the normal factors to estimate total directs and total indirects. The factors above are for installing the packing materials for the absorbers only.

Table 4.23: Estimated Total Installed Costs for carbon steel and stainless steel vertical process vessels and sieve tray internals, field installed. (Column diameter greater than 10 ft.)

Item	CS Unit (1990 \$)	SS Unit (1990 \$)
Vertical Process Vessel (VPV) (E' or E'')	241,564	886,540
Process Piping [0.7(0.595)](E' or E'')	100,611	369,244
Other Directs (1.999 – 0.4165)E	347,523	347,523
Total Directs	689,698	1,603,307
Indirects (1.14E)	250,349	250,349
Bare Module Costs (BMC)	940,038	1,853,656
Contingency & Fees (18% of BMC)	169,207	333,658
VPC Total Installed Costs (TIC)	1,109,245	2,187,314
Internals[a] (Sieve Trays) (E' or E'')	60,314	102,534
Field Labor, 0.58(E_i)	34,982	34,982
Total Direct Costs	95,296	137,516
Indirects 0.75E_i	45,236	45,236
Bare Module Costs (BMC)	140,532	182,752
Contingency & Fees (18% of BMC)	25,296	32,895
Total Installed Costs for Trays	165,828	215,647
Total Installed Costs for VPV and Trays	**1,275,073**	**2,402,961**

Contingency and fees are 30% of the Bare Module Cost in either case. For the initial installation, assume that M&S = 1070. In either case, the column is expected to operate for at least five years without replacement.

The company earns 15% p.a. Based on net present cost, should the company install a Raching ring column or use high efficiency packing? Based on venture cost, which option should be selected? Base your conclusions on total installed costs in either case. Estimate the Net Present Cost (NPC) and Venture Cost (VC) in either case. (See Chapter 8 for more details about NPC and VC.)

Solution: 4.9 *Option 1*

$C_3 = 33.143/110 + 0.127 = 0.4283,\ E = 2000(6.0/0.4283)^{1.047}(1070/273.1) = 124,273$

$TIC(VPV) = \$124,273(4.139)(1.30) = \$668,676$

Field Materials and Labor for Internals, $M_i + L_i = 3.1416(6/2)^2(100)(7)(1070/273.1) = \$77,545$

Indirects for Internals: $= 0.34(M_i + L_i) = 26,365$

Bare Module Cost $= 77,545 + 26,365 = \$103,910$

$TIC(Internals) = 1.3(103,910) = \$135,083$

Option 1 TIC $= 668,676 + 135,083 = \$803,759$

$NPC = 803,759,\ VC = 803,759 + 0.15(803,759) = 924,323$

Option 2

$C_3 = 33.143/90 + 0.127 = 0.78986,\ E = 2000(4.0/0.78986)^{1.047}(1070/273.1) = 42,827$

$TIC(VPV) = \$42,827(4.139)(1.30) = \$230,437$

Internals, $E = 3.1416(4/2)^2(40)(120)(1070/273.1) = \$236,327$

Other Directs $= 0.05E = 11,816$, *Indirects* $= 0.08E = 18,906$

Bare Module Cost $= 236,327 + 11,816 + 18,906 = \$267,050$

$TIC(Internals) = 1.3(267,050) = \$347,165$

Option 2 TIC $= 230,437 + 347,165 = \$577,602$

$NPC = 577,602,\ VC = 577,602 + 0.15(577,602) = 664,242$

Based on either Net Present Cost or Venture Cost, the high-efficiency packing should be installed.

4.4 Total Chemical Plant Costs: Predesign

Estimation of the total chemical plant cost requires an evaluation of the other three modules — industrial buildings, site development, and offsite facilities. In general, it is preferable to itemize these requirements for the chemical process which they support. This analysis requires detailed information about the site. Guthrie [35, 36] provides additional guidelines.

Chemical plants are frequently divided into Areas. Figure 4.4 is a typical layout. Major areas of the chemical plant include: (1) the battery limits (or chemical process area), (2) the tank farm, (3) the utilities area, and (4) the services area.

A complete cost picture can be organized by dividing the cost elements into modules. Five of the modules represent direct costs:

1. chemical processing,

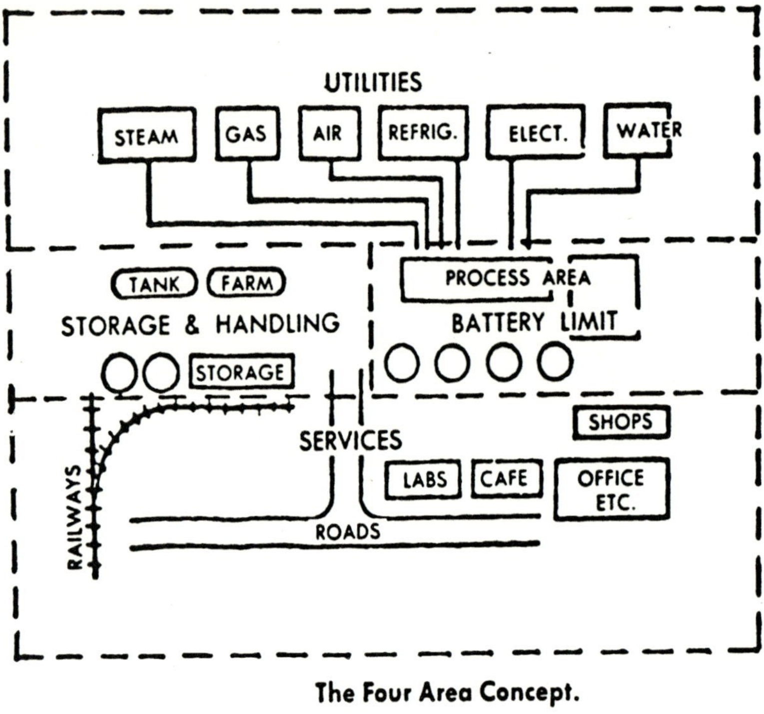

Figure 4.4: The four area concept of chemical plant layout. Adapted with permission from Miller [65]. Courtesy of McGraw-Hill Publishing.

2. solids handling,
3. site development,
4. industrial buildings, and
5. offsite facilities.

Within each of these modules, there is a degree of similarity with respect to the relative amounts by which the various cost elements contribute to the module cost.

The sixth module, consisting of the indirect module costs, is a function of the five direct module costs for a complete plant, but the contribution of any direct module to the indirect costs can be estimated by factors.

A chemical plant can usually be divided into generic areas. The chemical process area is often referred to as the *battery limit.* However, the battery limit requires additional support facilities such as storage and handling areas for raw materials and finished products, electrical, steam, air and water systems, sewers and effluent treatment facilities, analytical laboratories, offices, roads, and security systems. The offsites supporting the battery limit typically include three general components:

1. the storage and handling areas,
2. the utilities, and
3. the other support services.

Although it is desirable to itemize those components of the plant besides the battery limit, for purpose of preliminary design they can be factored. Table 4.24 summarizes typical values for factors for the components which are expressed as a percentage of the total installed cost of the plant. As a first approximation during Preliminary Engineering, it is often sufficient to assume that the Total Installed Cost is twice the installed cost of the Battery Limits.

Heuristic: 9 *As a first approximation, assume that the Total Installed Cost for a grass-roots chemical plant, including all offsites, is roughly twice (200% of) the installed cost of the battery limits. Assume that the installed cost for an expansion to an existing plant is 133% of the installed cost of the battery limits.*

Note that Heuristic 9 is an approximation. It is helpful for preliminary design, but it can overestimate results. If the battery limits include large amounts of equipment that require special materials of construction, say due to corrosion problems, then assuming that the total installed costs are 200% of the installed cost of the battery limits may overestimate true costs, since much of the offsites will not require special materials of construction. For the sake of simplicity, however, the effects of special materials of construction on the total installed cost estimate for the entire plant are not corrected using Heuristic 9. While this approximation may inflate the preliminary estimates somewhate, it should also motivate engineers to correct these errors during Conceptual Engineering, if the design project continues to that level of detail.

Table 4.24: Percentage factors [76, 75, 3] for industrial buildings, site development, and offsites.

Item	Percentage Range of Total Installed Cost
Industrial buildings	1–6
Site development	2–5
Offsites	9–40
Totals	12–51

Example: 4.10 *Estimate the total installed cost of a chemical plant supporting a chemical process with an estimated installed battery limit cost of \$12,000,000. from Table 4.24, the total cost of the non-battery limit items are estimated to be 40% of the total installed cost.*

Solution: 4.10 *Let X equal the total installed cost, then:*

$$\begin{aligned} 0.4X + 12,000,000 &= X \\ X = 12,000,000/0.6 &= \$20,000,000 \end{aligned}$$

4.5 Total Chemical Plant Costs: Page Method

Page [74] offers a method similar to Guthrie's for estimating the total installed cost of a chemical plant based on the delivered cost of equipment. It is similar to Guthrie's method in that ratio factors are used to estimate direct and indirect costs from delivered equipment costs (i.e., fob purchased cost, E', plus shipping, taxes, and insurance). It is different in that: (1) factors are selected based on the type of chemical plant, rather than the individual type of equipment and (2) all costs are based on the actual delivered equipment costs, rather than low pressure carbon steel.

It is the author's opinion that Page's method should be modified to compute the most meaningful results. All direct costs should be based on the fob purchased cost of equivalent, low pressure carbon steel equipment, except for 70% of the piping costs which correspond to process piping and the process equipment itself. Secondly, shipping, taxes, and insurance should be added as an indirect cost.

Page defines six categories, or chemical plant types, from which to select ratio percentages to estimate direct costs:

1. Refinery Process Systems
2. Liquid Type Chemical Plants
3. Solid Type Chemical Plants
4. Liquid/Solid Type Chemical Plants
5. Chemical Plants–Liquid High Pressure Types 4000 to 5000 psi

6. Chemical Plants–Liquid High Alloy Types

The Refinery Process Systems are based upon the following types of refinery systems: (1) Atmosphere and vacuum crude distillation, (2) Saturates gas plant, (3) Naphtha hydrodesulphurization, (4) Distillate hydrotreater, (5) Flexicoker, (6) Heavy oil hydrodesulphurization, (7) Fluid catalytic cracker, (8) Unsaturates gas plant, (9) Cycle oils hydrodesulphurization, (10) Fluid catalytic cracker, (11) Unsaturates gas plant, (12) Cycle oils hydrodesulphurization, (13) C_5, C_6 isomerization, (14) Catalytic reformer, (15) Aromatics extractions, (16) Steam reforming hydrogen plant, (17) Butane alkylation, (18) Olifin poly plant and (19) H_2S recovery and sulphur manufacturing. Most of these operations operate at low to moderate pressure and equipment is largely manufactured of carbon steel. So special materials of construction and high-pressure operation are not particularly important in refinery operations.

The basis for the remaining categories are not defined by Page. Comparisons with Guthrie's methods suggest that using Page's procedures will yield higher estimates than Guthrie's especially for chemical plants operating at high pressures or containing significant amounts alloys or other special materials of construction. This conclusion follows from the fact that the suggested ratios in Tables 4.29 and 4.30 are about the same as those in Table 4.26. Using Guthrie's method, direct materials and labor costs are always factored from the low-pressure carbon steel cost of equipment, not actual costs.

Nonetheless, Page's method, especially in a modified form, can be helpful in providing a benchmark and, with Guthrie's method, will give a range of likely costs. Since both methods depend ultimately on the accuracy of the fob purchased equipment costs, the results can be skewed by inaccurate estimates of process equipment requirements and, particularly, by incorrectly selecting required materials of construction. Page indicates that the accuracy of this type of ratio estimate is no better than ±30%. It has the advantage of being somewhat easier to compute, compared to Guthrie's approach, since ratios are applied only once to the total delivered equipment costs. Page's method claims to estimate the entire plant cost, if all equipment (battery limits plus offsites) is included in the delivered equipment costs. All of Page's tables and graphs are based on costs as of July 1, 1984. Ratios (e.g. Tables 4.25–4.34), are presumed to depend primarily on the type of plant constructed and to be virtually time independent.

If the project is to be located outside of the continental forty-eight states, then other items should be considered. The actual costs can vary widely for different parts of the world, and must be evaluated individually for each foreign project.

1. Mobilization and demobilization of personnel and equipment
2. Foreign differential labor pay
3. Permits, licenses, fees, duties
4. Load out, tie down and material handling
5. Transportation and ocean or air freight

6. Camp, family quarters, field office, and storage facilities
7. Rest and recreation for personnel
8. Personnel and site security
9. Potential acts of war or terrorism

The following information is required to achieve a degree of accuracy to ±30%.

1. Plant capacity, product form, basic process and raw material
2. Knowledge of general specifications
3. Plant location (site conditions)
4. Equipment list (priced out) for battery limits plus utilities

4.5.1 Direct Costs

Tables 4.25 through 4.30 provide factors for estimating direct costs using Page's method. He defines three types of direct costs: (1) material, (2) field labor, and (3) subcontracts. The direct materials and field labor are essentially the same as defined by Guthrie and described in Section 4.1. In addition, Page includes a subcontracts category as a direct cost. Subcontract items include the total cost of labor, material, overhead and profit of the supplier, and vendor or subcontractor to complete specified work. Subcontractors are usually specialists in applying a particular construction art or technology.

Process Equipment The equipment costs, labor, material and subcontracts to intall the process equipment for operation. This is the delivered[2] equipment cost. It includes the fob purchased equipment price plus freight, insurance and taxes to ship from the equipment manufacturer's site to the plant construction site. In the continental forty-eight states shipping may be taken as 3 to 5% of the fob purchased equipment cost. Overseas shipping costs are higher.

Site Preparation All general preparation of the plant site for constuction. Includes such things as clearing, rough grading, stabilization and finish grading.

Site Improvements Includes some permanent improvements such as drainage ditches, culverts, storm drains, etc, railroads, fencing.

Concrete All earthwork forms, reinforcing steel, and embedded items for placing, finishing, and curing concrete for equipment foundations and structures.

Structural Steel All cost of procuring, fabricating, and erecting structural steel for framed structures, equipment supports, platforms, walkways, ladders, etc.

[2]The author recommends basing direct costs on fob purchased equipment costs as Guthrie advocates, and then adding freight, taxes and insurance costs as an indirect.

Buildings All building costs including earthworks, foundations, superstructure, HVAC, plumbing and utility piping to five feet outside of building, insulation, electrical, and painting.

Underground Piping All underground piping including fire water loops, water lines and special process lines.

Above Ground Piping Fabrication and erection of all above ground process pipe, valves, and fittings required for plant operations. Installation of relief and control valves, and orfice fittings for instrumentation.

Underground Electrical All labor and materials for underground power and lighting, ducts, conduits, wire, cable, etc.

Above Ground Electrical All items for the complete above ground power and lighting systems for the total project, excluding buildings.

Instrumentation Includes all material and labor costs for instrumentation required to monitor and control the process.

Insulation All labor and material costs for insulation, refractories, and linings for process equipment and piping. Building insultation is included with the buildings.

Painting Includes all preparation and protective coatings of surfaces with paint and/or epoxy for total project exluding buildings.

Paving Includes furnishing and placing concret, asphalt, or gravel for surfacing roads, sidewalks, parking areas, and process work areas.

Proratables Includes the labor and material costs for daily area cleanup, all required scaffolding and final test, startup, and pre-commissioning assistance.

4.5.2 Indirect Costs

Tables 4.31 through 4.33 provide factors for estimating construction equipment costs, overhead and indirects, both as ratios of direct field labor, and home office service costs as a ratio of total direct and indirect costs, respectively.

Construction Equipment

Construction equipment costs are grouped into three categories.

Rental or Purchase Cost of rental or purchase of construction equipment required to construct the project. Does not include small tools.

Service Labor All labor required to service and maintain the construction equipment. Does not include the equipment operation which is part of the Direct Labor.

Fuel, Oil, Grease, Supplies Includes the cost of all fuels and supplies to operate and maintain the construction equipment.

Home Office Costs

Home Office Costs are divided into three categories.

Engineering/Design Services Total labor and material costs for complete engineering and design of the project including engineering fee.

Construction Services Services such as estimating, scheduling, expediting, etc. as required to assist the project.

Project General Management Time that contractor's management staff applies to direct the project.

Example: 4.11 *The delivered costs of all project equipment is \$10 million. Estimate the Total Installed Cost of the chemical plant using Page's method. Assume that it is a liquid type chemical plant, operating at near ambient pressures, with negligible special materials of construction.*

Solution: 4.11 *Use average values from Tables 4.26, 4.31, 4.32, and 4.33.*
Total Direct Materials = \$10(2.277) = \$22.28 million
Total Direct Labor = \$10(0.772) = \$7.72 million
Total Subcontracts = \$10(0.345) = \$3.45 million
Total Directs = \$33.45 million including all process equipment.
Construction equipment = \$7.72(.31) = \$2.39 million
Overheads and Indirects = \$7.72(0.85) = \$6.56 million
Home Office Servies = 0.133(33.45+2.39+6.56) = \$5.64 million
Total Installed Cost = 33.45+2.39+6.56+5.64 = \$48.04 million.

How well do Guthrie's and Page's methods compare? For low pressure, carbon steel equipment, and ignoring shipping costs, Page's method estimates that direct costs are about 50% greater than Guthrie's method (also see Fig 4.3). Part of this difference is due to the fact that Page includes certain categories that are excluded by Guthrie (i.e., Site Preparation and Improvements, Buildings, Paving and Proratables). Of greater concern, Page's direct cost factors for a High Pressure Liquid Type and a High Alloy Liquid Type are both about the same as for a Liquid Type. Presumably the latter type is low pressure carbon steel. If so, then the direct costs for high alloy and high pressure plants should be a lower percentage of the total equipment costs, not about the same as indicated in Table 4.34. Using Page's factors in Table 4.34 will give much higher estimates than Guthrie's method where the direct costs, other than for the equipment and the process piping, are based strictly on low pressure carbon steel. The following example illustrates this point.

Example: 4.12 *Estimate the direct costs for constructing a liquid high alloy type chemical plant using Guthrie's and Page's methods. Ignore shipping costs. Estimate the low pressure carbon steel equipment costs using an average materials factor. The plant is 30% monel, 10% titanium, 30% 316 stainless steel, and 20% carbon steel. Use an average material factor from Table 4.13. The fob purchased equipment cost is \$20 million.*

Table 4.25: Refinery Process Systems. Ratio Percentages Direct Costs. Typical ranges for Ratio Estimating Factors to predict direct field materials and field labor [74]. Expressed as percentages of the delivered equipment cost.

Cost Category	Direct Materials	Subcontracts	Direct Labor
Process Equipment[a]	100.00	44.30–44.60	7.70–7.80
Site Preparation	0.02–0.05	0.25–0.30	2.75–2.85
Site Improvements	0.95–1.10	0.0	1.10–1.12
Concrete	4.25–4.75	0.10–0.12	12.70–12.90
Structural Steel	8.00–8.30	0.0	3.70–3.75
Buildings	1.10–1.25	2.20–2.40	1.70–1.75
Underground Piping	1.10–1.20	0.0	1.40–1.48
Above-ground Piping	32.00–35.00	0.75–0.80	21.30–22.40
Underground Electric	0.30–0.50	0.0	0.65–0.70
Above-ground Electric	11.50–12.00	0.0	6.10–6.25
Instrumentation	10.30–10.70	0.0	3.20–3.50
Insulation	4.50–5.00	0.0	7.40–7.50
Painting	1.50–1.70	0.0	3.90–4.00
Paving	0.40–0.60	0.0	0.60–0.70
Proratables	1.00–1.10	0.0	2.80–2.95
Total Direct Costs	176.9–183.3	47.6–48.2	77.0–79.7

[a]delivered cost

Table 4.26: Liquid Type Chemical Plants. Ratio Percentages Direct Costs. Typical ranges for Ratio Estimating Factors to predict direct field materials and field labor [74]. Expressed as percentages of the delivered equipment cost.

Cost Category	Direct Materials	Subcontracts	Direct Labor
Process Equipment[a]	100.00	24.00–25.50	3.75–5.50
Site Preparation	1.75–2.50	0.01	0.50–1.00
Site Improvements	2.75–3.50	0.60–0.75	1.50–2.25
Concrete	7.50–8.75	2.00–3.00	9.00–12.00
Structural Steel	11.50–12.50	0.00	4.00–6.00
Buildings	4.00–5.00	3.00–3.50	3.50–4.50
Underground Piping	4.00–5.00	0.01–0.02	2.75–3.25
Above-ground Piping	37.00–43.00	2.75–3.25	24.00–27.00
Underground Electric	0.40–0.60	0.0	0.20–0.40
Above-ground Electric	15.00–17.00	0.0	6.50–7.75
Instrumentation	22.00–25.00	0.0	3.25–4.50
Insulation	7.50–8.50	0.0	4.50–5.25
Painting	2.50–3.25	0.0	2.75–2.90
Paving	1.00–1.50	0.30–0.40	0.50–1.00
Proratables	1.00–1.50	0.0	2.00–2.40
Total Direct Costs	217.9–237.5	32.57–36.43	68.7–85.7

[a]delivered cost

Table 4.27: Solid Type Chemical Plants. Ratio Percentages Direct Costs. Typical ranges for Ratio Estimating Factors to predict direct field materials and field labor [74]. Expressed as percentages of the delivered equipment cost.

Cost Category	Direct Materials	Subcontracts	Direct Labor
Process Equipment[a]	100.00	25.00–26.00	3.75–4.75
Site Preparation	2.50–3.00	0.01	0.50–1.00
Site Improvements	3.50–3.75	0.80–0.90	2.25–3.00
Concrete	8.75–10.25	2.50–3.00	12.00–13.75
Structural Steel	12.50–14.30	0.00	5.50–6.25
Buildings	5.00–5.50	3.75–4.25	4.00–5.25
Underground Piping	4.75–5.50	0.01	3.00–3.25
Above-ground Piping	45.50–46.00	3.25–3.50	27.00–28.25
Underground Electric	0.50–0.70	0.0	0.30–0.50
Above-ground Electric	15.50–17.00	0.0	6.75–7.25
Instrumentation	12.25–14.00	0.0	1.50–2.50
Insulation	5.50–6.50	0.0	3.00–4.00
Painting	2.00–2.75	0.0	1.75–2.25
Paving	1.10–1.80	0.30–0.40	0.75–1.00
Proratables	1.40–1.75	0.0	2.25–2.50
Total Direct Costs	220.8–232.8	36.12–38.07	74.30–85.50

[a]delivered cost

Table 4.28: Liquid/Solid Type Chemical Plants. Ratio Percentages Direct Costs. Typical ranges for Ratio Estimating Factors to predict direct field materials and field labor [74]. Expressed as percentages of the delivered equipment cost.

Cost Category	Direct Materials	Subcontracts	Direct Labor
Process Equipment[a]	100.00	25.00–26.50	4.00–6.00
Site Preparation	2.25–3.00	0.01	0.60–1.25
Site Improvements	3.30–3.80	0.75–0.85	2.25–3.00
Concrete	9.25–10.50	2.75–3.50	12.50–14.00
Structural Steel	13.50–14.50	0.00	5.50–6.50
Buildings	5.00–5.50	3.75–4.25	4.75–5.25
Underground Piping	4.50–5.25	0.01	3.00–3.50
Above-ground Piping	43.00–45.00	3.00–3.50	25.50–28.00
Underground Electric	0.60–0.90	0.0	0.30–0.50
Above-ground Electric	17.00–18.00	0.0	7.15–8.00
Instrumentation	24.00–27.00	0.0	3.30–4.60
Insulation	8.00–9.00	0.0	4.50–5.50
Painting	3.00–3.50	0.0	2.80–3.10
Paving	1.25–1.75	0.30–0.50	0.75–1.25
Proratables	1.50–2.00	0.0	2.25–2.50
Total Direct Costs	237.2–249.7	35.57–39.12	79.15–92.95

[a]delivered cost

Table 4.29: Liquid High Pressure Type (4000 to 5000 psi) Chemical Plants. Ratio Percentages Direct Costs. Typical ranges for Ratio Estimating Factors to predict direct field materials and field labor [74]. Expressed as percentages of the delivered equipment cost.

Cost Category	Direct Materials	Subcontracts	Direct Labor
Process Equipment[a]	100.00	25.00–26.00	3.75–5.00
Site Preparation	1.70–2.25	0.01	0.25–0.50
Site Improvements	2.30–3.00	0.50–0.75	1.50–1.75
Concrete	6.50–7.25	1.75–2.00	9.25–10.00
Structural Steel	9.00–10.00	0.00	3.50–4.00
Buildings	3.25–4.00	2.50–3.00	3.25–3.75
Underground Piping	4.75–5.25	0.01	3.00–3.25
Above-ground Piping	45.25–47.00	3.25–3.50	27.00–28.00
Underground Electric	0.50–1.00	0.0	0.25–0.40
Above-ground Electric	16.50–17.25	0.0	6.75–7.25
Instrumentation	26.50–27.50	0.0	3.50–4.00
Insulation	8.00–8.50	0.0	4.50–5.00
Painting	2.75–3.25	0.0	2.75–3.00
Paving	0.75–1.25	0.02–0.04	0.50–0.70
Proratables	1.25–1.50	0.0	2.00–2.50
Total Direct Costs	229.0–239.0	33.04–35.31	71.75–79.10

[a]delivered cost

Table 4.30: Liquid High Alloy Type Chemical Plants. Ratio Percentages Direct Costs. Typical ranges for Ratio Estimating Factors to predict direct field materials and field labor [74]. Expressed as percentages of the delivered equipment cost.

Cost Category	Direct Materials	Subcontracts	Direct Labor
Process Equipment[a]	100.00	25.00–25.50	3.75–4.25
Site Preparation	1.50–2.00	0.01	0.25–0.50
Site Improvements	2.25–3.00	0.50–0.75	1.50–1.75
Concrete	6.25–7.00	1.50–2.00	8.75–10.00
Structural Steel	9.25–9.75	0.00	3.50–3.75
Buildings	3.25–4.00	2.50–3.00	3.25–3.50
Underground Piping	4.50–5.00	0.01	3.00–3.50
Above-ground Piping	44.25–45.00	3.00–3.25	26.00–27.50
Underground Electric	0.25–0.50	0.0	0.01–0.02
Above-ground Electric	11.00–12.00	0.0	4.50–5.25
Instrumentation	23.50–25.00	0.0	3.00–4.00
Insulation	6.75–7.25	0.0	3.50–4.00
Painting	2.50–3.00	0.0	2.25–2.50
Paving	0.75–1.00	0.25–0.50	0.50–0.70
Proratables	1.25–1.50	0.0	2.00–2.50
Total Direct Costs	217.3–226.0	32.77–35.02	65.76–73.72

[a]delivered cost

Table 4.31: Construction Equipment. Ratio Percentages as a Percentage of Direct Field Labor. Adapted from Page [74].

	Percentage of Direct Field Labor	
Cost Category	Range	Average
Rental or Purchase	12.0-18.0	15.0
Service Labor	2.0–6.0	4.0
Fuel[a], Oil, Grease, Supplies	7.0–15.0	12.0
Total	21.00–39.0	31.0

[a]Fuel percentage based on diesel costs of $1.25 per gallon.

Table 4.32: Overhead and Indirects. Ratio Percentages as a Percentage of Direct Field Labor. Adapted from Page [74].

Cost Category	Percentage of Direct Field Labor	
	Range	Average
Salaried Indirect	8.00–12.00	10.0
Office Hourly Indirects	4.0–8.0	6.0
Field Hourly Indirects	3.0–7.0	5.0
Temporary Construction Facilities[a]	10.0–20.0	15.0
Burdens and Benefits	25.0–28.0	27.0
Small Tools and Consumables	5.0–10.0	7.0
Other Indirects	10.0–20.0	15.0
Total	65.0-105.0	85.0

[a]Includes construction labor.

Table 4.33: Home Office Services. Ratio Percentages as a Percentage of Direct and Indirect Costs. Adapted from Page [74].

Cost Category	Percentage of Total Directs & Indirects	
	Range	Average
Engineering/Design Services	8.0–16.0	12.0
Construction Services	0.1–0.4	0.2
Project General Management	1.0–1.5	1.1
Total	9.1–17.9	13.3

Solution: 4.12 *Using Page's method, the direct factor, obtained from Table 4.34, is 3.25. So the total directs are:*
Total directs = 3.25($20) = $65 million.
Using a materials correction from Guthrie (see Table 4.13), then
F_m *= 0.3(3.20)+0.1(4.23)+0.3(2.25)+0.2(1) = 2.26*
$E = E'/F_m = 20/2.26 =$ *$8.85 million.*
Processing Piping factor, $f_t = 0.7(0.0475 + 0.44625) = 0.3456$.
Equipment and Process Piping = 1.3456(20) = $26.912 million
Other directs = (3.25-1.3456)(8.85) = $16.854
Total directs = 26.912 + 16.854 = $43.8 million which is over $21 million less than with Page.

Example: 4.13 *Estimate the indirects and total installed cost using the Guthrie-Page modified method for the previous example 4.12.*
Solution: 4.13 *Use average indirect factors from Tables 4.32 and 4.33. Assume that the indidrects are not affected by special materials of construction.*
CS Directs = 3.25(8.85)= $28.76 million
CS Direct Field Labor = 0.6984(8.85) = $6.18 million
Construction Equipment = 0.31(6.18) = $1.92
Overhead and Indirects = 0.85(6.18) = $5.25 million
Home Office Services = 0.133(28.76 + 5.25 + 1.92) = $4.78 million
Total Installed Cost (less shipping) = 43.8 + 5.25 + 1.92 + 4.78 = $55.75 million

As the preceding example shows, using actual (E') costs, rather than low pressure carbon steel costs (E) for equivalent equipment has a dramatic effect on the direct costs. It is the author's opinion that Guthrie's method is more correct because many of the direct costs are not affected by special materials of construction. Why, for example, do special materials affect the paving or site development costs? Yet using Page's method, as originally presented by Page, these costs are affected. Similarly, painting and insulation costs are not drastically affected by special materials. It therefore seems more reasonable to base Page's direct and indirect costs on fob purchased costs for equivalent low pressure carbon steel equipment. The two exceptions, as pointed out by Guthrie, are the actual process equipment and the process piping (70%) of the total piping (both above and underground).

For these reasons, the author recommends that capital costs be estimated using either Guthrie or by a hybrid calculation partly based on both Guthrie and Page. It is attractive to use Page's direct cost factors because they are more complete, but then to ratio the costs relative to the low pressure carbon steel equipment costs, except for the equipment and the process piping. Equipment costs should include auxilliaries as well as battery limits when using Page.

Also, Guthrie recommends that freight, insurance and taxes be computed as an indirect cost, rather than added to the fob purchased cost to compute a delivered equipment cost. The reason for this is simple. Shipping equipment from New Jersey to California doesn't affect the costs of installation in California. The installation costs will be the same whether the equipment is fabricated in New Jersey and shipped across country or fabricated in Los Angeles and shipped 10 blocks. It

Table 4.34: Average Direct Costs for all Chemical Plant Types, based on Page [74].

Chemical Plant Type	Materials	Labor	Subcontracts	Total
Refinery Process Sysems	180.10	78.35	47.90	306.35
Liquid Type Chemical Plants	227.70	77.20	34.50	339.40
Solid Type Chemical Plants	226.80	79.90	37.10	343.80
Liquid/Solid Type Chemical Plants	238.95	86.05	37.35	362.35
Liquid High Pressure Type Chemical Plants	234.00	75.43	34.18	343.61
Liquid High Alloy Type Chemical Plants	221.65	69.74	33.90	325.29
Averages	223.20	77.78	37.49	338.47

is therefore illogical to base other direct materials, labor, and subcontracts on delivered equipment prices. They are more correctly based on fob purchased costs. Direct costs are artificially inflated when based on ratios to delivered equipment cost.

4.6 Summary

Capital investment may be estimated using several different methods. The simplest predesign method uses only one or two factors, but often leads to large estimation errors. A factored estimate is based upon cost estimates for each line item in the equipment flowsheet being considered separately. A composite of these estimates may then be used to generate the contingency and fees for the battery limits and to estimate the auxiliary equipment and other site development costs.

The materials of construction, operating pressures, and design features of various pieces of equipment can all affect the cost estimate. Of these factors, the materials of construction are perhaps the most important and should be considered first.

List of symbols

a power rule for unit size

C_t^a actual cost for process piping (or tubing)

C_p^a actual fob purchased equipment cost

C_p^{cs} fob purchased equipment cost, low pressure, carbon steel basis ($C_p^{cs} = E$)

E_o fop purchased cost for reference size, S_o

E fob purchased equipment cost, low pressure carbon steel basis

E' actual fob purchased equipment cost

F_{bm}^{a} actual bare module factor, adjusted for pressure or materials of construction

F_{bm}^{cs} bare module factor, carbon steel, low pressure basis

F_m materials factor

F_p pressure factor

f_t tubing or piping factor

S actual equipment size

S_o reference equipment size at which the fob purchased cost is E_o

4.7 Problems

4.1 Estimate the purchase cost in mid-1981 of a 316 stainless steel vertical pressure vessel that is 120 ft tall with a 6 ft diameter. The operating pressure is 200 psia. Estimate the installed cost for this same item. Estimate the purchased and installed costs for this same item if it is made out of carbon steel.

4.2 Estimate the purchased cost of a 500 BHP process gas compressor and driver in mid-1981. A carbon steel, reciprocating, steam-driven unit is required. Estimate the installed cost for the same item. Prepare a table indicating the costs associated with piping, insulation, instrumentation, electrical connections, concrete, and paint for the unit. Also estimate the direct field labor costs and the indirect costs.

4.3 If the installed cost of a compressor is three times its purchased cost, and the purchased cost is correlated by:

$$E = 906.069(BHP)^{0.73532}$$

where E is the purchased cost in mid-1968 U.S. dollars and BHP is the brake horsepower. Then what is the installed cost of a unit requiring 150 BHP in the 4th quarter of 1983?

4.4 A carbon steel shell and tube heat exchanger requires 500 ft^2 of transfer area and will be operated at 350 psia. Using the cost correlations and the installed cost factors for single units, then answer the following questions. If the heat exchanger is a floating head unit, then what is the purchased cost in mid-1983 dollars? What is the estimated cost of process piping for the heat exchanger during the same time period? What is the estimated bare module cost in the same time period?

4.5 Estimate the total installed cost of an equipment item operated at low pressure and made of carbon steel which has a purchased cost of $70,000. The bare module factor is 2.178 and contingency and fees are 22% of the bare module cost.

4.6 A preliminary estimate is needed for a completely installed pumping system. A pipeline is to be used to deliver 250 gpm of water at 60^oF and steady state. The theoretical horsepower requirements (100% efficiency) are 10 hp. What is the optimal pipe diameter in inches? What is the Reynolds number in the optimum pipe? What is the C/H factor for this pumping system? (Guthrie's C/H factor is the volumetric flowrate in gpm times the total developed head in psi.) Estimate the installed cost of a single unit, centrifugal pump and driver system made of cast iron in mid-1968.

4.7 Estimate the installed cost of a 316 stainless steel centrifugal pump operating with a total developed head of 50 psi and pumping 400 gpm in mid-1979 dollars.

4.8 A single shell and tube heat exchanger is to be installed in an existing plant. It requires 150 ft^2 of transfer area.

If the unit operates at low pressure and is made of carbon steel, what is its estimated purchased cost in mid-1983 dollars?

4.9 If the unit in Problem 8 operates at low pressure and is made of carbon steel, what is its estimated installed cost in mid-1983 dollars? Let contingency and fees equal 30% of the estimated bare module cost.

4.10 If the unit in Problem 8 operates at 350 psia and the tubes are made of titanium, what is its estimated installed cost in mid-1983 dollars? Set contingency and fees at 30% of the bare module cost.

4.11 Estimate the installed cost of a vertical pressure vessel with internals, assuming contingency and fees at 18%. The unit is 150 ft high with a 7 ft diameter. It is constructed of solid monel and operates at a pressure of 250 psia. (Use the same pressure factor coefficients as for a horizontal pressure storage vessel). The internals are bubble cap trays with 18 inch spacing between trays.

Indicate the following items in your cost analysis: purchased cost, field materials and labor, and the indirect costs. Determine the total bare module cost and the installed cost. Convert your estimate to mid-year 1986 dollars.

4.12 A company must install a spherical storage tank. If it operates at 30 psia or less, the fob purchased cost is $300,000. However, the tank must operate at a pressure of 200 psia. What is your estimate of the total installed cost of the spherical storage tank?

4.13 Estimate the installed cost of a vertical process pressure vessel with internals, assuming contingency and fees at 18%. The unit is 200 ft high with a 10 ft diameter. It is constructed of solid monel and operates at a pressure of 250 psia. The internals are sieve trays with 24 inch spacing between trays.

Indicate the following items in your cost analysis: fob purchased costs of the vertical pressure vessel and the internals, process piping costs, all other directs, indirect costs. Determine the total bare module and the total installed costs. Convert your estimates, including the fob purchased equipment costs, to mid-year 1986 dollars. Adjust the fob internals cost for materials, but not for pressure. Why? How many actual trays are in the column?

4.14 Estimate the 1988 installed cost for a field fabricated process furnace operating at 1000 psia. The unit requires Hastelloy C tubing. Use $F_m = 2.89$. The fob purchased cost in 1988 is \$450,000. Show the major costs in tabulated form. Assume contingency and fees are 18%.

4.15 Estimate the battery limits total installed cost of a process for producing acetic acid from carbide at a 12 ton/day plant operating 320 days per year, 24 hours/day. The process involves the following steps: generation of acetylene, hydration to form acetaldehyde, adsorption of acetaldehyde, distillation, oxidation to acetic acid, and distillation of the acid. Assume the plant is constructed of 300-series stainless steel. The highest operating temperature is 450 K. The highest operating pressure is 300 kPa. Factor your estimate to current dollars using a Chemical Engineering Index of 460.

Now estimate the personnel requirements for this plant using the Stein correlation (see Chapter 6 and the fob purchased equipment cost for items within the battery limits. Hint: Study Figure 4.3 to estimate the fob purchased equipment cost from the Battery Limits Total Installed Cost.

4.16 A pumping station is needed to transfer 500,000 U.S. gal/day of Greasy Kid's Stuff (GKS) from one storage tank to another. The pumping station will house two pumps configured in parallel. One unit will be in continuous operation with a second on standby. The GKS-1 feed tank is 500 ft from the pumping station. The GKS-2 receiving tank is 3500 ft from the station. The inlet and discharge tank pressures are essentially the same at all times, but the discharge point into GKS-2 is elevated 400 ft above the pumping station. The GKS liquid level elevation in GKS-1 fluctuates with time from 40 to 20 ft above the pump. The pumps should be designed to carry the maximum estimated load.

Greasy Kid's Stuff is a nontoxic liquid. Its viscosity is constant at 3 cp and it has a specific gravity of 0.4. It is not corrosive, but has a slightly disgusting greenish color. Thus, the two pumps (centrifugal) can be fabricated of cast iron.

The pipeline connecting GKS-1 and GKS-2 is of normal piping complexity. Assume the average roughness is $\epsilon = 0.0018$ in and that the overall pump efficiency, $\epsilon_o = 80\%$.

Estimate

1. the optimal pipeline diameter,
2. the pipeline Reynolds number,

3. the average pipeline friction factor,
4. the total developed head across the pumps,
5. the required brake horsepower (at 80% efficiency) for each pump,
6. the f.o.b. purchased cost of each pump (use an M&S Index of 910),
7. the total installed cost of the two pumps in the pumphouse, and
8. the annual utility cost for continuous pump operation (one at a time only) if operation is 24 h/day and 360 days/yr with electricity costing \$0.1/kw-h.

4.9 The installed cost of a chemical plant was \$77.5 million in mid-1970. How much will the same plant cost in mid-1983 if its design basis production rate is 2.6 times that for the plant built in mid-1970? (Hint: Use the 6 tenths rule.)

Chapter 5

Materials Selection

As is clear from Chapter 4, the selection of materials of contruction for equipment has a major impact on capital investment. Materials of contruction (MOC) can also have profound effects on operational safety and process maintenance costs so it is important to consider materials selection during the preliminary design.

5.1 MOC Selection Heuristics

Heuristic methods as outlined here can be used during Preliminary Engineering, but materials specialists and experimental measurements (e.g., coupon testing) should be completed no later than during the early part of the Conceptual Engineering phase. In some instances, less expensive materials of construction can be adopted, but such decisions should be based on experimental measurements of corrosion rates, not heuristics.

Heuristic: 10 *For industrial-scale processes, start by assuming all equipment is carbon steel. Check for corrosion properties by reviewing the properties of chemicals in the process. Adjust to more expensive materials of construction as problems are identified.*

There are no simple, infalliable heuristics when it comes to materials selection. Because of the large impact materials have on construction costs, it is important to provide reasonable levels of overdesign during Preliminary Engineering if, for no other reason, than to alert management that a potential materials problem exists and experimental work is needed to correctly select materials. Qualitative features of alternative materials of construction are summarized in Tables 5.1 and 5.2. Additional heuristics are listed below.

Heuristic: 11 *Carbon steel may be acceptable if the process is essentially free from water and doesn't involve halogens.*

Heuristic: 12 *Stainless steel (316) may be acceptable if the process is contains water, nitric or sulfuric acids, and is free from halogens.*

Table 5.1: Materials of Construction (MOC). Adapted from Adapted from Chris Haslego, "Experience-Based Rules of Chemical Engineering," www.cheresources.com

Material	Advantages	Disadvantages
Carbon Steel	Relatively low cost, easy to fabricate, commonly used. Resists most alkaline environments well.	Very poor resistance to acids and stronger alkaline streams. More brittle than other materials, especially at low temperatures.
Stainless Steel	Relatively low cost, still easy to fabricate. Resists a wider variety of environments than carbon steel. Available is many different types.	No resistance to chlorides. Resistance decreases significantly at higher temperatures. .
254 SMO (Avesta)	Moderate cost, still easy to fabricate. Resistance is better over a wider range of concentrations and temperatures compared to stainless steel.	Little resistance to chlorides, and resistance at higher temperatures could be improved.
Inconel	Excellent halide resistance.	Expensive, hard to fabricate.
Titanium	Very good resistance to chlorides (widely used in sea water applications). Strength allows it to be fabricated at smaller thicknesses.	This material is moderately expensive, but fabrication is difficult. Expensive to weld.
Pd stabilized Titanium	Superior resistance to chlorides, even at higher temperatures. Is often used on sea water application where Titanium's resistance may not be acceptable.	Very expensive material and fabrication is again difficult and expensive.

Table 5.2: Materials of Construction (MOC). Adapted from Adapted from Chris Haslego, "Experience-Based Rules of Chemical Engineering," www.cheresources.com

Material	Advantages	Disadvantages
Nickel	Very good resistance to high temperature caustic streams.	Moderate to high expense. Difficult to weld.
Hastelloy Alloy	Wide range to choose from. Some have been specifically developed for acid services where other materials have failed.	Expensive alloys. Use must be justified. Most are easy to weld.
Graphite	One of the few materials capable of withstanding weak HCl streams.	Brittle, very expensive, and very difficult to fabricate. Some stream components have been know to diffusion through some types of graphites. Combustible.
Tantalum	Superior resistance to very harsh services where no other material is acceptable.	Extremely expensive, must be absolutely necessary.
Teflon	Excellent stability to virtually all low-temperature environments. Easy to fabricate. May be used as a liner.	Not suitable for large-scale applications. Teflon liners are easily scarred, exposing the underlying materials.
Fiberglass	Excellent resistance to many corrosive environments at low temperatures.	Subject to cracking. Relatively low in tensile strength. May be more suitable as a liner for large-scale applications.

Heuristic: 13 *Hastalloy or titanium may be reuired if the process requires the use of halogenated species.*

The impact of materials on costs is a good incentive to search for alternative chemical routes using "greener" chemicals whenever possible. Whenever halogens are involved (e.g., HBr, HCl, HF, or HI) an evaluation of green chemistry options is probably in order. Even if the process is nominally dry, say if working with F_2, but nominally not HF, it should be realized that water will inevitably work its way into the process despite best efforts to keep it out. Corrosion problems will therefore eventually occur.

Heuristic: 14 *Don't assume the best results based on lab data. Conditions are always more difficult to control in large-scale processes.*

It may be possible to work with certain chemicals, say under nitrogen and in a dry environment. Such conditions can be maintained quite well in a laboratory glove box. It may be possible to carry out certain processing steps without significant product loss or corrosion problems in the dry glove box, but it is much more difficult to duplicate such results in the plant. Mechanical equipment (e.g., pumps) contain seals that can leak. The plant piping is much more complex and, in fact, the glove box doesn't require the same piping since the entire box is under the nitrogen blanket. In the plant, however, blanketing with an intert gas requires the installation of a complex header system allowing gas flow into and out of each tank (during filling and draining operations), and there are many more opportunities for leakage and contamination by air or moisture.

Heuristic: 15 *High operating temperatures always increase corrosion rates. Favor ambient temperatures to reduce known corrosion problems. Ceramics may be required for high temperature operations. Recognize that such materials are often brittle and much more fragile than carbon steel equipment.*

Operating at high temperatures, say greater than 500 C, usually requires special materials of construction.

Heuristic: 16 *Corrosion problems may be controlled by using lined equipment, but linings are subject to tears and wear out.*

5.2 MOC Costs

Materials of construction that are more resistant to corrosion and other stresses also tend to be more expensive. Materials factors are provided by Guthrie and summarized in Tables 4.13 through 4.15. Tables 5.3 and 5.4 can also be used to generate approximate materials factors when none are available. It should be recognized that the approach illustrated below doesn't consider the fabrication costs which vary with the material since some are more difficult to machine and weld.

Table 5.3: New York spot nonferrous metal prices on November 3, 2005

Metal	Price
Aluminum	90.9 cents per lb., London Metal Exch.
Copper	197.00 cents Cathode full plate, U.S. destinations.
Copper	190.50 cents per lb., N.Y. Merc spot
Lead	$1011.0 per metric ton, London Metal Exch.
Zinc	73.93-74.43 cents lb., delivered.
Gold	$461.85 Handy & Harman (only daily quote).
Gold	$460.10 troy oz., NY Merc spot Thu.
Silver	$7.540 Handy & Harman (only daily quote).
Silver	$7.597 troy oz., N.Y. Merc spot Thu.
Mercury	$700.00 per 76 lb flask, N.Y.
Platinum	$934.00. troy oz., N.Y. (contract).
Platinum	$941.90 troy oz., N.Y. Merc spot Thu.

Example: 5.1 *Estimate the materials factor for stainless steel from the cost factors in Table 5.4.*
Solution: 5.1 $F_m = 2500/780 = 3.2$. *Gurthie gives a factors ranging from 1.93 to 3.20 in Table 4.14. Factors may be lower if the equipment is not 100% stainless steel and higher if the MOC is more difficult to machine and weld.*

Example: 5.2 *A specialty piece of equipment will be manufactured of stainless steel with gold plate. The estimated total mass is 300 lbs and the gold plate mass is 0.5% of the total. Estimate a materials factor for this item, relative to carbon steel. Assume 1% of gold is lost in the plating process and an 10% excess of stainless steel (or carbon steel) is required to fabricate the piece.*
Solution: 5.2 *From Table 5.3, gold metal costs $460.10 per troy oz. (12 troy oz = 1 lb)*
From Table 5.4, carbon and stainless steel are $780 and $2,500 per ton, respectively.
Required mass of gold $= (12/1)(0.005)(300)(1.01) = 18.18$ *troy oz*
Cost of gold $= (460.1)(18.18) = \$8,365$.
Cost of stainless $= 300(1.1)(2500/2000) =$ *$413*
Total materials cost $= 8,365 + 413 = \$8,778$.
Carbon steel cost$= (1.1)(780)(300/2000) =$ *$128.7.*
$F_m = 8778/128.7 = 68.21$.

Table 5.4: London Metal Exchange (LME) Official prices for 2 Nov 2005

Metal	**Cash Buyer Prices (US$/ton)**
Aluminum Alloy	1,605
Aluminum	1,976
Copper	4,090
Lead	988
Nickel	11,570
Steel, carbon[a]	780
Steel, stainless[a]	2,500
Tin	6,175
Zinc	1,531

[a]Approximate prices during same time frame, not traded on LME

Chapter 6

Operating and Product Costs

Important Operating Costs include raw materials, energy, and labor. Here shortcuts are provided to permit estimation during preliminary design.

6.1 Labor Costs

The cost of labor is one of the larger components in annual operating costs for chemical plants. A detailed estimate itemizes the personnel requirements based upon the plant layout, process complexity, and other requirements such as security and support services. For purposes of preliminary design, however, it is useful to factor personnel requirements as a function of the installed cost. Two short-cut methods are provided: the Stein and Wessel correlations. If correctly used, they give about the same answers.

6.1.1 Stein Correlation

Stein [97] offers such a correlation to estimate personnel requirements:

$$P = 7TIC(end - 78)^{0.88} \tag{6.1}$$

and

$$\text{TOWN SIZE} = 5.5P \tag{6.2}$$

$TIC(end - 78)$	=	the total installed cost of the chemical plant built in the U.S., average cost, in millions of end-1978 U.S. dollars
P	=	the estimated number of employees ($\pm 20\%$)

According to Stein, this relationship appears to be nearly independent of the type of plant for large projects. It was derived from data on 35 plants including oil refineries, ore plants, steel mills and nonferrous smelters. The plant total installed costs ranged from \$30 million to \$10 billion. The personnel requirements (number of employees) include plant maintenance and support workers, and plant staff and administration. They exclude sales, home office and corporate staff. The estimated personnel requirements are based on average U.S. worker productivity and, therefore, they must be adjusted for local productivity (*e.g.* for plants outside the U.S.). The projection also should be adjusted for local construction costs.

Example: 6.1 *Estimate the personnel requirements for a chemical plant if the total installed cost is $12,000,000 in end-1978 dollars, and the plant is constructed and operated in an average U.S. location.*
Solution: 6.1 *The total installed cost for the plant is $12,000,000. From Equation 6.1:*
$P = 7(12)^{0.88} = 62$ *employees*

Example: 6.2 *A 120,000 ton/year aluminum smelter project is being studied for a foreign country where construction costs are 70% higher than in the U.S. and where worker productivity is 60% of that in the U.S. If the U.S. smelter costs in end-1978 dollars are $3,000/ton of capacity for plant total installed costs, then how many employees are required to run the plant in the U.S. and in the foreign country?*
Solution: 6.2
$U.S.\ \$ = 120,000(3,000) = \$360\ M$
$P(U.S.) = 7(360)^{0.88} = 1243$ *employees*
$I(foreign) = 1.7(360) = \$612\ M$
$P(foreign) = (7/.6)(612)^{0.88} = 3306$ *employees.*

Example: 6.3 *The total installed cost for a new plant is estimated as $408 M in mid-1983 dollars. How many employees are required to run the plant?*
Solution: 6.3 *From Table 4.2, the end-1978 CE index is estimated by interpolation:*
$I_{CE}(end - 1978) = (218.8 + 238.7)/2 = 228.8$
The mid-1983 CE index is 316.9. Therefore, the end-1978 plant total installed cost is:
$I = 408(228.8/316.9) = \$294.6$ *in end-1978.*
$P = 7(294.6)^{0.88} = 1042$ *employees.*

6.1.2 Wessel Correlation

Labor requirements for chemical plants are a function of the plant capacity and the number of processing steps in the manufacturing operation. Wessel [117] provides a procedure for estimating labor costs based upon a correlation that predicts Y, man-hours/ton-processing-step as a function of the plant capacity, X, in tons/day. Using Y, the number of processing steps, and an assumed average hourly rate, then the operating labor costs per ton of product can be estimated.

Wessel's approach is as follows. First, divide the process into component steps and then count the steps. Second, compute the labor per ton of product, per processing step using Wessel's correlation. It can also be used to estimate upper a lower bounds on labor requirements to determine a probable range. Third, multiply the computed factor Y by the number of steps to obtain the man-hours per ton of product. His correlation takes the form:

$$Y = aX^{-0.78} \tag{6.3}$$

The most likely value for a is $a = 18.1$, but it could be as low as 11.7 or as high as 23.7, based upon variability in his analysis. Plant sizes in this correlation range from 2 to 1,500 tons per day. Processes include batch, continuous, and mixed operations in tonnage organics and inorganics, low-volume organics, and tonnage chemical intermediates. Accuracy of the method depends largely upon how the steps are defined. Small batch operations are more likely to obey the higher limit with $a \approx 23.7$. Larger scale, continuous processes may approach the lower limit with $a \approx 11.7$, especially for very large processes involving only gases or liquids and with a high degree of instrumentation.

Several examples taken from Wessel are given below.

Example: 6.4 *Estimate the labor requirements for producing acetic acid from carbide at a seven ton per day plant.*

Solution: 6.4 *This process involves six steps: (1) generation of acetylene, (2) hydration to form acetaldehyde, (3) absorption of acetaldehyde, (4) distillation, (5) oxidation to acetic acid and (6) distillation of the acid. Using an average value,* $a = 18.1$

$Y = 18.1(7)^{-0.78} = 4$ *man-hrs/ton-step*

Since there are six steps, the labor is $6(4) = 24$ *man-hours per ton of product. If product cost analysis is estimated per ton of product, then multiplying the man-hours/ton by the assumed average operating labor rate will give \$ Operating Labor/ton of product.*

Example: 6.5 *Estimate the man-hours to produce a ton of ethanol from molasses in a 100-ton-per-day plant. Estimate the likely upper and lower bounds on man-hours/ton of ethanol product.*

Solution: 6.5 *The production of ethanol from molasses requires five steps: (1) yeast preparation, (2) fermentation, (3) beer distillation, (4) aldehyde distillation, and (5) final ethanol rectification. Using a average value for a*

$Y = 18.1(100)^{-0.78} = 0.5$ *man-hours/ton-step*

Likely labor is: $0.5(5) = 2.5$ *man-hours/ton of ethanol product.*

Likely upper bound is $5(23.7(100)^{-0.78} = 3.3$ *man-hours/ton of ethanol produced*

Likely lower bound is $5(11.7(100)^{-0.78} = 1.6$ *man-hours/ton of ethanol product.*

Example: 6.6 *Estimate the operating labor to produce sodium dichromate from chromite ore at a 15-ton-per-day plant.*

Solution: 6.6 *This process involves nine steps: (1) milling, (2) mixing, (3) roasting, (4) leaching, (5) filtration, (6) evaporation, (7) crystallization, (8) centrifugation, and (9)drying. Using an average value for a:*

Man-hours $= 9(18.1)(15)^{-0.78} = 20$ *man-hours/ton of sodium dichromate produced.*

Example: 6.7 *A new chemical process will be built as a grass roots plant. The process involves seven functional units (i.e., major unit operations). The highest process temperature is 425° C and the highest pressure is 23 atm. The equipment will be constructed of 316 stainess steel. The plant will produce 50 million lb/yr of product. Estimate the total installed cost of the plant using a CE index of 400. Estimate the working capital for this plant. The plant will operate 320 days/year.*

Use the Stein and Wessel correlations to estimate the personnel requirements for this plant. Estimate the annual payroll ($/yr) and cents/lb of product for labor charges using an average labor rate of $30/person-hr.

Solution: 6.7 *Use the predesign method of Zevnik and Buchanan described in Section 2.5.2 to estimate the Total Installed Cost of the plant. Then factor this cost back to end-1978 dollars to estimate the labor requirements using Stein's correlation.*

$T = 425 + 273.15 = 698.15\ K,\ f_T = 0.017(698.15/100) - 0.05 = 0.068685$
$P^* = 23(101.3/100) = 23.299,\ f_P = 0.10log_{10}P^* = 0.136733,\ f_M = 0.20$
$f_T + f_P + f_M = 0.4054,\ C_F = (2)10^{0.4054} = 5.0869$
$I_F = 1.5(5 \times 10^7)^{0.6} C_F(400/102) = 1.24561 \times 10^6$
$TIC_{BL} = N_F I_F = \$8,719,274,\ TIC_{GR} = 2(TIC_{BL} = \$17,438,549$
$Working\ Capital = 0.18(TIC_{GR}) = \$3,138,939$
Using the Stein Correlation:
$I(end - 1978) = (\$17,438,549/10^6)(228.8/400) = 9.974,\ P = 7I^{0.88} = 52.98 \approx 53$ *workers*
$Annual\ Payroll = 53(\$30/hr)(40\ hr/week)(52\ week/yr) = \$3,307,200/yr$
$Labor\ Cost = \$3,307,2000/50times10^6 = \$0.06614/lb\ of\ product.$
Using the Wessel Correlation:
$Throughput = (50 \times 10^6\ lb/yr)/2000\ lb/ton = 25,000\ ton/yr\ or\ 25,000/320 = 78.125\ ton/day$
$X = 78.125,\ Y = 18.1(78.125)^{-0.78} = 0.604\ man\text{-}hr/ton\text{-}step$
$N = 7,\ NY = 4.23\ man\text{-}hr/ton,\ (25,000\ ton/yr)(4.23\ man{-}hr/ton) = 105,760\ man{-}hr/yr$
$Annual\ Payroll\ \$105,760\ man{-}hr/yr(\$30/man{-}hr) = \$3,172,800/yr$ *or $0.06345/lb of product*
Employees = (105,760 man-hr/yr)/[52 wk/yr(40 hr/wk)] = 51 employees versus 53 employees using Stein.

In some instances, labor is specified in terms of shift workers. In such cases overtime may be required.

Example: 6.8 *A new chemical plant will require 5 laborers, 2 operators, and 1 supervisor per shift. The plant operates with three shifts per day, 24 hours per day, 7 days per week. The average payroll for this labor is $15/hr-worker with 1.5 times for overtime. Estimate the annual payroll for the chemical plant. How much overtime is required if the plant operates with exactly four shifts?*

Solution: 6.8 *Four sets of shift workers are specified. Typically, one shift works from 7 am till 3 pm, a second shift works from 3 pm till 11 pm, and a third shift works from 11 pm till 7 am. A fourth shift is off duty while the other three are working and the 2nd shift works 6 days/week.*
The weekly overtime is:
$[(7\ day/week)(24\ h/day) - (4\ shifts)(40\ h/week)](8\ workers) = 64\ worker - hr/week\ overtime$
Annual overtime payroll is:

(52 weeks/yr)(64 worker-hr/week)(1.5)($15/hr) = $74,880/yr overtime. The regular annual payroll is:
$15/hr-worker)(32 workers)(40 hr/week)(52 week/yr)=$998,400/year
The total annual payroll including overtime is:
$74,880 + $998,400 = $1,073,280

6.2 Utility Costs

Using the Four Area Concept (see Figure 4.4), the utilities are separate from the battery limits where chemical manufacturing occurs. Utilities (e.g., steam, gas, air, refrigeration, etc.) are provided to the battery limits as needed and, as a design proceeds into the Conceptual and Detail Engineering stages, utility design and layout must be detailed. Initially, however, utility costs may be estimated during Preliminary Engineering using the concept of internal pricing.

As suggested in Figure 4.4, utiilties are manufactured in the Utilities Area and supplied as needed to the Battery Limits to support chemical manufacturing. All of the cost factors described in Section 3 are applicable to utility manufacture, so both capital and operating expenses are incurred. These expenses result in an actual manufacturing cost (see Table 3.6) which is the actual cost to supply the utility to the Battery Limits. This manufacturing cost is also referred to an an"internal price" since it is the internal cost or price of the utility at that plant site.

Here procedures are described to estimate the internal prices of utilities based on historical data at large chemical plants on the US Gulf coast. In order to use these methods, the utility requirements must be estimated using energy balances for the Battery Limits. This information is then used with internal price factors, expressed as $/million BTU transferred to or from the Battery Limits, to estimate annual costs and appropriate product charges. This approach realistically estimates true utility costs while avoiding the need to detail the Utilities Area during Preliminary Design.

Lumping the utility charges into an internal cost for the utility tends to simplify the analysis of other chemical processing systems that are using the auxiliaries. That is, the lumped costs for auxiliaries are included in the system analysis in lieu of a detailed evaluation of the auxiliary operation as part of the chemical process optimization. On the other hand, the auxiliary investment term, I_A, may be added onto the fixed investment term, I_P, in Equation 3.9 if this approach simplifies the overall analysis.

A complete plant design includes not only the chemical process (*i.e.* the battery limits), but also the support areas. Generally, the support areas, particularly the utilities, can represent significant cost items – especially when high pressure steam and refrigeration are required. Moreover, these utilities can play an important role in determining the most economical operating conditions for the process because significant economic interactions often occur between the battery limits and the service areas.

In general, the support areas should be included in process optimization, but

Table 6.1: Typical internal prices for utilities in large U.S. chemical plants (mid-1980 dollars). These are the available temperatures except for Cooling Water which is available at 80 °F.

Utility	Temperature (°F)	Cost ($\$$/10^3 lb)	Cost ($\$$/10^6 Btu)
Steam (935 psig)	538	11.34	17.18
Steam (400 psig)	448	10.68	13.75
Steam (150 psig)	366	8.89	10.34
Steam (40 psig)	267	7.14	7.65
Steam (10 psig)	239	5.34	5.60
Cooling water (ΔT from 80 to 100 °F)	100	0.003	0.15
Refrigerant (propylene)	40	0.455	2.86
Refrigerant (propylene)	0	1.30	7.51
Refrigerant (propylene)	-40	2.06	11.19
Refrigerant (ethylene)	-75	2.40	13.97
Refrigerant (ethylene)	-110	3.47	18.35

inclusion via detailed capital and operating cost models often lead to models which are difficult to use due to their complexity. As an alternative, a system of internal pricing can be used to include the effects of the utility systems on the process, but with greater simplicity.

When internal pricing is used to account for the utility costs, then the utility charges are assigned to the process as an operating cost only. That is, a utility rate schedule (see Table 6.1, for example) is assumed, and the process is charged according to the amounts of the utilities that it requires. Typically, these rates are expressed as \$/M Btu transferred or \$/M lbs of heat exchange medium (*e.g.* steam, cooling water or refrigerant).

In general, the value of a utility, per million Btus transferred, is a function of the temperature of that utility. As conditions become more extreme, the utility becomes more expensive as indicated in Table 6.1 and Figure 6.1. As can be seen from Figure 6.1, a discontinuity is associated with cooling water because it is much less expensive than refrigeration and steam. Usually, a particular type of utility can be represented by a continuous function, but discontinuities occur with different utility types.

Additional examples of how internal price estimates can be used are given by Breuer [12] and Ulrich [107].

The M&S index should not be used to factor internal prices from one year to another. The U.S. Department of Commerce publishes an Energy Index that is appropriate for time factoring. Alternatively, one can use the price of crude oil to factor utility costs over time. Crude oil prices are published in the Wall Street Journal.

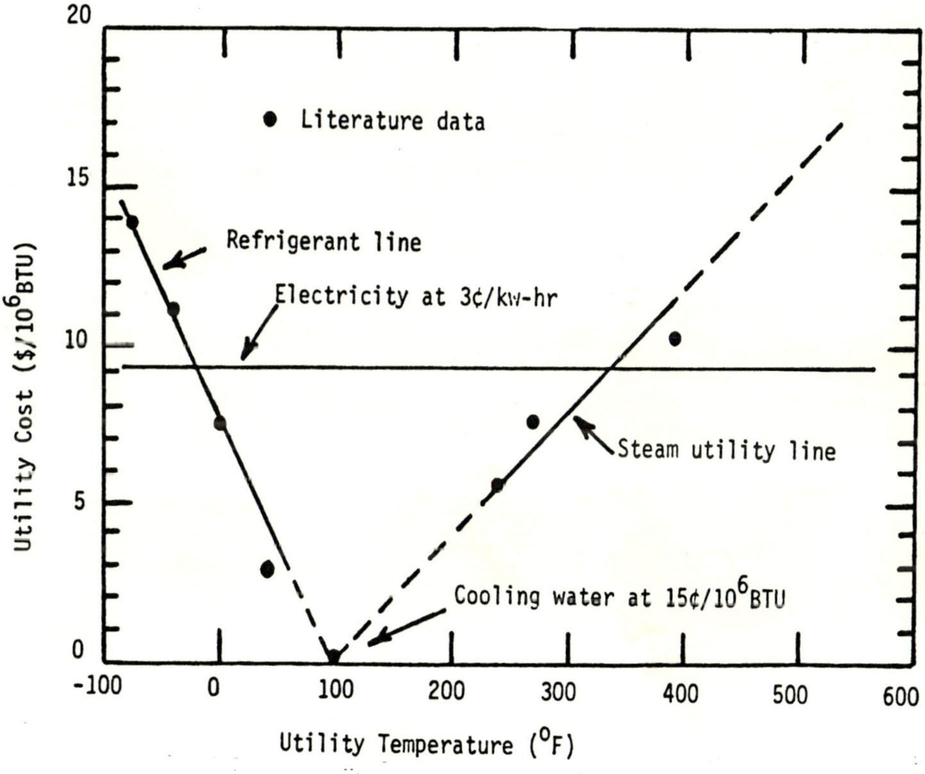

Figure 6.1: Typical utility cost spline for large chemical plants as a function of the utility temperature (mid-1980 $).

Table 6.2: Major equipment for steam generation. Basis: located in Alabama, producing 200,000 lb/hr of 50 psia steam, operating 7,200 hr/yr, December 1994 time frame, using M&S = 1004.

Item	Size		Factors		Cost($1,000)	
	Units	Value	SH[a]	P	fob	BM[b]
Boiler, 50 psia	klb/hr	200	0.2[c]	1	670	1,227
Elec Dist, Gen. Pur	kw	85				29.3
Water Dist, Gen. Pur	gpm	100				13.4
Cost Totals:					670	1,270
Contingency & Fees 18% of BMC						229
Total Installed Cost (TIC)						1,498
Working Capital 15% of TIC						225
Total Capital Investment (TCI)						**1,723**

[a]Superheat
[b]Bare Module Cost
[c]300 ^oF Superheat

6.2.1 Estimating Internal Prices

An internal price is essentially a manufacturing cost plus an additional fee to insure that the producer receives an expected rate of return on investment. In order to estimate an internal price, it is necessary to estimate the total capital investment for the steam generating facility. This information is then used to estimate the fixed costs. Utilities include fuel purchases as well as other utilities such as electricity and water that are required to produce the steam. Labor and overhead should also be included. A typical estimate is shown in Tables 6.2 and 6.3 which list an estimate for a facility producing 200,000 lb/hr of 50 psia steam, operating 7,200 hrs/year. As indicated in Table 6.2, the facility is located in Alabama and the cost estimate is in December 1994 dollars. Similar calculations were used to generate the data in Table 6.1.

6.2.2 Factoring Utility Prices

Because of its widespread use, crude oil and its price per barrel tends to determine the costs of other energy resources. As a first approximation, it is reasonable to use average crude oil prices to factor energy costs (e.g., utility internal prices) from one time period to another. Tables 6.4 and 6.5 summarize average domestic first purchase prices for crude oil in the U.S. market and prices in the state of Illinois, respectively. Table 6.6 summarizes energy costs for several types of energy in the U.S. markets from 1980 through 1996. Generally speaking, it is better to obtain current estimates for energy from an appropriate vendor. In some cases quotes can be obtained from a simple phone call or from data on the internet. The data in Table 6.6 are representative. Similarly, those in Table 6.4 can be used to estimate

Table 6.3: Steam manufacturing cost and internal price estimate. Basis: See Table 6.2.

Item	Costs ($1,000)
Fixed Costs	
TCI Finance Charge, 5 Years at 10.00%	454.5
Taxes & Insurance, 5% of TIC	74.9
Maintenance, 5% of TIC	74.9
Subtotal:	604.3
Utilities	
Electricity ($0.044/kw-hr)	7.5
Fuel Purchases $6.20/10^6 Btu	13,703.4
Water ($2.00/1000 gal)	2.0
Subtotal:	13,712.9
Labor(1 worker/day at $12/worker-hour)	105.1
Overhead (50 % of Labor and Maintenance)	90.0
Steam Manufacturing Cost:	14,512.4
	$/10^3 lb steam
Manufacturing Cost:	10.08
Internal Price: (30% Return on TCI[a])	10.45

[a]Annual Return before Taxes.

Table 6.4: Summary of Crude Oil Prices in U.S. $/barrel, based on Average U.S. Domestic First Purchase Prices, provided by the U.S. Department of Energy, Energy Information Administration (See http://www.eia.doe.gov/fueloverview.html).

1973	3.89	1990	20.03
1974	6.87	1991	16.54
1975	7.67	1992	15.99
1976	8.19	1993	14.25
1977	8.57	1994	13.19
1978	9.00	1995	14.62
1979	12.64	1996	18.46
1980	21.59	1997	17.23
1981	31.77	1998	10.88
1982	28.52	1999	
1983	26.19	2000	
1984	25.88	2001	
1985	24.09	2002	
1986	12.51	2003	
1987	15.40	2004	
1988	12.58	2005	
1989	15.86	2006	

internal prices by factoring from the data in Table 6.1.

Example: 6.9 *Estimate an internal price for 40 psig steam in 1998.*
Solution: 6.9 *From Table 6.1, 40 psig steam in mid-1980 dollars cost* $\$7.65/10^6$ *Btu supplied to the process. Using crude oil prices as cost indices,*
Cost in 1998 $= (10.88/21.59)(7.65) = \$3.86/10^6$ *Btu*

6.3 Summary

Although it is preferable to itemize labor requirements, preliminary estimates can be obtained using a simple correlation. Utility costs can be estimated using internal prices instead of detailing the fixed investment and actual operating costs to manufacture them on site.

6.4 Problems

6.1 A large oil company is considering the construction of a tar sands conversion project which is estimated to cost $1.7 G. Because of adverse climatic conditions, the average worker productivity is about 70% of that in the U.S. How many employees are required to operate the plant? If the project is located in a remote area and the company builds a new town for the workers to live in, how many people will live in the new town initially?

Table 6.5: Summary of Crude Oil Prices in Illinois \$/barrel, based on History of Crude Oil Prices, provided by the Illinois Oil and Gas Association, Mt Vernon, IL. (See http://www.ioga.com/Special/crudeoil_history.htm).

Year	Price	Year	Price	Year	Price
1946	1.59	1973	4.27	1990	23.19
1947	2.37	1974	5.20	1991	20.19
1953	3.02	1975	5.20	1992	19.25
1954	2.90	1976	11.42	1993	16.74
1955	3.00	1977	11.16	1994	15.66
1956	2.90	1978	12.32	1995	16.75
1957	3.08	1979	13.08	1996	20.46
1960	2.92	1980	14.35	1997	18.97
1963	3.00	1981	15.13	1998	11.91
1965	3.10	1982	31.55	1999	16.55
1967	3.15	1983	29.00	2000	27.40
1968	3.20	1984	27.50	2001	25.44
1969	3.30	1985	26.50	2002	
1970	3.60	1986	14.64	2003	
		1987	17.50	2004	
		1988	14.87	2005	
		1989	18.33	2006	

Table 6.6: U.S. Energy Prices: 1980 to 1996. Provided by the U.S. Census Bureau[111]

Product	Unit	1980	1985	1990	1991	1992	1993	1994	1995	1996
Crude Oil	\$/bbl	21.6	24.1	20.0	16.5	16.0	14.3	13.2	14.6	18.5
Motor gasoline	Cents/gal	122.1	119.6	121.7	119.6	119.0	117.3	117.4	120.5	128.8
Natural gas[a]	$\$/10^3$ ft^3	3.7	6.1	5.8	5.8	5.9	6.2	6.4	6.1	6.3
Heating oil[a]	Cents/gal	161.3	134.3	113.6	104.7	93.4	88.8	84.3	80.6	90.1
Coal	\$/ton[b]	28.8	34.5	30.5	30.0	29.4	28.6	28.0	27.0	26.5
Electricity	Cents/kw-hr	4.7	6.4	6.6	6.7	6.8	6.9	6.9	6.9	6.9
Uranium[c]	\$/lb		31.4	15.7	13.7	13.5	13.1	10.3	11.1	13.8

[a]Residential

[b]Short ton

[c]Domestic purchases

6.2 The installed cost of a new battery limits plant is 156 million dollars in the first quarter of 1982 when the M&S index was 739 and the CE index was 311.4. This cost estimate is based on the assumption that the plant is located in the northeastern part of the U.S. where the workers are 20% more efficient than the U.S. average. Estimate the personnel requirements for this new plant in the northeast.

6.3 A chemical plant has an installed cost of $1750 million in mid-1983. How many employees are required to run this plant based on average U.S. worker productivity?

Chapter 7

Cash Flow Analysis

Here we examine ways that economic inputs (i.e., benefits or receipts) and outputs (i.e., costs or disbursements) can be used to determine the profitability of alternative investments. Since larger and more expensive investment projects usually require investment over longer time perods than smaller projects, they often require longer project lives to become economically attractive. Comparison of such projects with smaller projects that occur over shorter time periods requires cash flow analysis. Ultimately, the time value of money must be considered in comparative analysis to arrive at correct funding conclusions because time has a monetary value. But initially we consider alternatives in their simplest terms and ignore the effects of time on the value of money.

7.1 Cash Flow

Investment projects can usually be characterized as a series of cash flows that are either positive (receipts) or negative (payments). In fact payments and receipts occur on a daily basis, but for the purposes of project analysis, they are often grouped into annual expenses and receipts, and may be represented in a bar chart as shown in Figure 7.1. In this example there is an initial cash expenditure of \$2,000 in the first year, and income of \$500 annually for four years thereafter. So the net cash flow is zero, if money in the fifth year is worth as much as in the first year. But is that assumption really true?

In order to determine the value of a potential investment, it is necessary to estimate the stream of disbursements and receipts for the proposal over its estimated life, or at least for a defined planning horizon. For private corporations, cash flows are estimated from expected gross revenues and expenses. Since the gross profits (or revenues) are taxable, cash flows should be based on after-tax receipts. In engineering analysis, cash flows are often evaluated on an annual basis, particularly in design.

When alternative investments are compared, it is usually assumed that the future expenses and gross profits are know precisely, but this is seldom actually

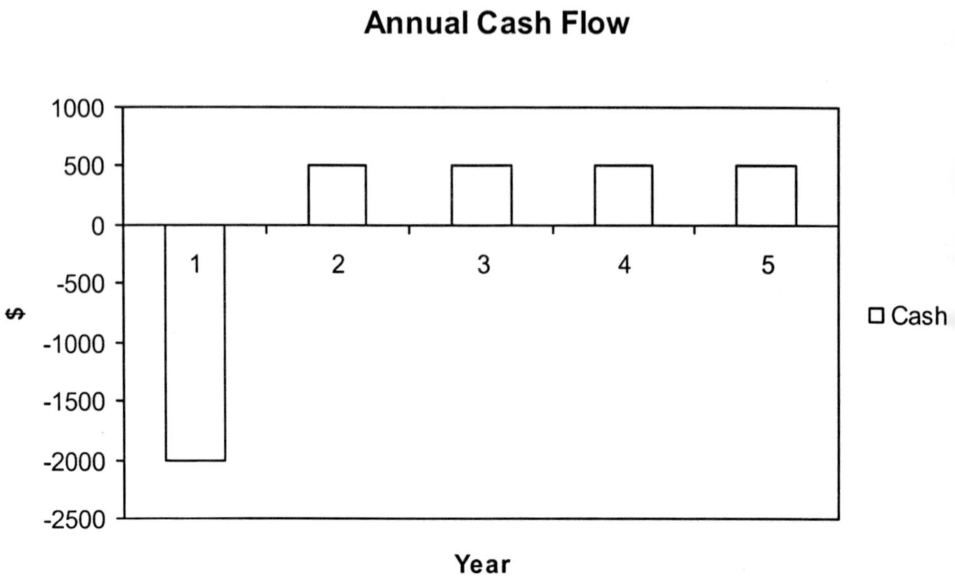

Figure 7.1: Investment cash flows

true. Given this assumption, it is then possible to estimate cash flows for each year of a project life and eventually determine the net present worth of an investment when the time value of money is taken into consideration.

Example: 7.1 *An office computer and in-house printer system costs $7,000 if purchased today. It is expected to be used over a five year period. After that time, it will have a salvage value of $250. The annual operating and maintenance cost of this equipment is $600 per year, but it will enable the production of reports without going to an outside printer with an estimated savings of $2,200 per year. Assuming the receipts and disbursements are made at the end of each year, except the initial cost which is paid at the end of this year (year 0), describe this project in terms of its anticipated cash flows.*

Solution: 7.1 *The cash flows may be represented using a table to compute the net benefit or loss each year of the project as shown below. Based on cash flow analysis, this project will save $1,250, compared to the alternative of sending all reports to outside printers, but the initial expenditure will not be completely recovered until the end of the fifth year.*

Investment profitability can be measured in several ways. Inevitably, an investor wants to know how profitable an investment is compared to the alternative that was foregone. If an investor is considering either investing in the stock market or placing the money in a savings account, the issues revolve around relative safety (risk) for

Table 7.1: Estimated cash flows.

Year	Annual Expense	Annual Benefit	Annual Cash Flow	Accumulated Cash Flow
0	$7,000	$ 0	-$7,000	-$7,000
1	600	2,200	1,600	- 5,400
2	600	2,200	1,600	- 3,800
3	600	2,200	1,600	- 2,200
4	600	2,200	1,600	- 600
5	600	2,450	1,850	1,250

the principal invested and earnings. In the former case, stock investments yield dividends and they may appreciate. On the other hand, stock values may decline. A savings acount, at least in a U.S. bank insured by the Federal Deposit Insurance Corporation (FDIC), is considered relatively safe, but the interest earnings may be much less than on the stock market. So an investor may compromise and invest some funds in both places. She may reduce her risk by investing in blue chip stocks, diversifying (i.e., not putting all of her funds into a single stock), and placing some funds in a savings account (e.g., a money market account which earns higher interest rates than a regular account). In this way she can realize a degree of safety, but also higher earnings.

Differences in tax liabilities are also considered by investors. A company may need to raise capital. Funds may be raised by issuing bonds or by selling stock. In the first case, bonds earn interest which is tax deductible as an operating expense and the funds are repaid to the investors after a fixed number of years. In the case of public stocks, dividends are paid by corporations on net earnings after taxes and the stock holders are not repaid unless the company chooses to buy back stock issued. Sometimes companies choose to do this in order to control the market price of their stock or to maintain control of the company. Comparing these two means of raising capital, interest paid on bonds is deductible from taxable income so it has a smaller effect on retained earnings than stock dividends which are not tax deductible.

When we consider alternatives, it becomes apparent that any investor is faced with a variety of options that have different levels of risk, projected returns, and tax liabilities. Liquidity, the ability to convert investments from one form to another, is also usually an issue. Stock investments are considered to be equivalent to cash because stocks can be sold readily on the market, but we don't want to be forced to sell stocks when their market price is low to maintain liquidity (i.e., to pay actual expenses). So it is also important to maintain cash reserves to cover anticipated shortfalls without having to sell stocks under adversity.

In addition, there are also tax differences between different investments. For example, interest paid by a corporation to bond holders is tax deductible. On the other hand, dividends paid to stock holders are not tax deductible. Stock dividends must be paid out of net profits after taxes. So there is a tax advantage for the corporation to issue bonds, rather than sell stock, because the interest paid on bonds reduces corporate taxes.

Heuristic: 17 *Don't invest more on the stock market than you are prepared to lose.*

Heuristic: 18 *Only invest in blue chip stocks that have earned dividends for the past five years, and that are rated at least "B" or higher by Standard and Poor.*

Heuristic: 19 *Always maintain cash reserves equal to at least three months of anticipated expenses.*

7.2 Time Value of Money

Money available today is worth more than funds available tomorrow because money has a time value. Several factors are at play. One is economic inflation which decreases the value of money over time. A second factor is the ability of the money holder to invest funds and use them to earn more money. In order to consider the former case, we discount future values by the inflation rate. In the latter case, we compute the future value of a present sum using an appropriate interest rate.

Given an initial principal, P, it can be invested in a savings acount to earn interest. Interest may be calculated in several ways: simple interest or, more commonly, as compound interest. For simple interest calculations, the amount of interest paid is based solely on the initial investment. Given a simple interest rate, i_s, then the amount of interest earned in any year is Pi_s, After n years of investment, the amount earned is nPi_s and the total value of the future sum F after n years earning simple interest is:

$$F = P + nPi_s = P(1 + ni_s) \tag{7.1}$$

This is the earnings rate of the original investment P if the earned interest is continually removed from the account, but most commonly the earned interest is held in the account and it also earns interest. Simple interest is sometimes paid on short-maturity securities (e.g., U.S. Treasury bills).

Discrete compound interest refers to interest that is compounded at regular (or discrete) intervals. Discrete compound annual interest i_a is most commonly used to determine the future value of a present investment. If the intervals are annual, then the interest is discrete annual compound interest. Usually other forms of interest are compared with this form as the compartive basis.

Initially we have P as our investment. In the first year, P earns Pi_a in interest. In the second year the account value is $P + Pi_a$, the principal plus the interest earned. If this amount remains in the account, then at the end of the second year, the account value is $P(1 + i_a) + P(1 + i_a)i_a$ or $P(1 + i_a)(1 + i_a)$. Similarly, at the end of the third year, the account value is $P(1+i_a)^3$. After n years of compounding the value of the account is

$$F_n = P(1 + i_a)^n \tag{7.2}$$

which is usually expressed on an annualized or per annum (p.a.) basis. By turning this equation around, we calculate the present value of a future sum:

$$P = \frac{F_n}{(1 + i_a)^n} \tag{7.3}$$

which is known as discounting.

Several other definitions of interest rates are in common usage. The effective interest rate, i_{eff}, is the interest rate when the compounding period is exactly one year. The nominal interest rate, i_n, is the interest rate when the compounding period is not equal to one year (e.g., it could be a monthly rate).

Example: 7.2 *What is the future value in ten years of $1000 invested today in a savings account if it earns 4% p.a. compound interest?*
Solution: 7.2 $F = P(1 + i_a)^n = 1000(1 + 0.04)^{10} = \$1480.24$
In this case simple interest would yield a future value of
$F = P + nPi_s = 1000(1 + 10(0.04)) = \$1400$.

Example: 7.3 *How much needs to be invested in a savings account paying 5% interest p.a. to have $8,000 in the account after six years?*
Solution: 7.3 *In this case we discount the future sum*
$P = F/(1 + i_a)^n = 8000/(1 + 0.05)^6 = \$5,969.72$
So we will have $8,000 in the account after six years if we invest $5,969.72 today.

Example: 7.4 *What will be the value of a savings account after ten years if $100 is invested at the beginning of each year and the account earns 4% p.a.?*
Solution: 7.4 *After ten years, the first $100 is worth* $100(1.04)^{10}$
The second $100 invested is worth $100(1.04)^9$ *and so forth. So the value of the account at the end of ten years is:*
$F = 100[1.04 + 1.04^2 + 1.04^3 + \cdots + 1.04^{10}]$
but this is a geometric series of the form $ar, ar^2, ar^3, \cdots, ar^n$ *which can be written as:*

$$S_n = ar + ar^2 + \cdots + ar^n \tag{7.4}$$

Multiplying both sides of Equation 7.4 by r and subtracting Equation 7.4 from the product gives:

$$rS_n - S_n = -ar + ar^2 - ar^2 + \cdots - ar^n + ar^n + ar^{n+1} \tag{7.5}$$

$$(r - 1)S_n = a(r^{n+1} - r) \tag{7.6}$$

$$F_n = S_n = a\left[\frac{r^{n+1} - r}{r - 1}\right] = a\left[\frac{(1 + i_a)^{n+1} - (1 + i_a)}{i_a}\right] \tag{7.7}$$

with $r = 1 + i_a$, *and* $a = \$100$.
So $F = 100[(1.04^{11} - 1.04)/0.04] = \$1248.64$ *Using the Excel financial function FV, the same answer is obtained with =FV(0.04,10,-100,0,1).*

Example: 7.5 *What will be the value of a savings account after ten years if $100 is invested at the end of each year and the account earns 4% p.a.?*

Solution: 7.5 *After ten years, the first \$100 is worth* $100(1.04)^9$
The second \$100 invested is worth $100(1.04)^8$ *and so forth. So the value of the account at the end of ten years is:*
$F = 100[1 + 1.04 + 1.04^2 + \cdots + 1.04^9]$
but this is a geometric series of the form $a, ar, ar^2, \cdots, ar^{n-1}$ *which can be written as:*

$$S_n = a + ar + \cdots + ar^{n-1} \tag{7.8}$$

Multiplying both sides of Equation 7.8 by r and subtracting Equation 7.8 from the product gives:

$$rS_n - S_n = -a + ar - ar + \cdots - ar^{n-1} + ar^{n-1} + ar^n \tag{7.9}$$
$$(r-1)S_n = a(r^n - 1) \tag{7.10}$$
$$F_n = S_n = a\left[\frac{r^n - 1}{r - 1}\right] = a\left[\frac{(1+i_a)^n - 1}{i_a}\right] \tag{7.11}$$

with $r = 1 + i_a$, *and* $a = \$100$.
So $F = 100[(1.04^{10} - 1)/0.04] = \$1200.61$ *Using the Excel financial function FV, the same answer is obtained with =FV(0.04,10,-100,0,0).*

The term $a(r^n - 1)/(r - 1)$ is also known as the discrete uniform-series compound-amount factor. If we wish to compute the present value, P, of the future sum, S_n, then we must discount S_n back to the present time so:

$$P = \frac{S_n}{(1+i_a)^n} = a\left[\frac{(1+i_a)^n - 1}{i_a(1+i_a)^n}\right] \tag{7.12}$$

Example: 7.6 *What discrete annual compound interest rate doubles the money in an account within eight years?*
Solution: 7.6 $2P = P(1 + i_a)^8$ *so* $i_a = 0.0905$ *or 9.05% p.a.*

The compounding period is not always annual, and any regular period may be used. Quarterly compounding is common in savings accounts, but the nominal annual interest rate (number of periods/year)(interest rate/period) is not equal to the discrete annual compound interest rate.

Example: 7.7 *The nominal annual interest rate is simply the number of periods per year times the discrete interest rate per period. Calculate the nominal annual interest rate and the discrete annual compound interest rate if interest is compounded quarterly and the rate is 5% per quarter.*
Solution: 7.7 *The nominal annual rate is* $4(0.05) = 0.20$ *or 20%*
To calculate the discrete annual compound interest rate: $P(1 + i_a) = P(1 + 0.05)^4$ *and* $i_a = 0.215506$ *or 21.55% p.a.*

Continuous compounding occurs when the time interval for discrete compounding approaches zero. In this case, the future value F, is continuously increasing in an amount proportional to $Fi_c dt$ so we can write:

$$\begin{aligned} dF &= Fi_c dt && (7.13) \\ \int_P^F \frac{dF}{F} &= i_c \int_0^n dt && (7.14) \\ \ln \frac{F}{P} &= in && (7.15) \\ & && (7.16) \end{aligned}$$

So for continuous compounding, we have

$$F = Pe^{i_c n} \tag{7.17}$$

By equating future sums, we can derive relationships between interest rates that are discretely compounded at nonannual periods:

$$\begin{aligned} F_n = P(1+i_a)^n &= P\left(1+\frac{i}{p}\right)^{pn} && (7.18) \\ i_a &= \left(1+\frac{i}{p}\right)^p - 1 && (7.19) \end{aligned}$$

Similarly, to compute an effective interest equivalent to a nominal, continuous rate:

$$\begin{aligned} F_n = P(1+i_a)^n &= Pe^{i_c n} && (7.20) \\ i_a &= e^{i_c} - 1 && (7.21) \\ & && (7.22) \end{aligned}$$

Inflation is another factor affecting the time value of money. Because of inflation, a dollar tomorrow is worth less than one today. If we define an effective annual inflation rate, i_f, then the present value of future dollars is simply:

$$P = \frac{F_n}{(1+i_f)^n} \tag{7.23}$$

So for a savings account earning i_a p.a. and experiencing inflation at a rate i_f, the future value is:

$$F'_n = P\left(\frac{1+i_a}{1+i_f}\right)^n \tag{7.24}$$

where F'_n indicates the future worth of a present sum in current dollars, and the value of savings account actually decreases if the inflation rate is greater than the earnings rate of the account. From Equation 7.24, it should be clear why inflation is so damaging to people living off of fixed income in savings accounts!

The present value of a future sum in inflated dollars is given by:

$$P = \frac{F_n}{(1+i_a)^n} \tag{7.25}$$

However, current dollars are worth more than inflated dollars, so a second discount factor must be included to correct for inflation. Thus in current dollars the present worth is not P, but P':

$$P' = \frac{F_n}{[(1+i_a)(1+i_f)]^n} \tag{7.26}$$

Example: 7.8 *The same investment may be made in two alternative projects. One project will yield \$1500 five years from now. The second project will yield \$3500 in 15 years. The investor has an annual earnings rate of 20%. Which investment is more attractive (i.e., has the greater present worth?*

Solution: 7.8 *For the first project* $P = 1500/(1.20)^5 = \$602.82$
For the second project $P = 3500/(1.20)^{15} = \$227.17$
So the first project is more profitable, even though it earns less money, because it has the greater present value.

7.3 Net Present Worth

The history of a project may be described as a series of cash flows. Receipts flow into the projects and disbursements flow out. Since money has a time value, cash flows over a period of time must be adjusted by an appropriate discount factor to correct for this effect. If each future cash flow is discounted back to its value today, one obtains the present worth (or present value) of each future cash flow. If all disbursements and receipts for a project are discounted back to today and added together, one then obtains the net present worth, P, of the project.

Consider Example 7.1. In the earlier analysis, the time value of money was not considered. Suppose that the company considering the purchase has an earnings rate of 20% p.a. Then the cash flows alone do not determine the value of the investment to the company, but these cash flows, CF_j, must be discounted first. Taking the initial purchase time as time zero, then the Net Present Worth, P, in general is

$$P = \sum_{j=1}^{n} \frac{CF_j}{(1+i)^j} \tag{7.27}$$

and for our specific example

$$P = -7,000 + \sum_{i=1}^{4} \frac{1,600}{1.20^i} + \frac{1,850}{1.20^5} = -2,115 \tag{7.28}$$

The annual discounted cash flows and the year by year net present worth are shown in Table 7.2.

So the net present worth of this project is $-\$2,115$. This means that if the company can really earn 20% p.a. over the next five years, then it will be better off going to an outside printer.

Consider this same problem in terms of net present cost. That is, compute the costs of each option, but discounted by the 20% earnings rate of the company. In

Table 7.2: Estimated discounted cash flows and net present worth.

Year	Annual Expense	Annual Benefit	Annual Cash Flow	Discounted Cash Flow	Net Present Worth
0	$7,000	$ 0	-$7,000	-$7,000	-$7,000
1	600	2,200	1,600	1,333	- 5,667
2	600	2,200	1,600	1,111	- 4,556
3	600	2,200	1,600	926	- 3,630
4	600	2,200	1,600	772	- 2,858
5	600	2,450	1,850	743	-2,115

the first case the company pays $2,200 p.a. for an outside printer. The Net Present Cost (NPC) in this case is:
$NPC(1) = 2200(1/1.2 + 1/1.2^2 + 1/1.2^3 + 1/1.2^4 + 1/1.2^5) = \$6,579$
In the second case, the NPC is:
$NPC(2) = 7000+600(1/1.2+1/1.2^2+1/1.2^3+1/1.2^4+1/1.2^5)-250/1.2^5 = \$8,694$
The diffence in net present cost is $NPC(1) - NPC(2) = -\$2,115$ which is also the net present worth of buying the in-house printing system. The conclusion is that the company is better off send its work to an outside printer because the $7,000 for the in-house system can earn more money elsewhere.

Consider this problem in a slightly different way. In the first year, the discounted rate of return is $\$1,600/[(1.2)7,000] = 0.1905$ or 19.05%. This is slightly less than the current earnings rate of the company which is 20% so the alternatives offering the higher earnings rate are favored.

Suppose the in-house printing system is operated for more than five years. How many years would it have to be held in order for the discreted, discounted cash flows to equal the present investment of $7,000? This problem may be answered using a spreadsheet and computing the net present worth for each year as a separate row. An analytical solution, for the case of constant annual cash flow, may be obtained by recognizing that successive terms in the discounted cash flow sum are related to each other by a common ratio.

This question can also be answered by treating it as a uniform cash flow problem, if the salvage value at the end of the time period is ignored. A geometric progression is a succession of terms where each, except the first, is related to the preceding term by a common ratio, r. Geometric progressions have the form $a, ar, ar^2, \ldots, ar^{n-1}$. One can show that the sum of a geometric progression, if $|r| < 1$, is given by:

$$S_n = \sum_{j=1}^{n} ar^{j-1} = \frac{a}{1-r} - \frac{ar^n}{1-r} = \frac{a(r^n - 1)}{r - 1} \tag{7.29}$$

With the substitution, $r = 1/(1+i)$,

$$P = \sum_{j=1}^{n} \frac{A}{(1+i)^j} = \frac{A}{i}\left[\frac{(1+i)^n - 1}{(1+i)^n}\right] \tag{7.30}$$

where P is the present worth of the sum of discounted future annual, uniform cash flow A, which are the same for each year. The total cash flow received or paid is therefore $CF = nA$, but the present worth of the future cash flow is less due to the discounting factor $1/(1+i)^j$ which is applied to cash received in the jth year.

Returning to our printing problem, consider it as being a disbursement of \$7,000 at $t = 0$, and a receipt of \$1,600 thereafter (i.e., ignore the salvage value). Then to compute the time to break even:

$$\begin{aligned} P &= \frac{A}{i}\left[\frac{(1+i)^n - 1}{(1+i)^n}\right] \\ 7000 &= \frac{1600}{0.2}\frac{1.2^n - 1}{1.2^n} \\ \frac{7000(0.2)}{1600} &= 0.875 = 1 - \frac{1}{1.2^n} \\ 1.2^n &= \frac{1}{0.125} \\ n &= 11.405 \text{ years} \end{aligned}$$

Relationships similar to Equation 7.30 can be derived for other types of predictable future cash flows (e.g., ramps), but cash flows are most commonly not uniform in project analysis. Thus the simplest approach is to set up an appropriate table in a spreadsheet, discount as required, and then sum the discounted cash flows to compute the present worth of the project.

Example: 7.9 *A corporation earns 25% p.a. It is considering a 10 year project with the following cash flows.*

Table 7.3: Estimated Expenses and Receipts

Year	0	1	2	3	4	5	6	7	8	9
Expense (k\$)	-50	-70	-30	-20	-20	-25	-25	-30	-30	-30
Receipt (k\$)	0	0	40	100	100	120	120	140	140	140

Solution: 7.9 *Cash flows are computed for each year and discounted by* $1/1.25^n$ *to give the table below. As can be seen, the Net Present Worth is about \$86,000 for this project*

7.4 Measures of Profitability

Different measures of profitability are in common use. The most rigorously correct measure is venture profit as discussed in Chapter 8. That measure of profitability combines the present worth and the rate of return criteria into a single measure. However, different organizations may prefer different measures for a variety of reasons and it is important to be aware of how the most common measures are computed.

Table 7.4: Estimated Discounted Cash Flows Per Annum and Net Present Worth

Year	(k$) Expense	(k$) Receipt	Actual Cash Flow	Discounted Cash Flow	Net Present Worth
0	-50	0	-50	-50	-50
1	-70	0	-70	-56	-106
2	-30	40	10	6	-100
3	-20	100	80	41	-59
4	-20	100	80	33	-26
5	-25	120	95	31	5
6	-25	120	95	25	30
7	-30	140	110	23	53
8	-30	140	110	18	72
9	-30	140	110	15	86

7.4.1 Internal Rate of Return

Net Present Worth is one key measure of profitability. Another important concept uses discounting. By adjusting the discount factor to a series of cash flows, some positive and some negative, one can compute a discount factor such that the net present worth of the discounted cash flows is zero. This discount factor is the Internal Rate of Return. Solving:

$$P = 0 = \sum_{j=1}^{n} \frac{CF_j}{(1+i)^j} \tag{7.31}$$

gives a value of i known as the Internal Rate of Return (IRR). Note that negative or positive cash flows can occur in any years. It is not necessary that all investment, for example, occur in year one or year zero.

Applying equation 7.31 to the actual cash flows in Table 7.4 gives an IRR=0.4269 or 42.69% p.a. using the solver function in Excel. This is much larger than the 25% the corporation usually earns, so the investment is attractive based on Internal Rate of Return.

7.4.2 Profitability Index

The Profitability Index is another method of using discounted cash flows to evaluate investments. In this case, discounting is often based on the corporate earnings rate. Attractiveness over a project life is determined by computing a numerator and denominator where all discounted cash flows that are positive are added into the numerator. All discounted cash flows that are negative appear in the denominator. The ratio is the Profitability Index which should be greater than unity before an investment is recommended.

$$N = \sum_{j=1}^{n} \frac{CF_j}{(1+i)^j},\ CF_j > 0.0 \tag{7.32}$$

$$D = \sum_{j=1}^{n} \frac{CF_j}{(1+i)^j}, \; CF_j < 0.0 \quad (7.33)$$

$$PI = N/|D| \quad (7.34)$$

Applying equation 7.34 to the discounted cash flows in Table 7.4 gives a Profitability Index, PI=1.81. The numerator is, $N = 192$. The denominator is $D = -106$.

Example: 7.10 *Two alternative investments are being considered with cash flows as shown below. The company considering these investments has an earnings rate of 20% p.a. Calculate the Profitability Index and Net Present Value for each. What do you recommend that they do? The company could choose to invest in neither, only one, or both.*

Year	*Cash Flow (10^6)*	
	Investment I	*Investment II*
0	*-10.0*	*-5.0*
1	*-2.0*	*-4.0*
2	*3.0*	*-2.0*
3	*5.0*	*4.0*
4	*8.0*	*5.0*
5	*9.0*	*8.0*
6		*8.0*

Solution: 7.10 *Calculate the net present worth and profitability index for each alternative.*
$N_1 = 3/1.2^2 + 5/1.2^3 + 8/1.2^4 + 9/1.2^5 = 12.4517$
$N_2 = 4/1.2^3 + 5/1.2^4 + \cdots + 8/1.2^5 + 8/1.2^6 = 10.6202$
$|D_1| = 10 + 2/1.2 = 11.6667$
$|D_2| = 5 + 4/1.2 + 2/1.2^2 = 9.722$
$PI_1 = 12.4517/11.6667 = 1.067$, $NPV_1 = 12.4517 - 11.6667 = 0.7851$
$PI_2 = 10.6202/9.7222 = 1.092$, $NPV_2 = 10.6202 - 9.722 = 0.898$
Both have PI values greater than one. Both have positive net present values. Recommend both.

Example: 7.11 *The township of Lake Woebegon normally earns 15% p.a. through several local monopolies. It has an opportunity to rent land it owns for a parking lot during a regional hockey tournament, a one time event, but must upgrade the land first at an estimated cost of $2,700. If it makes this expenditure, it expects to realize $5,100 from the rental, payable at the end of the year. Is this investment interesting to the township?*

Solution: 7.11 *This problem may be evaluated in terms of net future value, net present value, and internal rate of return. Net Future Value (NFV)* $= 5100 - 2700(1+.15) = \$1,995$
Net Present Value (NPV) $= 5100/1.15 - 2700 = \$1,735$
Internal Rate of Return (IRR) $= (5100 - 2700)/2700 = 0.8888 = 88.88\%$
Lake Woebegon will earn over 80% return for the $2,700 investment. This is more than the normal 15% p.a. earned by its monopolies, so the money should be invested, even though the receipts are only realized for a single year.

7.5 Summary

7.6 Problems

7.1 A paper company is planning a $30 million expansion of an existing mill. One third of the capital will come from corporate reserves, but the balance will be raised from outside sources. The combined state and federal income taxes are 55%. Several plans are being considered to raise these funds.

Plan A One third of the funds are raised from selling bonds at 10% interest. One third is raised from selling stock which pays 5% dividends p.a.

Plan B Two thirds of the funds are raised from selling stock which pays 5% dividends p.a.

Which plan do you recommend? Hint: Interest paid on bonds is tax deductible, but dividends paid to stockholders are not. Calculate the retained earnings in either case. For discussion, consider the differences between bonds and stocks in terms of the corporation buying them back (i.e., stocks are traded on the market, but bonds are eventually redeemed).

7.2 What annual amount is required to pay back a $15,000 loan in eight years if interest is 8% p.a.?

7.3 A chemical manufacturer wishes to purchase a new distillation tower and must borrow $150,000. Banks are willing to loan money at 8% p.a. for a five year period. Three payback options are being considered: (1) lump-sum payment of loan plus interest at the end of five years, (2) five uniform payments over five years and (3) a sliding scale payback where each payment is twice the previous one. Calculate the cash flows and total payments for these three options. Which one is least expensive?

7.4 A pump in corrosive service must be replaced every three years. The pump costs $12,000 to replace. If money earns 12% p.a., estimate the amount that must be initially invested to pay for six pump replacements over the 21 year life of the plant.

Chapter 8

Economic Objectives

Frequently, the goal of design is to either increase profits or decrease costs. In such cases one must understanding the most important economic factors in order to know how to approach the problem. Although alternative goals, such as improving safety or reducing the environmental pollution, may be the primary objectives, in many instances these problems can also be restated in terms of their effects on either costs or profits. Consequently, these economic goals are quite general and valuable even when the design effort is focused primarily on peripheral issues such as waste management.

8.1 Major Economic Factors

A number of alternative criteria are also available, and Peters and Timmerhaus [76] provide an excellent review. Rudd and Watson [86] focus more heavily on the venture profit criterion as it is discussed here. Happel and Jordan [37] also emphasize the importance of venture profit and cost. Basically, the venture profit criterion results from a consideration of net profit and the real goals of any profit-making entity in a competitive environment.

8.2 Justification for Venture Analysis

Most companies want to increase their net profits. Estimating corporate profits for a complex organization, however, is often a challenging task in itself. Moreover, these difficult calculations are normally beyond the scope of planned activities for an engineer who has just been asked to design a new heat exchanger for one particular application.

On the other hand, that same engineer may find it useful to study the economic characteristics of the proposed heat exchanger design using an internal corporate profitability measure. In this case, a criterion is defined which includes terms for describing both the capital requirements and the estimated operating costs for the new heat exchanger, but it also includes one or more additional economic terms which reflect the estimated impacts of the proposed heat exchanger investment on the overall corporate economic picture.

One such economic criterion is the venture profit. Another one is the venture cost. Using these criteria, a minimum rate of return on capital investment (ROI) is required for all corporate ventures. Although these criteria may be modified for specific applications, the internal ROI requirement is usually specified by the firm's management based upon the alternative corporate investment opportunities. Modifications to the required ROI may include specified increases to account for increased risks for some of the options. These risks may be due to uncertainties in the design, a lack of adequate pilot plant data, or novel processing features.

Most organizations which have money to invest will consider the profitability of making investments outside of the normal corporate activities as well as spending the money internally (for example, on a new heat exchanger). For another example, suppose that a company has a million dollars to invest and it identifies two alternative ventures. Suppose that the first venture requires an investment of the entire sum and yields a net profit of \$100,000 annually. On the other hand, suppose that a second investment is available which requires only \$200,000 in capital investment and it returns \$40,000 annually. Also, part or all of the million dollars can be invested in money markets where it earns 12% annually and the investment is renewable each year. How should the company invest?

If only the two internal investments are considered without regard to the possibility of investing some of the funds in money markets, then one might conclude that the first investment is better than the second since it yields \$100,000 rather than \$40,000 per annum. On the other hand, investment of the entire sum in the money markets will yield \$120,000 in the first year. At the end of that time the funds could be withdrawn and reinvested elsewhere or else renewed.

A third possibility is to invest \$200,000 in the second venture and \$800,000 in the money markets. This arrangement results in \$136,000 net profits the first year and the option to reinvest the \$800,000 at the end of that period.

In this simple example, investment in the first option yields 10% return on investment (ROI), the second possibility yields 12% ROI, and the third combination yields 13.6%. Clearly, the third option is best because it yields the greatest ROI, the largest net profits, and it permits a degree of flexibility that does not exist with the first option.

With respect to chemical process design, however, the choices are generally not so obvious for several reasons. First of all, the typical design engineer who is working for a large corporation will not be aware of the total spectrum of investment opportunities that are available to the company. In general, neither the availability of funds, nor the interest rates which the company can realize for the various market investments which are available to it, will be well known to the engineer. Yet the engineer must recommend a level of investment — frequently the sizes and configurations for processing equipment. Also, this equipment may require utilities to operate. If so, then what utility cost reductions are required in order to justify increases in equipment capital investment?

A second problem is the fact that a continuum of alternatives is available in the sense that the equipment size can usually be varied incremental over some

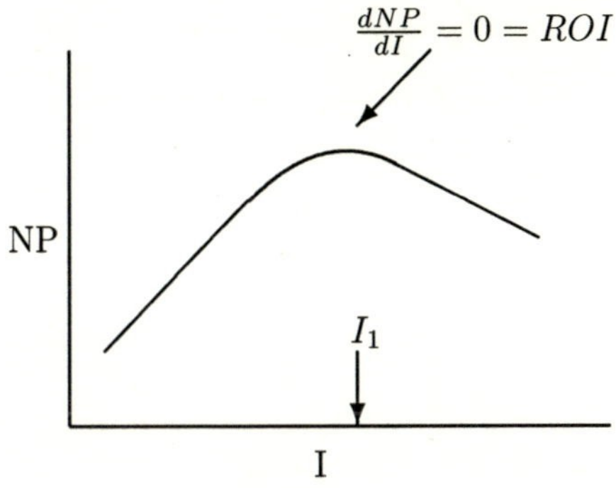

Figure 8.1: Net profits as a function of the capital investment.

permitted range of values. One aspect of this problem has already been mentioned. That is, how reversible should a process be in order to reduce operating costs at the expense of additional capital investment?

For another example, how large should a given plant be? In general, these situations result in profitability curves similar to that one depicted in Figure 8.1.

In this figure, the typical effects of incremental investment on the net profits associated with a single venture are shown.

As the level of capital investment increases in Figure 8.1, the net profits also increase until an investment level, I_1, is reached and the net profit, NP_1, associated with that investment reaches its maximum value. Clearly, investment levels greater than I_1 should be avoided since they will actually reduce the net profits. However, should the engineer recommend investment at the level I_1 which maximizes the net profit?

An engineer who makes such a recommendation will have failed to recognize the possibility of alternative investments. Figure 8.1 only considers the profitability of the single investment. It does not consider the corporate profitability and, therefore, maximizing the profits in Figure 8.1 alone is similar to recommending the million dollar investment mentioned in the first example. That case only yields 10% ROI when 13.6% is possible.

If Figure 8.1 is to be used as a basis for optimizing the process whose economic behavior it reflects, then an economic criterion must be developed which considers both the net profits for the particular process and also specifies that at least some minimum ROI be achieved. After all, if the investment level I_1 is chosen in Figure 8.1, then the company will lose money. At this investment level, the incremental ROI is 0% and the company is better off investing some portion of I_1 in outside markets.

An improved economic criterion results from recommending that the corporation

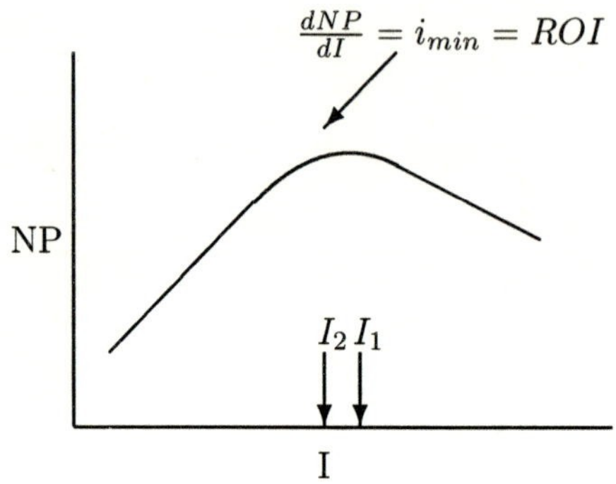

Figure 8.2: Effect of investment on venture profits.

maximize the net profits associated with a particular investment, *subject to the constraint that a minimum incremental ROI is also achieved.* This situation is illustrated in Figure 8.2 where an investment level, I_2, is specified. In this case, the engineer recommends investment until the incremental ROI, shown as i_{min}, is reached. The difference in capital investment levels, $(I_1 - I_2)$, is then available for alternative investments with greater than zero ROIs.

A suitable goal for local chemical process optimization is achieved by maximizing the venture profit for the process. This term is defined as:

$$VP = NP - i_{min} f(\boldsymbol{r}, \boldsymbol{I}) \tag{8.1}$$

Where $f(\boldsymbol{r}, \boldsymbol{I})$ is defined in section 8.4. The venture profit, VP, includes consideration of both the net profit for the process and the incremental return on investment. When the venture profit is maximized (or the venture cost is minimized) then

$$\frac{dVP}{dI} = 0$$

but at that investment level:

$$\frac{dNP}{dI} = i_{min}$$

which is apparent from differentiating Equation 8.1 Hence, maximizing the venture profit is equivalent to maximizing the net profit, but with the added constraint that the ROI associated with the process under consideration be equal to the minimum ROI for that type of investment within the corporation.

Example: 8.1 *Calculate the incremental return on investment (ROI) from implementing the heat exchanger network discussed in Example 13.1, versus supplying all heating and cooling requirments with a silicone oil heating medium and refrigeration. Refrigerant is available at 40 °F. It costs \$2.861/10⁶ Btu removed* *from the process. The refrigerant temperature rise is 20 °F when matched with a process stream. The silicone oil heating medium is available at 640 °F at a cost of \$20.00/10⁶ Btu added to the process. The oil temperature drop is 20 °F when it is used as a heating medium. Assume that the overall heat transfer coefficient for any heat exchanger in the network, or any heat exchanger supplying utility, is 100 Btu/hr-ft²-°F. Also assume that floating-head heat exchangers, constructed of carbon steel,will be used, regardless of the required sizes. Also assume that that plant operates 24 hours/day and 320 days/year.*

Solution: 8.1 *Two cases need to be computed. In the first case, all heat transfer requirements are provided by utilities. In the second case, only the minium requirments are provided by utilities, but additional heat exchangers are required. So an incremental savings (from reduced annual utility costs) result from an incremental investment in additional heat exchangers. So the incremental ROI in this case is given by:*

$$ROI = \left(\frac{Ann\ Utility\ wo\ Matching - Ann\ Utility\ w\ Matching}{TIC\ w\ Matching - TIC\ wo\ Matching}\right) \times 100\% \qquad (8.2)$$

Detailed analysis of this problem leads to the conclusion that heat exchanger matching is highly beneficial.

8.3 Venture Cost Analysis

Venture cost is defined as the net profit with the sales term deleted:

$$VC = C(1-t) + gI - (eI_S + dID + lI_L)t + i_{min}f(\boldsymbol{r}, \boldsymbol{I}) \qquad (8.3)$$

Although many of the terms can fluctuate for any project in its initial stages of design, heuristics show that, as long as C_Q and I_P are considered, operating costs and structures vary little with the other coefficients. As mentioned earlier, Rudd and Watson suggested a rough calculation of venture cost as :

$$VC = .5(\text{operating expenses}) + .25I \qquad (8.4)$$

Using the data in Table 8.1, the following equation results:

$$VC = .52(\text{operating expenses}) + .25I_P \qquad (8.5)$$

From equation 7.12 we have the present value of a constant annuity:

$$P = \frac{S_n}{(1+i_a)^n} = A\left[\frac{(1+i_a)^n - 1}{i_a(1+i_a)^n}\right] \qquad (8.6)$$

Applying l'Hôpital's Rule and taking the limit as $n \to \infty$ gives $P = A/i_a = A/i$ and the present value of a constant annuity continuing forever is simply the value of the annuity divided by an appropriate discount factor.

Table 8.1: Factors leading to the coefficients in the venture cost equation.

Item	Rate
Income tax	48% of net taxable income
Experienced return	12% of investment
Working capital	18% of fixed investment
Project life	20 years
Risk factor	8%, an average risk
Minimum rate of return	20% of investment
Annual property taxes	4% of fixed investment
Annual property insurance	1% of fixed investment
Annual maintenance	5% of fixed investment
Plant operating time	8500 hours/year

Consider the simplified case where we have an investment opportunity that consists of an initial, one-time investment followed by an annual operating cost that continues indefinitely. Then the net present cost is simply:

$$NPC = I_o + \frac{A}{i} \tag{8.7}$$

Multiplying both sides of equation 8.7 by i gives an expression that is analogous to a venture cost:

$$VC = i(NPC) = A + iI_o \tag{8.8}$$

from which it is clear that minimizing venture cost is equivalent to minimizing net present cost where the annual operating expense is constant and continues indefinitely.

Example: 8.2 *A company must install a new distillation column. The company requires 15% p.a. return on all new investment. They can install a smaller column that requires more reflux or a larger column requiring less reflux to do the same separation. The annual fixed charges are 10% of the Total Installed Cost of the column. The annual energy costs for steam in each case are shown below. All other costs for the column are about the same and can be ignored. Based on the information below, what is the optimal reflux ratio?*

Reflux Ratio	*1.10*	*1.15*	*1.20*	*1.30*	*1.40*
Total Installed Cost ($1000)	*1700*	*1600*	*1500*	*1200*	*800*
Energy Cost ($1000/yr)	*260*	*280*	*300*	*400*	*500*

Solution: 8.2 *Minimizing the Venture Cost,* $VC = A + iI_o$*,* $i = 0.1 + 0.15$*, gives:*

$VC_1 = 260 + 0.25(1700) = 685$
$VC_2 = 280 + 0.25(1600) = 680$
$VC_3 = 300 + 0.25(1500) = 675$
$VC_4 = 400 + 0.25(1200) = 700$

Table 8.2: Estimated discounted cash flows for first nine years of project. All values in \$1000. $i = 0.15$.

Year	Gross Profits	Amortization	Depreciation	Net Taxable Income	Income Taxes	Actual Cash Flow	Discounted Cash Flow
1						-550	-478
2						-2800	-2117
3						-6100	-4011
4	5300	190	1170	3940	1655	3645	2084
5	5300	190	1716	3394	1425	3875	1926
6	5300	190	1638	3472	1458	3842	1661
7	5300	190	1638	3472	1458	3842	1444
8	5300	190	1638	3472	1458	3842	1256
9	5300			5300	2226	3074	874

$VC_5 = 500 + 0.25(800) = 700$
So option three is preferred. The same conclusion results from calculating incremental return on investment, but case five ($I_o = 800$) must be used as the base or initial case.

Example: 8.3 *A company spent \$550,000 on research and development during the first year of a project. In the second year they spent \$700,000 on land and \$2.1 million on construction. During the third year they spent \$5.7 million to complete the plant installation. Start-up during the third year cost \$400,000. In the fourth year, they went into full production and continued in full production for the next five years thereafter. During each year of full production, the operating costs were \$8.7 million/year and the company realized \$14 million/year in sales.*

Management wants to recover all capital expenditures in three years of operation. They are allowed to use 5-year ACRS depreciation on all fixed assets and amortization. The company presently earns 15% pa.

Estimate the gross profits, cash flow, net taxable income, income tax (at 42% of the net taxable income), capital recovery and net profits during the first 9 years of this project. What is the internal rate of return for this project over a nine year project life (i.e, counting years 1 through 9)? What is the Profitability Index, the Net Present Worth, and the Venture Worth for this project?

Solution: 8.3 *Applying the appropriate formulas yields Table 8.2. The Net Present Worth is the sum of the discounted cash flows which in this case is \$2,639,000. The Internal Rate of Return is obtained by adjusting the discount factor until the sum of the discounted cash flows is zero. Using the solver function in Excel gives IRR=0.269163 or about 26.92%p.a. return.*

Net Profit and Venture Profit must be calculated in order to estimate Venture Worth. When calculating Net and Venture Profit, Capital Recovery is subtracted in the appropriate years, but capital outlay is not deducted during the years in which they occur. Otherwise, capital expenditures are incorrectly deducted twice. Performing these calculations gives Table 8.3. The venture worth is the sum of the discounted venture profits, \$989,000.

Table 8.3: Estimated discounted venture profits for first nine years of project. All values in \$1000. $i = 0.15$.

Year	Actual Cash Flow	Capital Recovery	Net Profits	Venture Profits	Discounted Venture Profits
1	-550				
2	-2800				
3	-6100				
4	3645	3150	495	-922	-527
5	3875	3150	725	-693	-345
6	3842	3150	692	-726	-314
7	3842		3842	2424	911
8	3842		3842	2424	792
9	3074		3074	1657	471

Example: 8.4 *Use the data in Example 3.5 to determine the venture profit internal rate of return with $i_{min} = 0.125$. Capital recovery is uniform over four years of operation.*

Solution: 8.4 *The internal rate of return based on venture profits (with $i_{min} = 0.125$) is determined from:*
$\sum_{j=1}^{5} VP_j/(1+i)^j = 0$, *and by trial-and-error,* $i = 0.189$.
Since the internal rate of return is greater than i_{min}, this venture should be accepted.

8.4 Methods of Risk Compensation

Whenever differences in financial risk exist between alternative investments, then these differences should be accounted for by penalizing those options involving the greater levels of risk. Basically, the penalty should amount to a type of insurance where the additional charge is proportional to the required investment and increases with the risk. Estimation of the risk is complicated, but it should consider the economic consequences of design failure, unexpected project termination, equipment salvage values, and the uncertainties of the technical and economic design bases. The risk penalty should also be a function of the minimum acceptable rate of return when risk is not considered since the penalty factor should increase whenever interest rates, for example, increase. One method of risk compensation is given by:

$$i_{min} = i_e(1+r) \tag{8.9}$$

where: r = the relative risk penalty factor which falls into one of several categories adapted from Rudd and Watson [86]:

1. High risk ($2 \leq r \leq 20$). Ventures in this category involve untested flowsheets with novel design features or they are based on unproven market conditions.

2. Fair risk ($1 \leq r \leq 2$). Ventures in this category include partially tested flowsheets with some new aspects that have not been completely evaluated.

3. Average risk ($.5 \leq r \leq 1$). Ventures here include the modification of existing processes to unproven conditions, but where some data exists and the probability of success is at least 80%.

4. Good risk ($.1 \leq r \leq 0.5$). Expansion projects of well known processes where an established market exists.

5. Excellent risk ($0 \leq r \leq 0.1$). Optimization of existing processes to reduce costs in a stable environment.

Note that each investment carries its own risk factor, where r_i is the risk factor of the ith investment.

Clearly, some subjective judgment is involved in setting risk penalty factors. The general categories listed above, however, can be used for purposes of initial estimation whenever risk is considered important.

Risks also vary with the type of investment. In general:

$$\begin{aligned} f(\boldsymbol{r}, \boldsymbol{I}) &= (1 + r_p)I_p + (1 + r_A)I_A + (1 + r_W)I_W \\ &\quad + (1 + r_L)I_L + (1 + r_S)I_S \end{aligned} \tag{8.10}$$

where: I_A = auxiliary investment
I_P = process battery limits investment
I_W = working capital investment
I_L = labor investment
I_S = start-up investment

8.5 Summary

8.6 Problems

8.1 The venture cost is simply the venture profit equation with the sales term deleted and a sign change. Derive Equation 8.5 from the information given in Table 8.1.

8.2 How are net profit and venture profit related to each other?

8.3 How is return on investment related to net profit and venture profit?

8.4 Four alternative investment opportunities are summarized below in terms of capital investment, annual income, and annual expenses. Which of these investments, if any, should be recommended if the minimum rate of return on the investment in 10%? Which should be recommended if the minimum rate of return on investment is 7% annually?

8.5 Four alternative investment opportunities are summarized below in terms of their total capital investment and annual cash flows. The company wants to recover the investment in two years (ignoring inflation) and expects 20% return on the investment before taxes. Which of these investments, if any, should be recommended? They all have the same level of risk.

	Alternative			
	I	II	III	IV
I(millions)	1.2	4.2	0.6	7.3
CF(millions)	1.1	3.1	0.2	1.8

8.6 Four alternative investments are summarized below in terms of capital investment, I, and annual cash flows, CF. The company considering these investments wants to recover its investment (if any is made) in 5 years. They are willing to ignore inflation and use straight line recovery. Also, the company expects to achieve at least 25% return on investment each year.

Investment	I	II	III	IV
I(millions)	22.5	6.4	1.8	11.5
CF(millions/yr)	8.3	3.3	0.9	4.5

Each investment is separate from the others. The company could invest in only one, or it may choose to invest in more than one. They have the same risk factors.

Which of the four investments have the highest rate of return?

8.7 In problem 6, which of the investments, if any, meet the company's criteria and should, therefore, be recommended?

8.8 What is the venture profit in the first year for the following investment? Use the data tabulated below.

Income tax rate	38%
Minimum annual rate of return	12%
Capital recovery period (straight line)	3 years
Annual gross sales (millions)	62
First year depreciation	15%
Annual operating costs (millions)	27
Total investment (millions)	37

8.9 What is the optimal investment policy? A corporation is considering six alternative investments. Their total capital investment and annual cash flows are tabulated below. The company wants to recover all capital investment in 5 years, using straight line ignoring any inflation. They expect to receive a 25% return on investment in the first year. All investments have the same risk factors which are incorporated into the required rate of return. The company can invest up to 12 million. What do you recommend?

	Alternatives					
	I	II	III	IV	V	VI
Investment (millions)	12	4.7	1.8	9.8	2.8	7.4
Cash Flow (millions)	6.9	2.1	1.2	4.3	1.9	5.3

8.10 Estimate first year gross profits, net taxable income, annual cash flow, net profits, venture profits, and rate of return for the following investment. The corporation pays 40% of its net taxable income in income taxes. It requires a minimum rate of return of 25% on any new venture and capital must be recovered in three years (straight line). The projected gross sales are 75 million/yr. The total investment is 45 million and the estimated annual operating costs are 22 million. The company may use ACRS depreciation (i.e., 15% in the first year).

8.11 A chemical company is operating a new process which required several years to develop, construct and start-up. In the first project year, the company rolled out $800,000 for research and development. In the second project year they spent another $1.6 million for pilot testing (a development cost). During the second year they also laid out $300,000 on land. In the third year they dropped an additional $0.7 million on pilot testing and a cool $4.8 million on plant construction. In the fourth year, they scraped the barrel and kissed off an additional $5.2 million to complete the construction and $0.9 million on start-up costs including extensive operator training.

Fortunately, they realized gross revenues of $24 million in the fourth project year while their annual operating expenses were $7.2 million.

Management wants to recover all capital expenditures within three years of operation. They are using 5-year ACRS depreciation on all fixed assets and amortization where allowed. Ignoring the time value of money, estimate the fourth year net profits if the corporate income tax rate is 52%.

Part II

Preliminary Design

Chapter 9

Engineering Estimates

The preliminary definition of a base case flowsheet is an important first step in the life of a design project. The base case is important as a reference point from which to lauch the examination of alternatives, and it is often used to evaluate economic feasibility.

If the product or products that will be produced are well known, then work can begin on selecting an optimal process to manufacture them. However, in some instances a marketing study is required to determine which products will be produced, based on current and past product prices and raw material costs. Prior to developing a base case, profit upper bound (PUB) analysis (see Chapter 12) is often used to determine whether or not a particular reaction path is at all attractive and to screen prospective products that may be attractive to produce and market. The initial base case flowsheet results from such preliminary analysis.

Once the products that will be produced are decided upon, then a manufacturing process may be selected and a flowsheet devised. The steps that are required to invent a flowsheet are summarized in Table 1.7, and they are straightforward, but sometimes tedious. Actually, the most difficult step is often the reliable calculation of energy and material balances for a process. Simulators can help, but may not converge reliably. Also, their results are only meaningful if the underlying physical property package accurately describes stream behavior. Eventually, experimental data are essential to verify and complete the final design of unit operations, particularly those which are more expensive. Each of these steps in the Preliminary Design of a chemical process is discussed in more detail below.

9.1 Process Flow Diagrams

Developing a Process Flow Diagram (or PFD) is one of the most important steps in a design project. Ultimately, the PFD determines the resulting problems in calculating energy and material balances. It also ties the entire design project together throughout its life. In the early phases of a project, the PFD may change considerably as the design evolves, but at some point in the project it is necessary

Table 9.1: Typical elements in an engineering style process flow diagram.Adapted from *A Guide to Chemical Engineering Process Design & Economics* [107] with permission from John Wiley & Sons.

Essential Elements	Equipment symbols Equipment numbers Process stream flow lines Utilities designations
Optional Elements	Energy exchange rates Temperature and Pressure designations on flow lines Volumetric and/or molar flow rates Equipment names[a] Major instruments Physical properties of process streams

[a]Because they are defined on the equipment list

to *freeze* the PFD so that the detailed design tasks may be completed.

Initially, only a few engineers and chemists are interested in and use the PFD. Toward the end of the project, it will be used by many individuals both in management and the technical divisions. As such, the PFD is a key focal point that hold the project together.

An effective PFD conveys a considerable amount of information with clarity and simplicity. In order to do so, the PFD should use a style that is generally accepted by design engineers, and should include a certain number of elements. Some of these essential are essential while others are optional as is indicated in Table 9.1.

The general style of a PFD should be as follows:

1. Stream flow is from left to right. If room is unavailable on a drawing to complete the flow, use a stream break symbol from Table 9.2.

2. Flowsheets are normally oriented horizontally with equipment placement mimicking those in the actual plant to a certain extent. Use some discretion to layout the equipment items on the PFD so that there is adequate room for the streams and their designations.

3. Process streams are designated with heavy solid lines. If streams cross each other, then usually the vertical line is broken. Do not allow process stream lines to touch each other unless they are connected.

4. Each process line, or process line segment, should terminate with an arrowhead indicating its direction of flow.

5. Packaged units, consisting of several unit operations, are designated by enclosing the units of the package by a dotted box.

6. Designate each equipment item or packaged unit on the flowsheet with an appropriate letter *and* number.

Table 9.2: Flowsheet symbols reproduced from Ulrich [107] with permission from John Wiley & Sons.

Symbol	Definition	Symbol	Definition
	Process Streams		*Utility Streams*
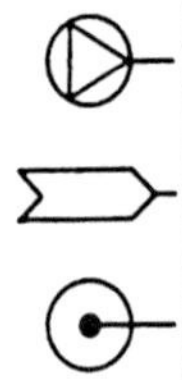	Raw material influx (Identify stream by name)	A-	Compressed air [A-12, e.g., would denote 12 barg]
		CTW	Cooling water (tower)
		CTWR	Cooling water return (tower)
		CW	Cooling water (natural source)
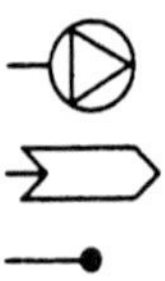	Finished product effluent (Identify stream by name)	CWD	Cooling water discharge (natural receiver)
		E−	Electricity [E−220, e.g., would denote 220 volts]
		HO-	Hot oil [HO-300, e.g., would denote 300°C]
		HW-	Hot water [HW-150, e.g., would denote 150°C]
	Designates a skip or break in a process line on the sheet. "B" is a match letter to identify the stream at another place on the same or an adjoining sheet	HWR	Hot water return
		R	Refrigerant [R-10, e.g., would denote -10°C]
	Stream designation for material balance point	S-	Steam [S-15, e.g., would denote 15 bara pressure]
		SC	Steam condensate
	Temperature (°C)	WD	Demineralized water
	Pressure [barg (0.1 MPa gage)]	WM	Municipal water
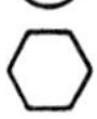	Pressure (kPa gage)	WP	Process water
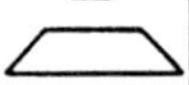	Pressure (kPa absolute)		
	Gas flow (std m^3/s; ideal gas at 273 K, 1 atm)		
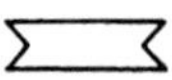	Liquid flow (liters/s)		
	Mass flow (kg/s)		
	Molar flow (kgmol/s)		

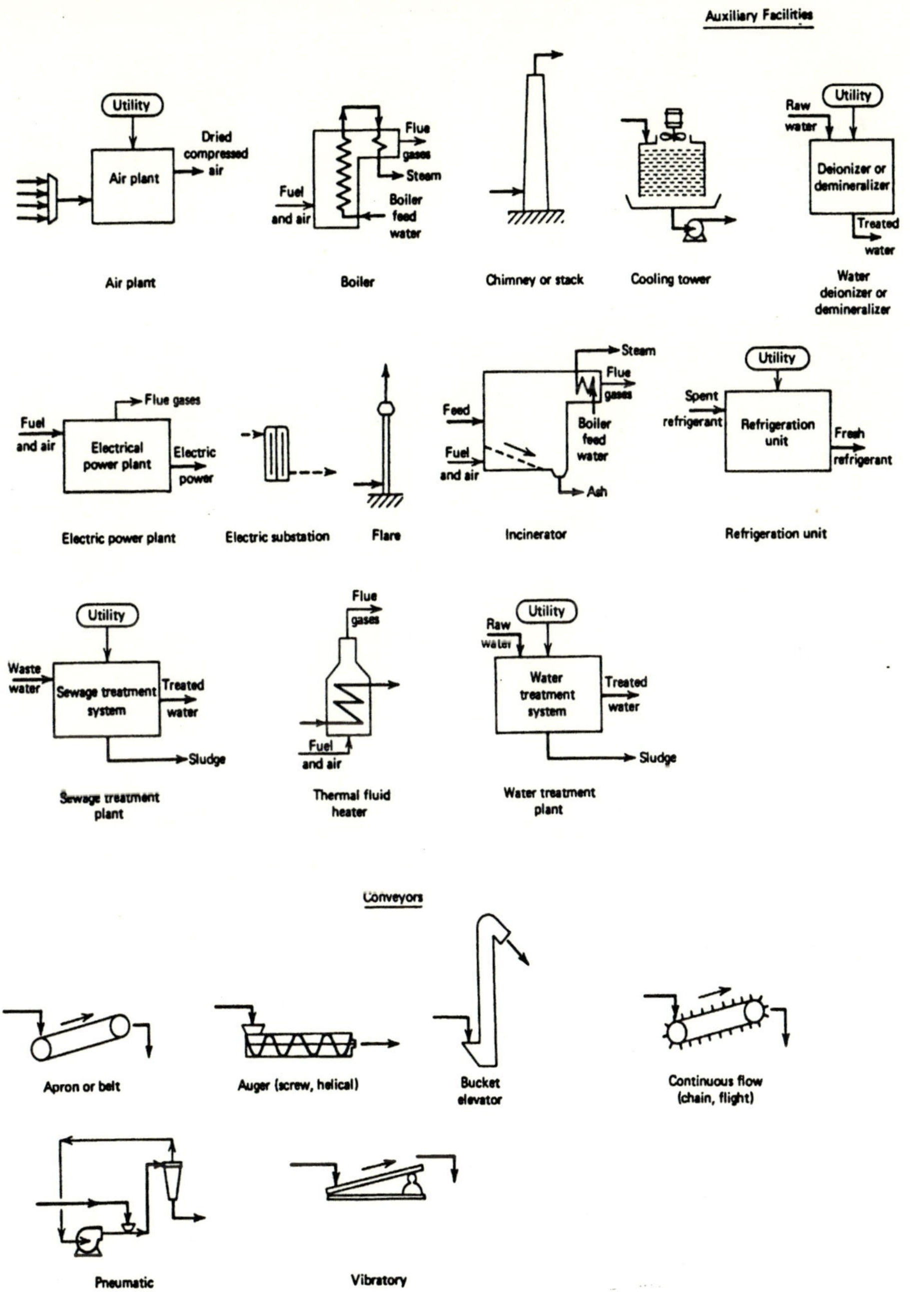

Figure 9.1: Equipment symbols for flowcharts. Reproduced from Ulrich [107] with permission from John Wiley & Sons.

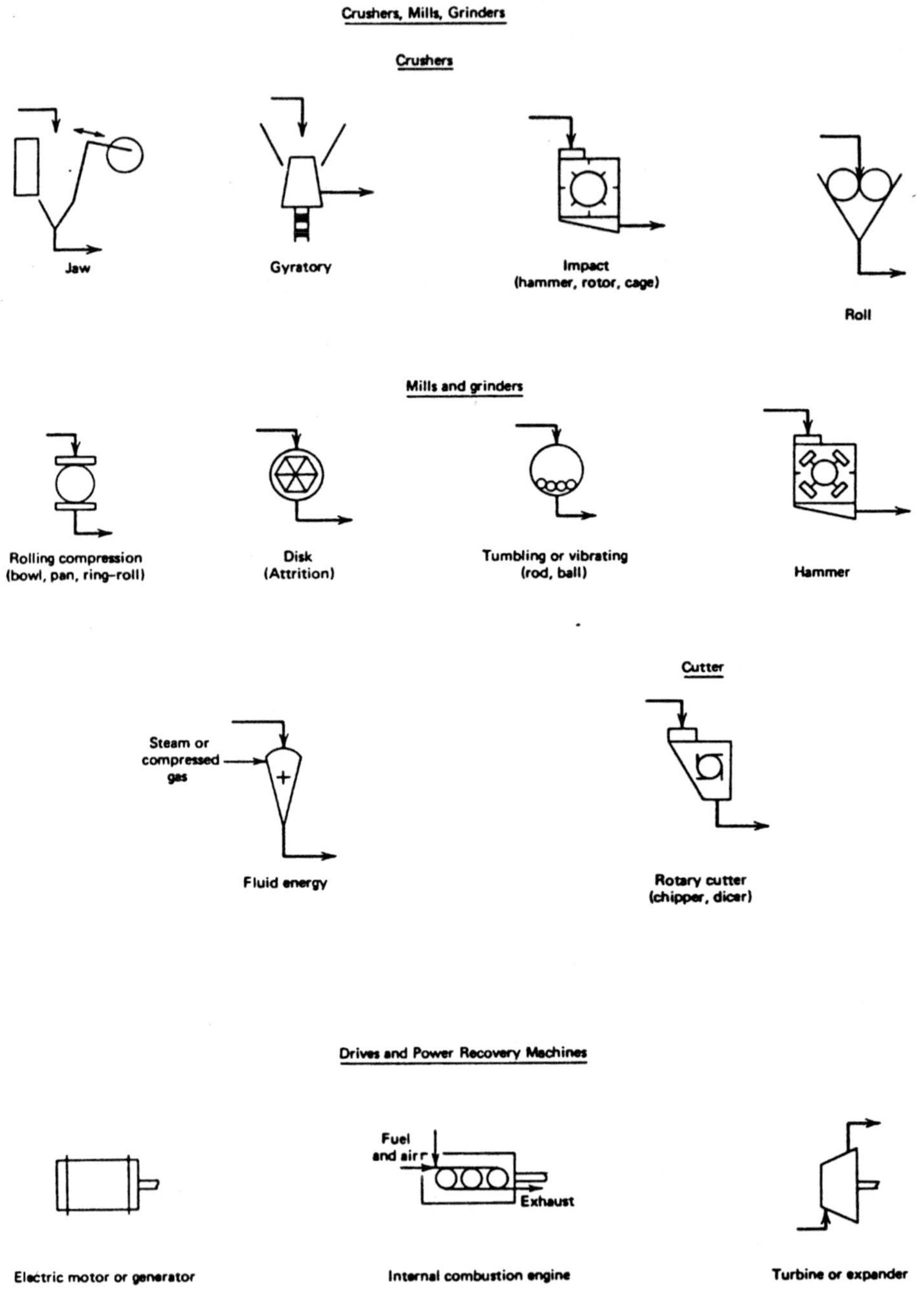

Figure 9.2: Equipment symbols for flowcharts. Reproduced from Ulrich [107] with permission from John Wiley & Sons.

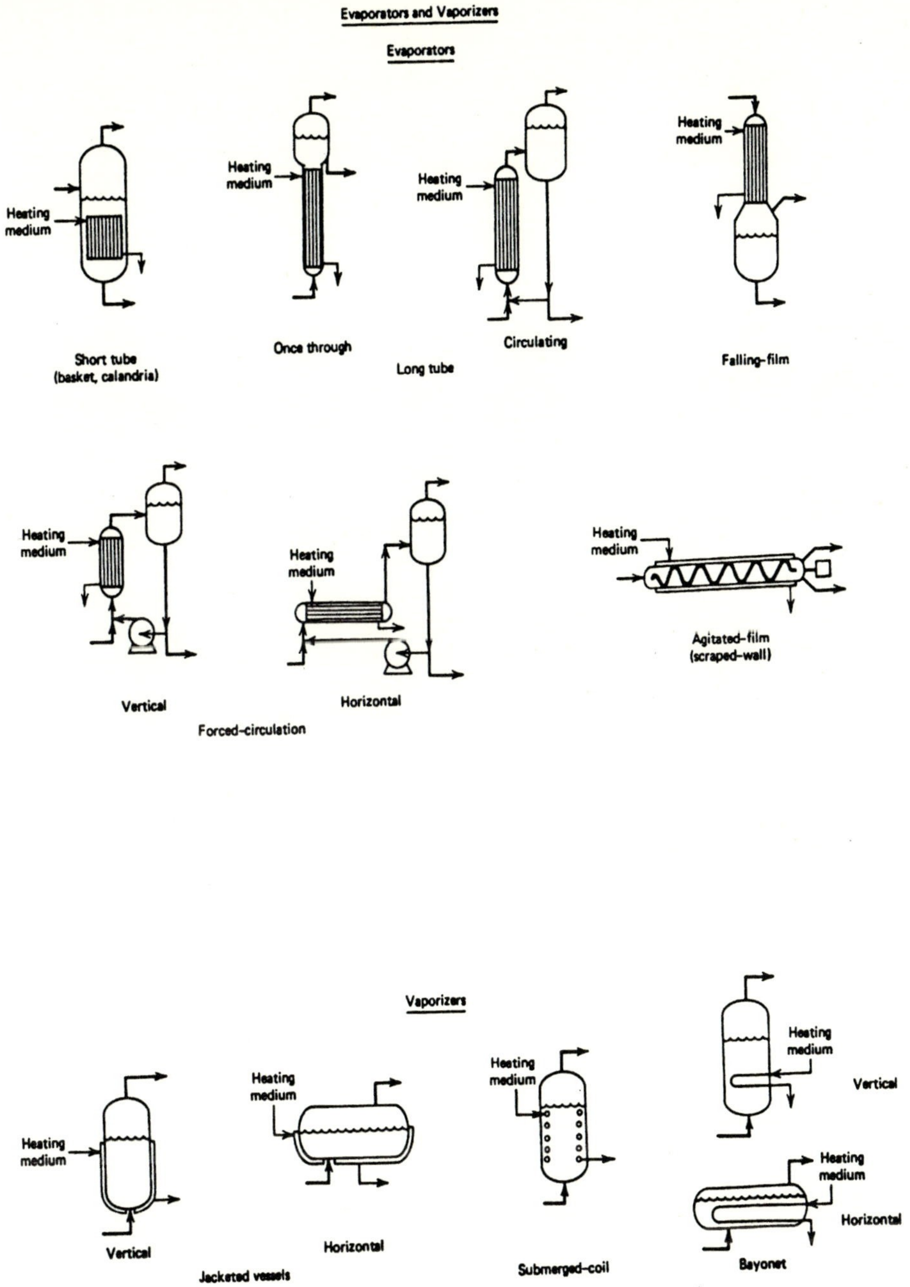

Figure 9.3: Equipment symbols for flowcharts. Reproduced from Ulrich [107] with permission from John Wiley & Sons.

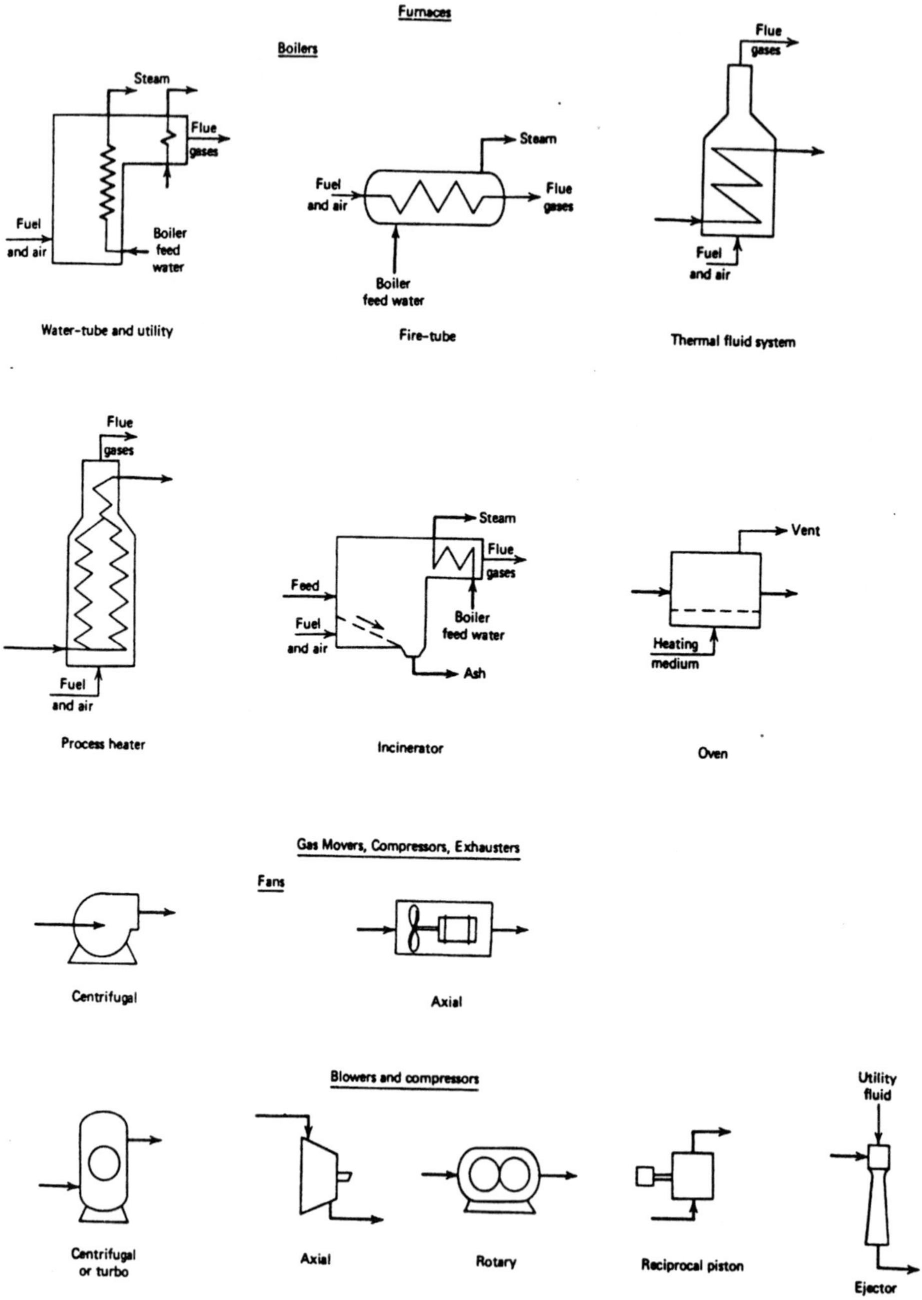

Figure 9.4: Equipment symbols for flowcharts. Reproduced from Ulrich [107] with permission from John Wiley & Sons.

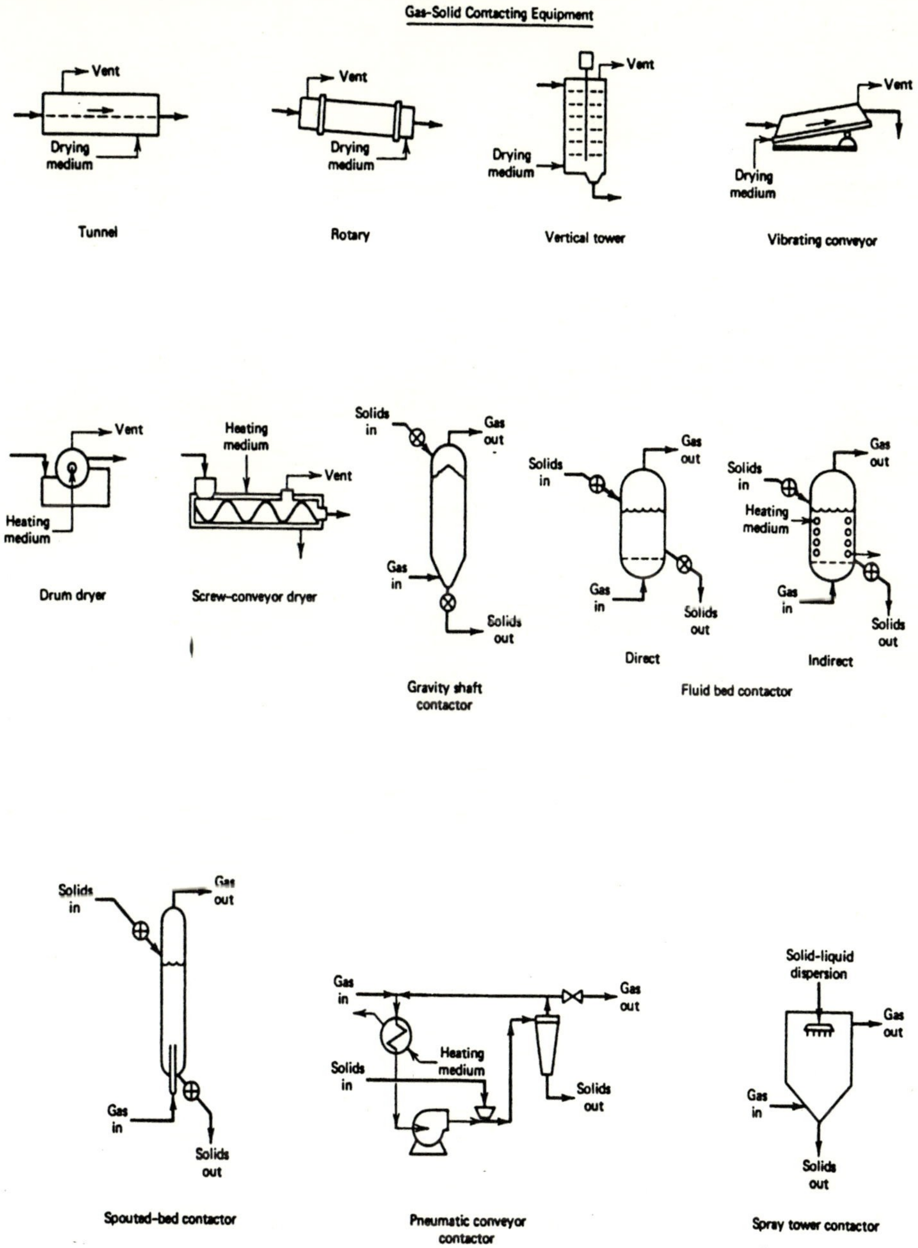

Figure 9.5: Equipment symbols for flowcharts. Reproduced from Ulrich [107] with permission from John Wiley & Sons.

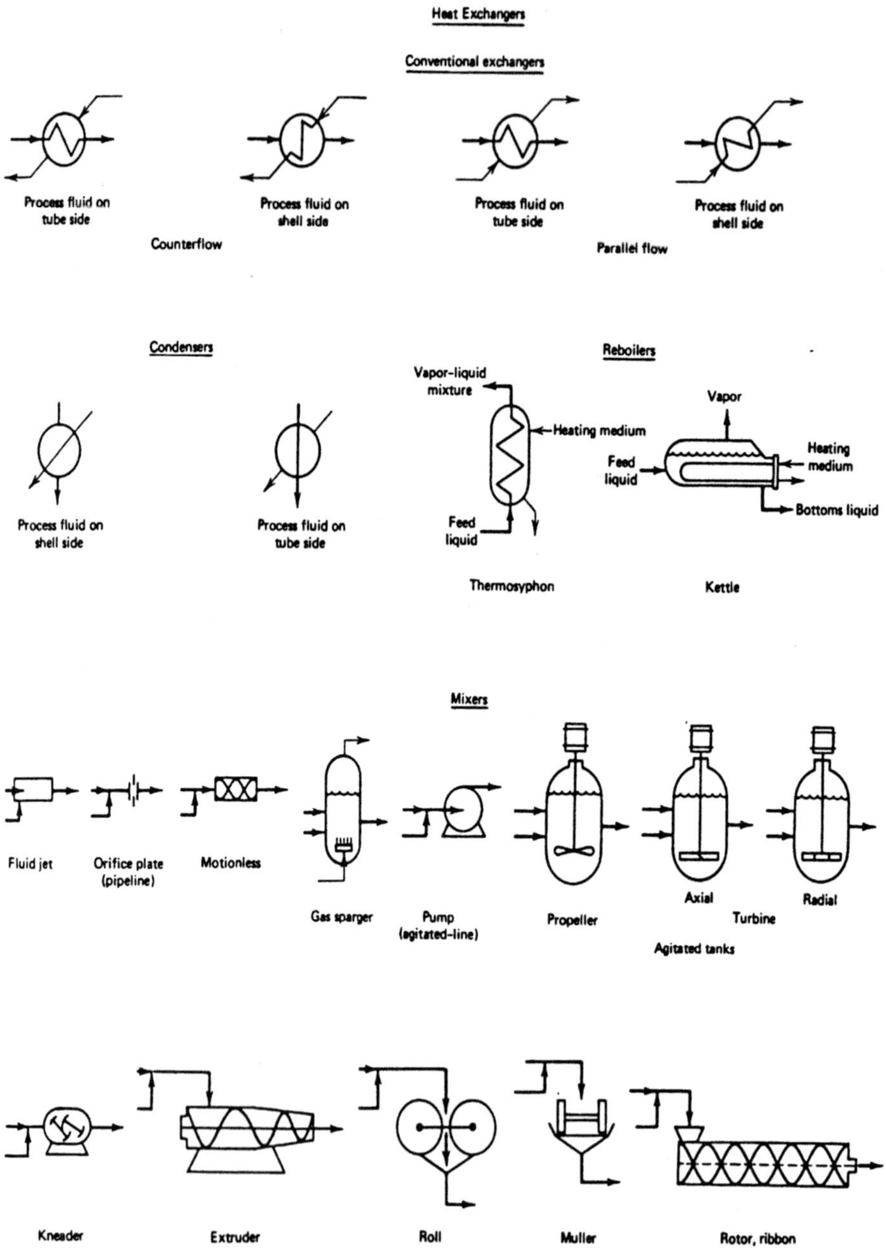

Figure 9.6: Equipment symbols for flowcharts. Reproduced from Ulrich [107] with permission from John Wiley & Sons.

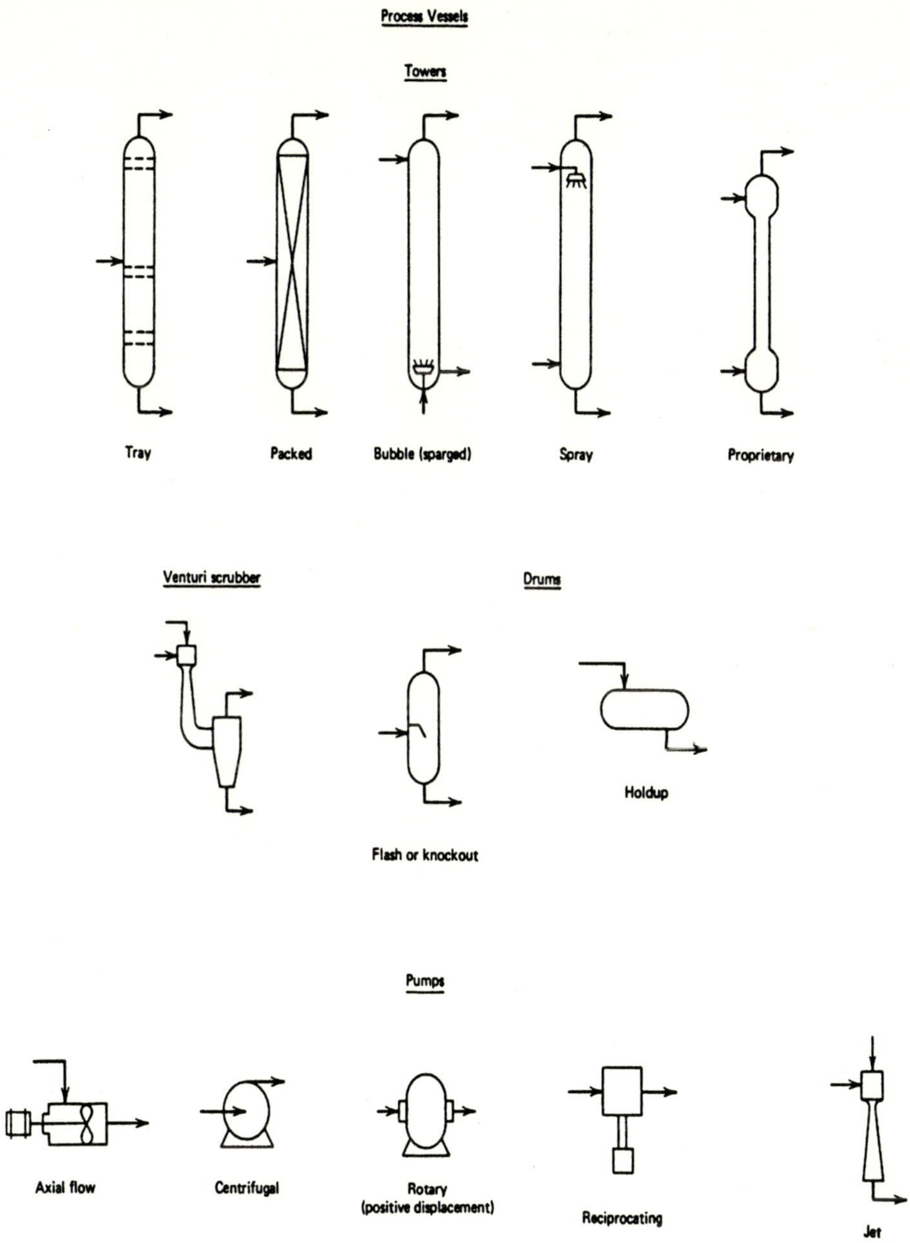

Figure 9.7: Equipment symbols for flowcharts. Reproduced from Ulrich [107] with permission from John Wiley & Sons.

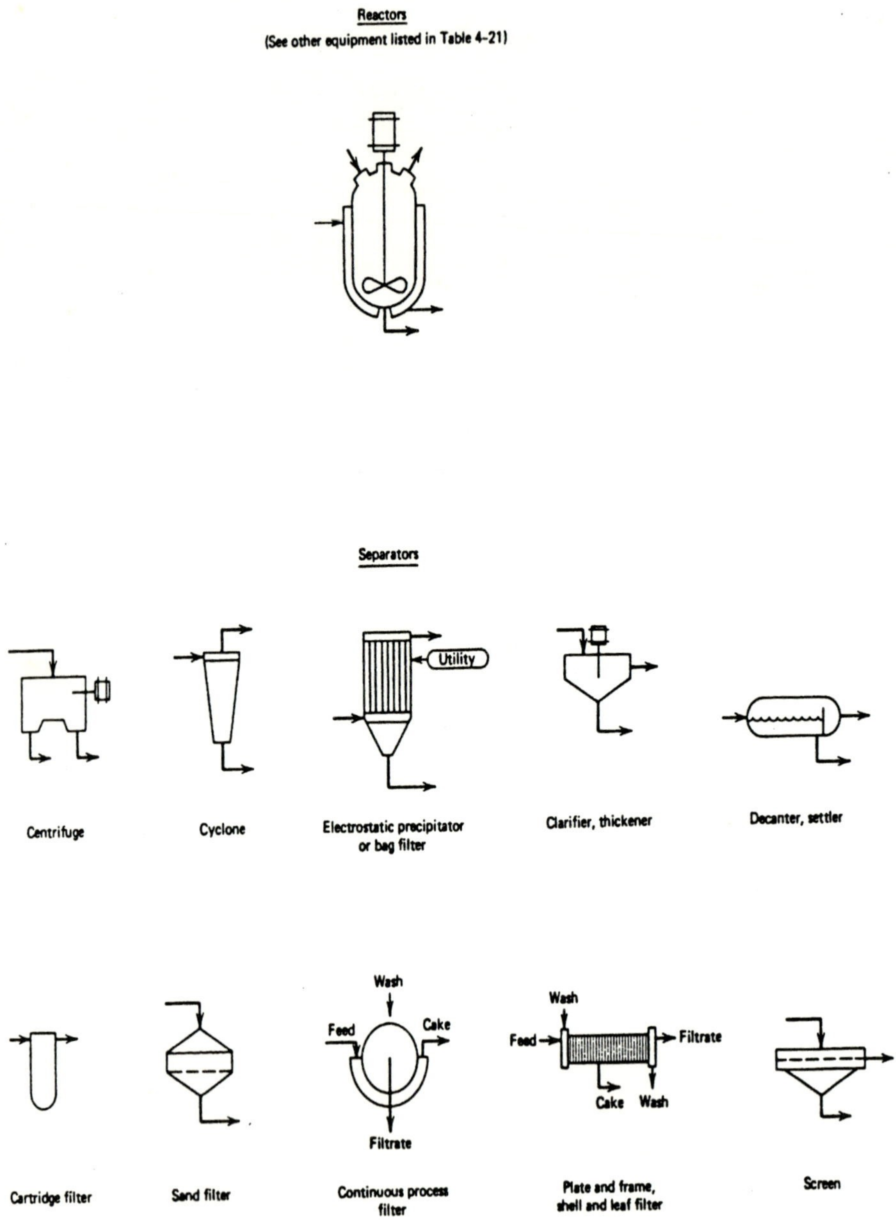

Figure 9.8: Equipment symbols for flowcharts. Reproduced from Ulrich [107] with permission from John Wiley & Sons.

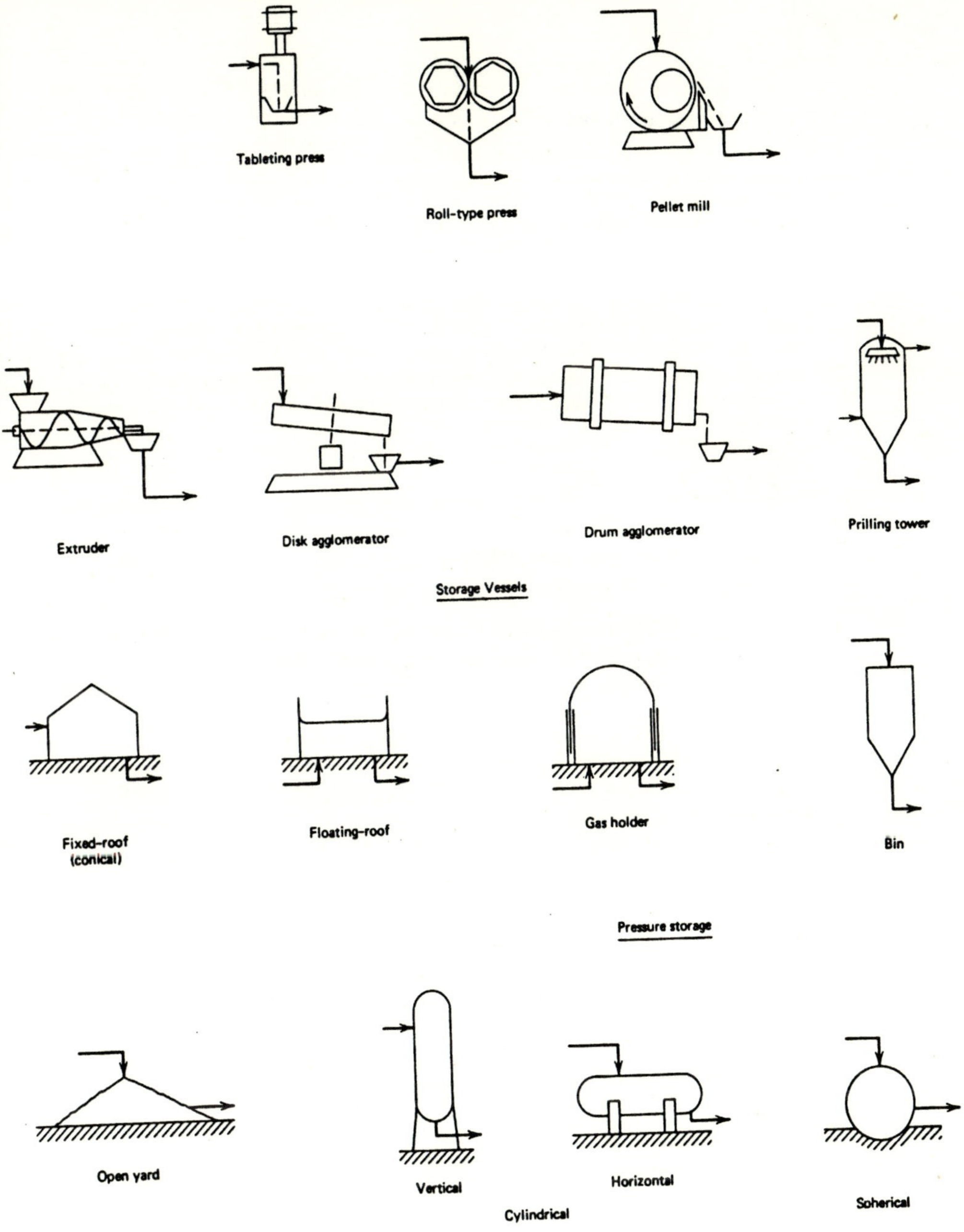

Figure 9.9: Equipment symbols for flowcharts. Reproduced from Ulrich [107] with permission from John Wiley & Sons.

Table 9.3: Equipment designation procedures for flowcharts. Reproduced from Ulrich [107] with permission from John Wiley & Sons.

Equipment Numbering System

1. Number each process area, starting with 100, 200, 300, etc.
2. Number major pieces of process equipment in each area, starting with 110, 120, 130, etc.
3. Number supporting pieces of equipment associated with a major process unit by starting with the next higher number than the major process unit, e.g., 111, 112, 113, 114 for supporting pieces associated with 110.
4. Attach a prefix letter to each equipment number from the following list to designate type. The equipment list might then appear thus:

F-110	B-120	D-210
G-111	G-121	E-211
E-112	G-122	G-212
E-113		E-213

5. The first nine numbers in each area are reserved for equipment servicing the entire area such as a packaged refrigerating unit (e.g., P-105). However, any packaged unit may be broken into its components and numbered as in steps 2 to 4.
6. Use letters following the equipment number to denote duplicates or spares (e.g., G-111A, B denotes two identical pumps in G-111 service).

Equipment Lettering System

A. Auxiliary facilities.
B. Gas–solids contacting equipment (calciners, dryers, kilns).
C. Crushers, mills, grinders.
D. Process vessels (distillation towers, absorption columns, scrubbers, strippers, spray towers).
E. Heat exchangers (coolers, condensers, heaters, reboilers).
F. Storage vessels (tanks, drums, receivers, bins, hoppers, silos).
G. Gas movers (fans, compressors, vacuum pumps, vacuum ejectors).
H. Separators (bag filters, rotary filters, cartridge filters, centrifuges, cyclones, settlers, precipitators, classifiers, extractors).
J. Conveyors (bucket elevators, augers, belts, pneumatic conveyors).
K. Instruments (control valves, transmitters, indicators, recorders, analyzers).
L. Pumps.
M. Agitators, mixers.
N. Motors, drives, turbines.
P. Package units (refrigeration, air units, steam generators, cooling towers, etc.).
Q. Furnaces, process heaters.
R. Reactors.
S. Size-enlargement equipment.
V. Vaporizers and evaporators.
X. Miscellaneous.

9.2 Material and Energy Balance Style

Initially, design projects only involve a few individuals, but successful projects eventually involve many people. Participants in a design project must communicate with each other, and the material and energy balance styles are of paramount importance in making these communications effective. Some of the essential features of process flow diagrams (PFDs), major equipment lists, energy and material balance tables, and cost analyses are summarized in Table 9.4.

Computer simulators (e.g., HYSYS or PROII) can generate PFDs with some of the features listed in Table 9.4. Some editing of the unit operation labels and stream labels is required and, in addition, computer simulators may not explicitly itemize all of the major equipment items that are needed for a proper cost estimate. Figure 9.10 is an example of a PFD produced by HYSYS for ethylene glycol manufacture. Note that the streams can be both numbered and labeled, but manual sorting of the material stream table is required to place the streams in the proper order (i.e., numerically increasing order). The unit operation labels must be edited to conform with the guidelines in Table 9.3. Equally importantly, the condenser and reboiler are not labeled as heat exchangers, but they should be included on the major equipment list. Also missing is the reflux accumulator drum for the distillation column, D-300, the reflux, feed, and reboiler pumps and, possibly, a feed preheater for D-300. Thus, this PFD is helpful, but incomplete from the standpoint of economic analysis.

9.3 Types of Cost Estimates

Cost estimates may be categorized by type and according to their level of detail. Two sets of similar estimate definitions are summarized in Table 9.5. The top two rows in Table 9.5 describe estimate types that can hardly be considered as engineering estimates because they do not require a PFD, a major equipment list, or energy and material balances. Approximations for completing either Order of Magnitude or Screening estimates are presented in Section 2.5.2, but these methods should be used with caution. Desai [26] (also see Table 9.6), expects that Order of Magnitude and Study estimates, both predesign or factored estimates, have percentage errors in the range of 40-50% and 25-40%, respectively.

The Preliminary Estimate, row three in Table 9.5, results from the Preliminary Design Phase of a project. It has been discussed in more detail in Section 1.1. According to Desai, a correctly executed Preliminary Design estimate should have an error of 25-40%. What Desai refers to as a Project Control estimate, but is similar to a Conceptual Design estimate, should have an error of 10-15%. A Detailed or Definitive Cost estimate should be within 5-10% error. As greater levels of detail are applied to a project cost estimate, the uncertainty in the estimate should decrease. Usually, the cost estimates increase as more is known about the design.

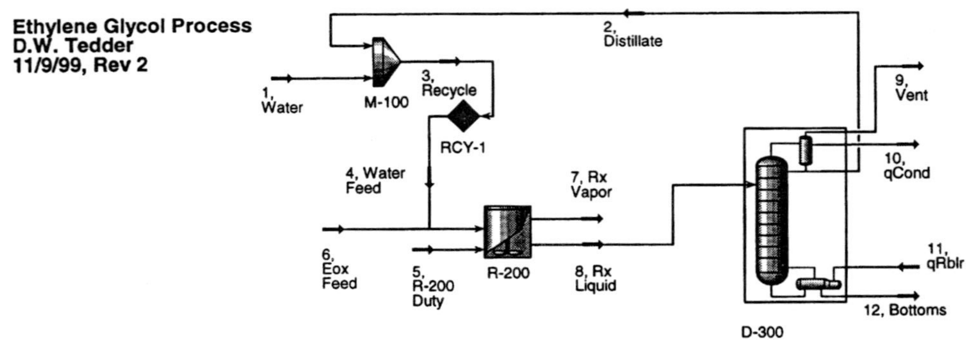

Material Streams										
		1, Water	2, Distillate	3, Recycle	4, Water Feed	6, Eox Feed	7, Rx Vapor	8, Rx Liquid	9, Vent	12, Bottoms
Vapour Fraction		0.0000	0.0000	0.0000	0.0000	1.0000	1.0000	0.0000	1.0000	0.0000
Temperature	C	25.00	41.30	33.68	33.66	25.00	65.00	65.00	41.30	197.4
Pressure	kPa	101.3	101.3	101.3	101.3	101.3	101.3	101.3	101.3	101.3
Molar Flow	kgmole/h	47.06	52.94	100.0	100.0	100.0	86.44	83.19	2.115e-07	30.24
Mass Flow	kg/h	847.7	1029	1877	1877	4405	3376	2906	8.909e-06	1877
Liquid Volume Flow	m3/h	0.8494	1.048	1.897	1.897	4.994	3.786	2.738	1.006e-08	1.690
Heat Flow	kJ/h	-1.344e+07	-1.445e+07	-2.789e+07	-2.789e+07	-5.267e+06	-7.582e+06	-2.792e+07	-0.01394	-1.277e+07
Comp Mole Frac (C2Oxide)		0.0000	0.0544	0.0288	0.0288	1.0000	0.8055	0.0346	0.9257	0.0000
Comp Mole Frac (H2O)		1.0000	0.9455	0.9711	0.9712	0.0000	0.1930	0.6017	0.0743	0.0000
Comp Mole Frac (EGlycol)		0.0000	0.0001	0.0001	0.0001	0.0000	0.0015	0.3636	0.0000	1.0000

Figure 9.10: Process Flow Diagram as produced by HYSYS. What essential features from Table 9.4 are missing?

Table 9.4: Steps in Preliminary Design.

1. Develop a process flow diagram (PFD). The essential features include:
   (a) Show major equipment (graphic representation and alpha-numeric designation). The alpha-numeric designation (e.g., R-210) should give information about the type of equipment and its location (also see Table 9.3). Using the scheme in Table 9.3, Reactor R-210, is located in the 200 Area of the plant. It may be supported by other equipment (e.g., L-211 and E-212).
   (b) Show process streams (arrow and number with diamond touching arrow). Try to avoid a "spagetti-like" appearance. Ideally, each stream number should be placed inside a diamond with one corner of the diamond touching the arrow that designates the stream.
   (c) Identify PFD (name of PFD, design group, date, revision number in box in lower right hand side.) It's important to clearly name all PFDs and date them as they change frequently, especially in the early stages of design.
2. Develop a major equipment list. The essential features include:
   (a) Use alpha-numeric designations from your PFD.
   (b) Provide qualitative descriptions. Eventually add size and materials of construction information.
3. Complete material balance tables. The essential features include:
   (a) Stream numbers (increasing from left to right)
   (b) Stream name [reference equipment, indicate type of stream (e.g., feed, distillate, etc.)]
   (c) Provide rows for T, P, density, enthalpy, total and component mass or molar rates (see Fig 9.10)
   (d) Each stream is a column, not a row
4. Size equipment and determine materials of construction. Use short-cut methods in initial estimates.
5. Complete cost analysis. Essential estimates include:
   (a) Total capital investment
   (b) Gross profits
   (c) Annual return on investment
6. List your most important assumptions and identify the most important topics to evaluate further. Where should research funds be spent? Which parts of the flowsheet are the most important, economically speaking?

Table 9.5: Comparison of different cost estimate definitions from the literature. Also compare with Figure 1.1

Desai[a]	Clark & Lorenzoni[b]	Basis
Order of Magnitude	Screening	A ball park estimate based on past plant costs for similar facilities. Costs are escalated and factored for size effects.
Study	Planning	The above estimate is further adjusted for a particular project using location, labor productivity, site development, availability of construction materials and scheduling considerations.
Preliminary	Preliminary	This estimate is based on a process flow diagram, a detailed equipment list with sizing and materials specifications, overall plot plans, local cost surveys of labor productivity, the availability of skilled labor and construction materials, and a site development survey.
Project Control[c]	Semi-definitive	This estimate requires all of the documents that are required for a Preliminary Estimate as well as preliminary bids on most major equipment items.
Detailed	Definitive	All of the above are required for this estimate plus firm quotations and delivery dates on all major equipment items, all construction drawings, and a definitive economic environment including all subcontract costs.

[a]Chemical Engineering (July 27, 1981) 65.

[b]Applied Cost Engineering, Marcel Dekker, New York (1985).

[c]or Conceptual Design

Table 9.6: Range of uncertainty or error associated with various types of cost estimates. Taken from Desai, Chemical Engineering (July 27, 1981) 65.

Type of Estimate	Error, ±%
Order-of-magnitude	40–50
Study	25–40
Preliminary	15–25
Project Control	10–15
Detailed	5–10

9.4 Accuracy in Cost Estimation

It is desirable to estimate costs independently for as many items on a major equipment list as practicable. By including more items in the equipment list, the errors associated with individual items will also tend to cancel out, and if no *systematic* bias has been introduced, then the resulting estimate should be more reliable. As a minimum, a Preliminary Estimate should consider all major equipment in the Battery Limits (see Chapter 4). It is also preferable to explicitly consider the offsites, rather than estimating them from heuristics.

Systematic errors in an estimate can totally swamp the benefits of including more items in an equipment list. An important example of a systematic error is selecting the wrong materials of contruction for equipment. This mistake could result in grossly low capital cost estimates by an error of 100% or more. On the other hand, incorrectly estimating direct and indirect costs that are based on fob equipment manufactured of special materials of construction, will significantly inflate true costs.

Suppose a Preliminary Estimate is completed with the assumption that the process will be constructed from carbon steel, but in reality 316 stainless steel is required. Then the resulting error in the Total Installed Cost of the process will be much greater than 15-25% even though the cost estimate is on a Preliminary Design basis. The reasons for this are discussed in greater detail in Chapter 4, but they can be qualitatively understood simply by recognizing that 316 stainless steel costs three to four times as much as carbon steel.

Incorrectly choosing the materials of construction for a process is one example of a systematic error. Another example would be selecting simple distillation to perform a separation when, in reality, the mixture forms an azeotrope and cannot be separated to the desired purity as assumed. In this latter instance an infeasible distillation sequence would be specified and a systematic error would have occurred. In such cases, the expected errors are much larger than the 15-25% error that is expected for a Preliminary Estimate without systematic bias.

If major items have been omitted from the equipment list and are actually required, then the estimate will be low due to systematic errors of omission. In most cases, initial cost estimates tend to be low, rather than high, for these and similar reasons. Systematic errors most frequently result in low estimates.

Heuristic: 20 *Conceptual process design is similar to dating. The less you know about a process or person, the more attractive they appear. As you continue to learn more about either one, it eventually becomes painfully clear that reality is much more complex and expensive than you could ever have imagined.*

9.5 Successful Teamwork

Most design projects are carried out by teams of engineers and scientists. During the Preliminary Engineering stage, the team size is typically small, say six or fewer participants. During Detailed Engineering, the team is much larger. At all stages, effective design means effective teamwork. The best engineers recognize this fact of life and work consistently to make it happen.

Effective teamwork must be cultivated. It does not happen by itself. Teamwork requires basic skills that, once recognized, can be improved, regardless of how severe an individual's past history of ineptitude may be. It is relatively easy to become effective team players once we decide to do so.

Stancell [95] has provided two lists of suggestions which are included here with permission. The first describes the Elements of Teamwork. The second focuses on the Elements of Leadership. Both are important in order for a design group to function effectively. Also, both are essential to maximizing your potential and success beyond the design team, in all that you do, whether it is a research career, a business career, academic career, or in your personal life.

Design groups that spend the most time on projects do not necessarily produce the best reports. The operating efficiency of the group is also an important factor that can be improved by recognizing certain inherent characteristics of an effective team.

1. Team members must rely on each other to accomplish goals. That means the group is interdependent and positive interactions must be cultivated within the group to become effective.

2. Some or all of the work is done by members working together. As a consequence, team activities require face-to-face interactions. Members must respect each other's positions.

3. Each group must have a leader, but all team members should practice leadership, decision-making, communication, and conflict management. It is not appropriate, for example, to allow two team members to square-off against each other. The other team members should be proactive and intervene to help dissident members resolve their conflicts.

4. Members must be held accountable for completing their fair share of the work and mastering all material. The group leader necessarily plays a key role in maintaining individual accountability. It is also helpful if the group leader maintains a log of group activities and leads periodic reviews within the group on each individual assignment during each group meeting. These should be friendly reviews that are designed to help the group focus.

5. Good leadership involves regular assessment of group performance. A good team periodically reviews its performance and looks for ways to operate more effectively.

Team effectiveness can be improved by making sure each member understands why he or she is completing a particular task. Each member of the team needs to have a vision of the total project and its end results (e.g., Table 9.10). In addition, team members should be in agreement that each task undertaken is helpful toward achieving the end results. Task selection and assignments should be carefully thought out. In some instances it is helpful to rotate assignments. This approach has the added benefits of promoting creative thought and helping to catch errors.

Members of effective groups ask lots of questions. When in doubt, a question should be asked. Is a task really required? Is there an easier way to answer the same question? Are the assumptions underlying the model being considered really meaningful? Is another member of the group more adept at completing a particular task? Do certain members of the group have preferences for particular tasks?

Group creativity can be improved by observing a few rules. First, no idea should be immediately discounted, regardless of how "off-the-wall" it may appear at first glance. No member should ever be ridiculed for advancing a novel idea. If someone is verbally abused after forwarding an idea, it is probably the last idea the group will obtain from that member. It is therefore better to consider the idea fairly, and then to ask questions about it. Second, weaknesses in a poor idea will usually become apparent to the innovator after additional thought and, typically, the idea will be withdrawn by the one who initially suggested it, without being ridiculed by the group. Third, an effective group leader encourages members to provide ideas, keeps a record of those ideas that are submitted, and makes sure that each is given an honest evaluation. The team will be much more effective if all members are treated in an equitable manner. As a consequence, individual ideas must be respected by the group as well as encouraged by the group leader.

Typically, creativity will be greatest during the initial steps of a Preliminary Design. As Process Selection is completed, tasks become more detailed and defined. Opportunities for innovation decrease, but still exist. It is important that a design group identify "problem" areas as soon as possible. The sooner problems are identified the more quickly they can be addressed and resolved. A failure to obtain initial cost estimates early in a design can be detrimental to project development and should be avoided. Consequently, simplified design methods, leading to the identification of apparently expensive steps in a process, are particularly helpful.

Table 9.7: Elements of Teamwork, adapted from Arnold Stancell, Georgia Institute of Technology, Atlanta, GA (1999).

Japanese saying, "None of us is as smart as all of us."

---

1. Communicate and share what you know, quickly.
2. Listen and stay tuned in. Try to respond to points as they are raised.
3. Treat others as you would like to have them treat you.
4. Know the culture of your organization. How are things done around here? Do not try to "buck the system." Instead, try to learn from individuals who are successful in the organization. Ask a lot of questions.
5. Do not be offended if your ideas are not accepted. Maybe other ideas are better alone or in combination with your idea.
6. Do not ever lose your cool. Shakespeare once said, "A man conquers you when he angers you."
7. Do not take things personally. Do not personalize issues. This response implies ineptitude or mistrust of others.
8. Do not bring old baggage. Learn from past mistakes and experiences (both good and bad), but do not be burdened by them (i.e., do not be thin-skinned).
9. Remember that your boss is part of your team. You want your boss to be successful.
10. Have some fun too. A proper level of humor is always welcome and expedites teamwork.

Table 9.8: Elements of Leadership, adapted from Arnold Stancell, Georgia Institute of Technology, Atlanta, GA (1999).

1. **Exercise Vision**
   (a) Use information, judgement and logic to see future trends and opportunities.
   (b) Look for parallels with earlier problems to better understand a new situation.
   (c) Consider scenarios to avoid surprises and overcome obstacles. Make contingency plans accordingly.
2. **Support Reasonable Risk-Taking**
   (a) Take and support calculated risk. Have a "can do" attitude. Do not dive for cover.
   (b) Encourage people to share their ideas, however different from the norm.
   (c) Have meetings to brainstorm new ways of achieving results.
3. **Communicate Effectively**
   (a) Be a good listener. Recognize that all points raised by others are worthy of a response.
   (b) Provide timely, candid "to the point" information to bosses, colleagues, and employees on priority issues. Avoid excessive detail.
   (c) Share bad news as well as good news.
   (d) Provide feedback to employees on the results of their efforts.

Table 9.9: Elements of Leadership, adapted from Arnold Stancell, Georgia Institute of Technology, Atlanta, GA (1999).

4. **Delegate Efficiently**
   (a) Assign important tasks to employees and give them room to be creative and make their own decisions.
   (b) Explain the rationale behind assignments including all the background information.
   (c) Provide guidance and feedback while allowing employees to be responsible for results.
   (d) Have high levels of expectation and trust for subordinates.
5. **Set High Standards and Expect Results**
   (a) Set clear goals and expectations. Develop performance measurements for yourself and your employees.
   (b) Emphasize results rather than activity.
   (c) Monitor progress toward goals. Schedule periodic review meetings.
6. **Reward Performance**
   (a) Creatively develop forms of non-financial rewards that are meaningful to employees.
   (b) Hire, promote, and transfer employees based on proven accomplishments.
   (c) Take prompt action against substandard performers.
7. **Exhibit High Integrity**
   (a) Speak plainly, but considerately.
   (b) Have high ethical standards which transcend the minimum legal requirements. Always do the right thing.

9.6 Time Management

Time management is an important part of any design project. The more complex the design, the more important time management becomes. Good leadership requires a clear understanding of major tasks and an ability to allocate tasks in an equitable manner that also increases group efficiency. Table 9.10 is a bar chart indicating some of the major tasks in a Preliminary Design. The deliverables are the three reports (both oral and written, usually). Other major tasks support the report production. One week is set aside for internal review of each report within the group before it is actually submitted to management. This is a key step to improve the quality of the report and avoid major omissions.

Every design requires extensive literature review, especially in the beginning. Profit Upper Bound (PUB) calculations may be used to select attractive processes for further evaluation. The Process Flow Diagram (PFD) generally evolves over a period of time. Energy and Material (E & M) Balances are required as part of the Preliminary Report, and they are often conveniently completed using a simulator (e.g., HYSYS or PROII). On the other hand, the design group can easily become overly involved in simulator calculations. It is therefore useful to focus more heavily on spreadsheet calculations to complement simulations and check underlying assumptions.

The Cost Estimation, Optimization, and Economic Analysis tasks are particularly important, but necessarily are among the last steps completed in a Preliminary Design. It is usually helpful to begin Cost Estimation as early as possible, particularly on "big-ticket" items. Typically, a cost problem is not recognized as such until a Preliminary Cost estimate has been obtained. Equipment Sizing is also critical in this analysis and necessarily precedes Cost Estimation.

9.7 Summary

Design and cost estimation is completed as a series of repetitive cycles in which subsequent cycles involve greater levels of detail. As more detail is added to the design and cost analysis, a higher degree of accuracy is achieved. By completing a project in levels of increasing detail, the chances of significant errors in the design are reduced since the calculations and design basis are repeatedly reviewed and checked.

Efficient teamwork is essential for effective design. Team members should remember to be respectful of other viewpoints, and to treat others as they themselves wish to be treated. Developing a time management plan can be very helpful in improving efficiency, and in making sure everyone knows the direction in which the project is headed.

9.8 Problems

9.1 An air liquefaction plant produces liquid nitrogen and oxygen products. The process compresses and chills air which is then distilled to produce the desired

Table 9.10: Time-Consuming Tasks in Preliminary Design (Semester Basis)

	Task Duration (Weeks)														
Task	**1**	**2**	**3**	**4**	**5**	**6**	**7**	**8**	**9**	**10**	**11**	**12**	**13**	**14**	**15**
Report Preparation															
Process Selection			—	—	—										
Preliminary						—	—	—	—	—					
Final											—	—	—	—	—
Internal Review					—					—					—
Literature Survey	—	—	—												
PUB Calculations		—	—	—											
PFD Development		—	—	—	—	—	—	—							
Process Selection	—	—	—												
E & M Balances[a]					—	—	—	—	—						
Equipment Sizing							—	—	—	—	—				
Cost Estimation									—	—	—	—	—		
Optimization									—	—	—	—	—		
Economic Analysis									—	—	—	—	—		

[a]May be completed using spreadsheets or simulators (e.g., HYSYS or ASPEN)

products. The compression and chilling operations may utilize reversing heat exchangers, subcoolers, and expansion turbines as configured by Wolff, Eyre, and Grenier [119]. Draw a process flow diagram for such a plant.

9.2 Air liquefaction may be carried out using liquid natural gas (LNG) as described by Yamanouchi and Nagasawa [120]. Draw a process flow diagram for the chilling operation they describe.

9.3 Sulfuric acid may be produced [16] by burning sulfur in a furnace to produce a mixture of SO_2 and SO_3. The furnace gases are passed through primary and secondary reactors, an economizer, and then scrubbed with dilute sulfuric acid to produce more sulfuric acid from the SO_3 and to absorb the SO_2. The bottoms effluent from the absorber is then sent to the top of a stripper column where fresh air feed to the furnace is introduced into the bottom of the column. In this stripper, the SO_2 is desorbed and recycled to the furnace and reactors while the dilute sulfuric acid is discharged as a product from the bottom of the stripper. Gaseous effluents from the absorber are passed through a tail gas reheater and used to heat the primary and secondary reactors before being expanded and discharged to a stack. Draw the process flow diagram.

9.4 In the Wacker process [4], ethylene is reacted with pure oxygen to form acetaldehyde via:

$$C_2H_4 + 2CuCl_2(aq) + H_2O \xrightarrow{PdCl_2} CH_3CHO + 2HCl + Cu_2Cl_2 + \text{heat}$$

$$Cu_2Cl_2 + 2HCl + \frac{1}{2}O_2 \rightarrow 2CuCl_2 + H_2O$$

The reactor effluents are transferred to a flash separator from which the hot gases are subsequently quenched and then distilled to produce acetaldehyde, crotonaldehyde, and higher boiling polymeric species. The liquid bottoms from the flash separator are recycled to the reactor. Draw a process flow diagram.

9.5 Why does chemical process design occur in steps as shown in Table **??**?

Chapter 10

Safety

Safety is a paramount concern in process design. As professionals, engineers must be concerned for the welfare of everyone on a plant site. Unplanned events (e.g., spills, explosions, fires, etc.) may be avoided altogether by thoughtful planning, but prudent design requires us to assume that such events will occur. Simple guidelines can often be used in the conceptual design stage to determine whether or not such hazards exists. Two important parameters are the Lower Explosion Limit (LEL) and the Upper Explosion Limit (UEL) that define conditions leading to combustible mixtures. The LEL and UEL are also referred to as the Lower Flammability Limit (LFL) and Upper Flammability Limit (UFL), respectively. Air mixtures with flammable mole factions y between LEL and UEL are considered combustible and may be explosive.

The Merck Index [14] provides experimental LEL and UEL values for many substances. This information[1] is also available from Material Safety Data Sheets (MSDS) which can be found on the Internet.

10.1 Lower Explosion Limits (LEL)

The lower explosion limit (LEL) is the minimum concentration of a flammable vapor in air that will propagate a flame if ignited. A flash point is the lowest temperature at which a substance will provide sufficient vapor to burn in air at atmospheric pressure. Gooding [33] points out that for many species, the LEL occurs at about 50% of the stoichiometric concentration that yields complete combustion, and the flash point is that temperature where the pure component vapor pressure is high enough to reach the LEL. Jones [49] puts the LEL at 55% of the stoichiometric concentration. Pressure has little effect on the LEL except at pressures less than 50 mm Hg where flames are unable to propagate [25].

[1]Using an Internet search engine and the entry "MSDS" will generate helpful data bases. One of the larger ones is that maintained by Cornell at http://msds.pdc.cornell.edu/issearch/msdssrch.htm.

Example: 10.1 *What are the LEL and flash point values for ethanol?*
Solution: 10.1 *The combustion equation for ethanol is:*

$$C_2H_5OH + 3O_2 \rightarrow 2CO_2 + 3H_2O$$

Air contains 1 mole of oxygen for 4.76 moles of air. In order to burn one mole of ethanol, 3(4.76) *moles of air are required. Therefore, the stoichiometric ethanol mole fraction at complete combustion is:*
$y_s = 1/[1 + 3(4.76)] = 0.0654$
Since the LEL is about 50% of y_s*:*
$LEL \approx 0.0654/2 = 0.0327$ *and* $p_{LEL} = y_{LEL}P = 0.0327$ *atm*
For a pure liquid, $yP = p = xP^{sat} = P^{sat}$. *The pure component vapor pressure,* P^{sat}, *may be computed using the Antoine correlation or found in a suitable reference. From Perry's Handbook, the pure component [75] vapor pressure of ethanol is 0.0327 atm at 13 °C. Therefore, the flash point is about 13°C using a pure component vapor pressure correlation for ethanol. Steere's [96]* Handbook of Laboratory Safety *confirms this calculation.*

10.2 Upper Explosion Limits (UEL)

The Upper Explosion Limit (UEL) or Upper Flammability Limit (UFL) is the maximum concentration of a flammable substance that will support combustion. According to Jones [49] the UEL is about 350% of the stoichiometric concentration in volume percent fuel in fuel plus air mixtures.

According to Zabetakis [121], the UEL increases with pressure. This effect broadens the composition range over which a mixture is combustible. To estimate the UEL at pressures greater than 1 atm, use

$$UEL(P) = UEL + 20.6(\log_{10} P + 1) \tag{10.1}$$

where P is the system pressure in mega pascals, absolute and UEL is the upper explosion limit in volume% of fuel in air at 1 atm.

Example: 10.2 *Estimate the lower explosion limit in air, flash point in °C at 1 atm pressure, and the upper explosion limit in air for 1-butene (* C_4H_8 *) using the information provided below. The pure component vapor pressure for 1-butene may be estimated using:*

$$\log_{10} P^{sat} = 6.7929 - \frac{908.8}{t + 238.54}$$

where P^{sat} *is the vapor pressure in mm Hg and t is in °C.*
Solution: 10.2 *First determine the stoichiometric composition* y_s *for complete combustion:*

$$C_4H_8 + 6O_2 \rightarrow 4CO_2 + 4H_2O$$

So $y_s = 1/(1 + 6(4.76)) = 0.033898$
and $y_{LEL} = 0.5y_s = 0.01691$
with $y_{UEL} = 3.5y_s = 0.11864$

The flash point is that temperature at which the composition equals y_{LEL}. *Assuming Raoult's law, then with equilibrium between the vapor and liquid:*
$\hat{f}_i^v = \hat{f}_i^L$
$y_i P = P_i^{sat}$
In the vapor, $y_i = y_{LEL} = 0.01691$, *and* $P = 760$ *mm Hg so* $P_i^{sat} = 12.8518$ *mm Hg at the flash point. Using the vapor pressure correlation gives* $t_{fp} = t_{LEL} = -78.65$ ^oC.
In addition we can use the correlation to estimate the minimum temperature at which the UEL is reached. In this case, $y_i = y_{UEL} = 0.11864$, $P = 760$ *mm Hg and* $P_i^{sat} = 90.1164$ *mm Hg. Use the vapor pressure correlation to estimate* $t_{UEL} = -50.65$ ^oC. *So in the temperature range* $-78.65 < t < -50.65$ *1-butene forms a combustable mixture with air. Tanks storing 1-butene must be kept warmer than -50.65* ^oC *to avoid a combustible mixture. In this case the mixture in the tank is not combustible because it is oxygen deficient. Or the mixture is not combustible if the tank temperature is less than -78.65* ^oC *since in that case it is fuel deficient.*

Example: 10.3 *Use the data in Example 10.2 to estimate the lower and upper explosion limits of 1-butene at 5 atm pressure. Also estimate the lower and upper explosion temperatures at that pressure.*
Solution: 10.3 *From Example 10.2,* $y_{LEL} = 0.01691$. *At 1 atm pressure,* $y_{UEL} = 0.11864$ *or 11.864 vol%, but the UEL is a function of pressure so:*
$UEL(P) = UEL + 20.6(\log_{10} P + 1) = 11.864 + 20.6(\log_{10}(0.50625) + 1) = 26.38$ *vol%.*
So at 5 atm, $y_{UEL} = 0.2638$. *But the lower and upper flash temperatures are both affected by system pressure.*
At the lower flash temperature: $P_i^{sat} = y_i P = 0.01691(3800) = 64.258$ *mm Hg. From the vapor pressure correlation for 1-butene,* $t_{LEL} = -56.23$ ^oC
At the upper flash temperature: $P_i^{sat} = y_i P = 0.2638(3800) = 1002.44$ *mm Hg. From the vapor pressure correlation for 1-butene,* $t_{UEL} = 1.13$ ^oC
So at 5 atm, the tank temperature must be kept either below -56.2 ^oC *or above 1.13* ^oC *to avoid a combustible mixture.*

10.3 Explosion Limits for Mixtures

The Le Chatlier Equation [19] can be used to estimate LELs and UELs for gas mixtures.

$$\frac{1}{LEL_{mix}} = \sum_{i=1}^{n} \frac{y_i}{LEL_i} \tag{10.2}$$

$$\frac{1}{UEL_{mix}} = \sum_{i=1}^{n} \frac{y_i}{UEL_i} \tag{10.3}$$

where y_i is the mole fraction of species i among n species on a fuel-only (or air-free) basis.

Example: 10.4 *A mixture contains 1 vol% ethylene and 2 vol% butene in air. The LEL and UEL for ethylene are 2.7% and 36.0%, and for butene they are 1.69% and 11.86%, respectively.*

Solution: 10.4 *On a fuel-only basis, the ethylene mole fraction is 1/3 and 2/3 for butene. So*
$1/LEL_{mix} = 0.33/2.7 + 0.67/1.69$ *and* $LEL_{mix} = 1.93$ *vol%*
$1/UEL_{mix} = 0.33/36.0 + 0.67/11.86$ *and* $UEL_{mix} = 15.27$ *vol%*
Since the mixture is 3 vol% fuel, it is combustible.

Example: 10.5 *A solvent cleaning system is being designed. A washing system, containing spray nozzles and racks inside a large rectangular vessel, will be used to remove oily films from contaminated equipment. This washing system will enable the recycle and reuse of the contaminated equipment, but it entails the use of several potentially flammable solvents. In particular, it is proposed to use a mixture of acetone (* C_3H_6O *) and ethyl acetate (* $C_4H_8O_2$ *) as the washing solvent. An equimolar mixture will be used.*

The washing procedure is as follows. The cleaning device will be opened from the operating floor where workers are located. After opening, the racks will be pulled out and loaded with contaminated equipment. The racks will then be pushed back into the cleaning device. After closing, the racks and contaminated equipment will be cleaned by recirculating the solvent mixture through the spray nozzles, producing a fine, high-velocity mist, at a temperature of 30 °C. After washing, the cleaning device will be drained of excess solvent and opened. The racks will be pulled out and onto the operating floor so workers can removed the cleaned equipment for reuse. During the wash cycle the washer will be sealed. When it is opened, vapors will be withdrawn from the device using a fan to draw them away from the workers.

Are there any safety issues that need to be addressed before this system is operated? Calculate the upper and lower explosion limits for both species in the solvent mixture. Estimate the upper and lower explosion limits for the mixture. Is there any possibility of forming a combustible mixture during any part of the washer cycle?

List your assumptions. Complete the calculations, and then discuss any safety issues that you feel may be important. The pure component vapor pressures of acetone and ethyl acetate can be estimated using the Antoine equations:

$$\log_{10} P^{sat}_{acetone} = 7.11714 - \frac{1210.595}{T + 229.664} \quad (10.4)$$

$$\log_{10} P^{sat}_{ethylacetate} = 7.10179 - \frac{1244.95}{T + 217.88} \quad (10.5)$$

where T *is in* $^\circ C$ *and* P_i^{sat} *is in mm Hg.*

Solution: 10.5 *At 30* $^\circ C$, $P^{sat}_{acetone} = 285$ *mm Hg and* $P^{sat}_{ethylacetate} = 120$ *mm Hg. The stoichiometric composition for complete combution of either species is found from:*

$$C_3H_6O + 4\,O_2 = 3\,CO_2 + 3\,H_2O \quad (10.6)$$

$$C_4H_8O_2 + 5\,O_2 = 4\,CO_2 + 4\,H_2O \quad (10.7)$$

For acetone:
$y_s = 1/(1 + 4(4.76)) = 0.0499$, $y_{LEL} = y_s/2 = 0.02495$, $y_{UEL} = 3.5y_s = 0.17465$
For ethyl acetate:
$y_s = 1/(1 + 5(4.76)) = 0.04032$, $y_{LEL} = y_s/2 = 0.02016$, $y_{UEL} = 3.5y_s = 0.141$
The LEL_m *and* UEL_m *are based upon vapor mole fractions on an air-free basis. Assuming Raoult's law, then on an air-free basis with saturation:*
$y'_{ac} = 285/(285 + 120) = 0.704$
$y_{ea} = 120/(285 + 120) = 0.296$
So for the mixture:
$1/LEL_m = \sum y_i/LEL_i = 0.704/0.02495 + 0.296/0.02016$, $LEL_m = 0.023$
$1/UEL_m = \sum y_i/UEL_i = 0.704/0.17465 + 0.296/0.141$, $UEL_m = 0.163$
Assuming Raoult's Law, the vapor pressure exerted by the liquid mixture is:
$P^{sat} = 0.5(285) = 0.5(120) = 202.53$ *mm Hg and* $P^{sat}/P = 202.5/760 = 0.266$

Assume that the cleaning vessel is saturated with vapor at the end of the wash cycle when the excess solvent is drained from the tank. At that point, $y = 202.5/760 = 0.266 > UEL_{mix}$ *and the mixture is not combustible since there is insufficient oxygen to support combustion. Now the vessel is opened and the fan is turned on. At that point the solvent concentration in the vapor decreases until it reaches the* UEL_m*. If no ignition occurs, ventillation will continue to decrease the solvent concentration until it is less than* LEL_m*, but for a period of time the mixture will be combustiblc. So this design faces a serious safety problem in that during parts of the cleaning cycle the workers will be working near a combustible vapor and air mixture which, if ignited, could result in loss of life.*

The assumptions are: (1) equilibrium concentrations at the end of the wash cycle (actually it would be supersaturated due to the presence of fine droplets), (2) applicability of equations for estimating UEL and LEL, (3) applicability of Raoult's law.

10.4 Green Chemistry

Risks can be reduced in many ways. The risk of fire necessarily follows from the use of flammable solvents in manufacturing. The consequences of a solvent fire are proportional to the amount of solvent inventory at the plant site. Risks can therefore be eliminated by removing the need for the solvent, and reduced by optimizing the process so as to reduce the required solvent inventory at the plant site. Such modifications, particularly the former, are commonly referred to as green chemistry. That is, the development of manufacturing processes that produce required chemicals, but without using hazardous chemical intermediates. Secondarily, if the use of hazardous chemicals is unavoidable, then green chemistry seeks to reduce the inventory of the hazardous species on the plant site.

Green chemistry applies process synthesis techniques to the most important issue facing chemical manufacturers–elimination of long-term, or habitual, risks. The most dramatic benefits result whenever manufacturing chemistry can be substantially modified to eliminate altogether the use of toxic or hazardous substances. In this case, the flow sheet is completely revamped to manufacture via another safer chemical manufacturing route. The need for monitoring and controls to handling the toxic materials is thereby eliminated and the overall process is simplified, at least in the most favorable circumstances. This strategy also reduces or eliminates entirely risks from terrorist activities and thereby reduces overhead, the cost of insurance, and gains public acceptance.

More traditional methods of risk management focus on the application of engineering techniques to provide better monitoring and control capability and early warning of problems. If the hazardous chemical of concern cannot be eliminated via green chemistry, then such methods are unavoidable. However, they are generally not as attractive as green chemistry for risk reduction as the risk still remains. The use of more sophisticated control and risk analysis techniques can be helpful and appropriate, but they often fail to gain public acceptance of the perceived risks.

Fault tree analysis is one technique that has been applied to assess risk. This methodology can be very helpful to engineers who are required to design safety systems, but it typically fails to reassure the general public that a process is "safe," regardless of how small the calculated probabilities (and consequences) of failure may be. More commonly, fault trees only serve to muddle the public perceptions and to elevate general anxiety further because fault trees inevitably detail failure modes that had not been contempated prior to reviewing the tree. Fault tree analysis, instead of reassuring the public by calculating very low failure probabilities, tends to increase anxiety because the public didn't realize so many things could go wrong.

In contrast, the application of green chemistry to eliminate the use of hazardous materials enables greater public acceptance. Unfortunately, it is not always possible to eliminate such risks, particularly if the final product is toxic and/or hazardous (e.g., gasoline). In such cases, one has no recourse except to design system safety through a thorough hazards analysis.

Even if the use of a toxic chemical cannot be eliminated altogether, it may be possible to minimize its inventory on a plant site and, thereby, to reduce the consequences of any accident scenario involving that chemical. For that reason, processing situations that involve habitual recycle of toxic or hazardous reagents should be carefully evaluated. Whenever the conversion per pass is low, recycle and inventory tend to be large. So options where conversion per pass and recycle rates can be reduced should be sought. Such options also tend to reduce capital investment, particularly if working inventories are also reduced by such process modifications.

Heuristic: 21 *It is better to get rid of hazards than to control them. Hazardous materials always add to process complexity, but the best engineered processes are inherently simple.*

10.5 Preliminary HAZOP Analysis

Hazard and Operability (HAZOP) study is a procedure that is used to identify hazard in a chemical process facility [40, 53, 80]. Like the design process, HAZOP analysis is usually performed iteratively. It should be given consideration even during the preliminary design, and it can be particularly helpful in helping to make intelligent choices in chemistry and process design that avoid problems early on. As the design proceeds, HAZOP should be completed in greater detail and with more formality. Initially, a very superficial HAZOP analysis, say based on information gathered from MSDSs, can be helpful.

Conceptual analysis plays a key role in HAZOP as the engineer is required to imagine "what if" scenarios. In order to be effective in this role, it is important to be familiar with the chemistry and the types of things that can go wrong and there is really no substitute for experience in this regard. As a consequence, it usually pays off to find technical people who have first hand knowledge of chemical and physical behavior under circumstances that are similar to those imagined for the plant.

Example: 10.6 *A chemical company needs a new incinerator. It is trying to find some technology that is safe and that avoids typical emissions problems associated with conventional incinerators. One possibility is to use a molten salt bed consisting of sodium carbonate. The salt bed would be air sparged to provide excess oxygen. Perform a preliminary HAZOP analysis of this idea.*

Solution: 10.6 *The Merck Index [14] provides the following information about sodium carbonate.*

> *Anhydrous, Solvay soda. The technial grade (about 99% pure) is known as soda ash. Odorless, hygroscopic powder; alkaline taste; d 2.53. mp 851° but begins to lose* CO_2 *even at 400°. On exposure to air it will gradually absorb one mole water–about 15%. Sol in glycerol; in 3.5 parts water at room temp, 2.2 parts water at 35°. Insol in alcohol. Dec by acids with effervescence. Combines with water with evolution of heat. Its aq soln is strongly alkaline. pH 11.6. Keep well closed.* $\mathbf{LD}_{50}$ *(30 day) i.p. in mice: 116.6 mg/kg (Norde).*
>
> *Monohydrate, odorless, small crystalls or cryst powder; alkaline taste. Stable at ordinary temps and atmospheric conditions; dries out somewhat in warm, dry air or above 50°, becomes anydr at 100°.*

A molten bed of sodium carbonate will be operating at temperatures above 851° C. Corrosion is likely a problem under these conditions and ceramic lined materials will be required. Solidification occurs at temperatures below the melting point and would likely be a problem during operation, especially if any transfer lines containing molten salt are required to operate. Startup would also be difficult with this concept.

Since the operating bed temperature is well above the normal boiling point of water, wastes containing significant amounts of water may react explosively with the bed. At a very minimum, high moisture-bearing wastes will rapidly generate gases in the form of steam. At room temperature, the bed picks up moisture from air. It will loose this moisture as the bed is reheated to operating conditions. This behavior will increase startup time and decrease availability.

Air sparging to provide oxygen will likely result in particulate migration to the offgas system. With high air flowrates, bed particle migration to the offgas may be quite high and it is likely that other species (e.g., inorganic particulates) may migrate as well. Such mass movement would defeat the stated purpose of the molten salt incinerator.

Additional potential problems with this concept will become apparent as one continues to imagine how it may operate. Discussions with pilot plant operators confirm that the points listed above are all significant issues.

10.6 Problems

10.1 Suggest two alternative distillation sequences for separating a ternary mixture into three essentially pure product streams. What qualitative differences exist between these two sequences in terms of reboiler duties, condenser loads, and tower size requirements?

10.2 Estimate the lower explosion limit and flash point of n-butane. Use the pure component vapor pressure correlation:

$$\ln\left(\frac{P^s}{P_c}\right) = 5.742 - \frac{4126.4}{T + 409.5}$$

where T is in ^{o}F and $P_c = 550.7$ psia.

10.3 A pure species exhibits a normal boiling point temperature of 350°C and it has a critical temperature of 780°C. Estimate the latent heat of vaporization for this species at a temperature of 510°C.

10.4 Estimate the lower explosion limit (LEL) for 1-pentene (C_5H_{10}) in air at one atmosphere total pressure.

10.5 Estimate the flash temperature for a species whose lower explosion limit in air is 5 mole% at one atmosphere. The pure component vapor pressure for this species is well correlated using the formula:

$$\ln\frac{P_i^s}{P_{ci}} = 3.2 - \frac{4336.4}{T + 388.6}$$

where $P_{ci} = 586$ psia and T is in ^{o}F.

10.6 Estimate the lower explosion limit and flash point of ethylbenzene. Use the pure component vapor pressure correlation:

$$\ln P^s = 14.0045 - \frac{3279.47}{t + 213.2}$$

where t is in ^{o}C and P^s is kPa. (100 kPa = 0.986923 atm)

Chapter 11

Preliminary Equipment Design

As discussed in Section 9.4, a design project typically goes through multiple levels of analysis. The final design level is the most expensive, involves the greatest detail, and results in the highest precision and cost accuracy. On the other hand, the preliminary design is the least expensive, has the greatest uncertainty in the cost estimate, and frequently underestimates the true costs. While the final detailed engineering study defines a process down to the materials of construction that are required for the smallest bolts, the preliminary design usually only sizes and evaluates the major equipment items.

The intention of this chapter is to provide guidelines for the preliminary sizing of major equipment items. Energy and material balances are generally required before these sizing calculations can be completed.

11.1 Materials Handling Equipment

In most chemical processes, some combination of gases, liquids and solids must be transferred within the plant. These transfer operations may be carried out using a variety of unit operations such as pumps, conveyers, chutes, hoists, fans, blowers, etc. The specification of such unit operations is usually completed through knowledge of the required mass and/or volume rates through them and the specific mechanical energy input that they must provide per unit of mass throughput. The former quantities are defined by the flowsheet material balances, while the latter can be estimated using the mechanical energy balance equation:

$$\frac{\dot{w}_f}{\dot{m}} = \int_{p_1}^{p_2} \frac{dp}{\rho} + \frac{\Delta u^2}{2} + \frac{g\Delta z}{g_c} + \frac{u^2}{2g_c}\left(\frac{4fL}{D} + \sum_i K_i\right) \tag{11.1}$$

where $\dot{w}_f$ = the frictional work
$\dot{m}$ = the the mass rate (pumping rate)
p_i = the pressure at ith state
Δu = the change in fluid velocity
Δz = the change in elevation
g/g_c = gravity conversion factor
ρ = fluid density
f = friction factor
L = pumping path length
D = pipe diameter
K_i = resistance factor for ith fitting

11.1.1 Pumping Compressible Fluids

For preliminary design, it is usually sufficient to compute the pressure–enthalpy diagram for the fluid being processed and to estimate the polytropic head requirements. For pure fluids, the P–H diagram can often be found tabulated in a suitable reference. For mixtures, the necessary properties must often be computed.

Estimation of the polytropic head is relatively straightforward [64, 56]. Basically, the polytropic head, H_p, must be calculated between the compressor inlet conditions and the final discharge pressure:

$$H_p = ZRT\left(\frac{n}{n-1}\right)(r_p^{\frac{n-1}{n}} - 1) = ZRT\phi\left(r_p^{1/\phi} - 1\right) \tag{11.2}$$

where H_p = the polytropic head in $\text{lb}_f\text{-ft}/\text{lb}_m$ (lb-force-ft/lb-mass)
Z = the average mixture compressibility factor between the inlet and outlet conditions
R = $1545/MW$ in $\text{lb}_f\text{-ft}/\text{lb}_m\text{-}^o\text{R}$
T = the inlet temperature in ^{o}R
r_p = the compression ratio per stage $= P_2/P_1 = P_{out}/P_{in}$
ϕ = a compression parameter.

The last parameter, ϕ, is calculated from the relationship:

$$\phi = \frac{n}{n-1} = \left(\frac{k}{k-1}\right) n_p \tag{11.3}$$

where k = the ratio of the mole weighted specific heats, (C_p/C_v)
MW = the average molecular weight of the mixture
n_p = the polytropic efficiency (typically in the range of 0.65 to 0.85). (An assumed value of $n_p = 0.76$ is recommended [56]).

In order to estimate the mixture compressibility factor, suitable mixture rules [93] must be assumed and used to compute a mole average critical temperature and pressure for the mixture. Then, based upon the mixture reduced properties, generalized compressibility tables can be used to look up the Z-values for both the inlet and outlet conditions. The average Z-value is then used to compute the polytropic head.

To compute the k-value for the mixture, it is first necessary to estimate the C_p value for each species at the specified temperature. These data are then used to compute a mole average C_p value for the mixture which is used in Equation 11.4 to calculate the mixture k-value:

$$k = \frac{C_p}{C_p - 1.986} \tag{11.4}$$

where C_p is in BTU/lb-mole^{o}R.

To improve the accuracy, k-values should be computed at both the inlet and outlet conditions. The outlet temperature can be estimated from:

$$T_{out} = T_{in} r_p^{1/\phi} \tag{11.5}$$

where T_{out} and T_{in} are in absolute units. Hence some trial-and-error is required to estimate the outlet conditions and the average k-value.

In addition to the polytropic head, an initial design requires a description of the proposed compressor cycle on a pressure-enthalpy diagram. Figure 11.1 summarizes a typical cycle together with a flowsheet indicating the reference conditions in the cycle. This information is used to compute the recycle rates through the compressor network. Although the enthalpy change from point a to b in Figure 11.1 can be computed from the polytropic head, the remaining points are determined largely by the characteristics of the fluid which is being compressed.

Example: 11.1 *A vapor compressor has an efficiency of 0.76 and the vapor has a heat capacity of 6.0 Btu/lb-mole^{o}R. What is the value of ϕ? What is T_{out} if $T_{in} = 430^oF$ and $r_p = 3$? What is T_{out} if $n_p = 1.0$?*

Solution: 11.1

$k = C_p/(C_p - 1.986) = 6.0/(6.0 - 1.986) = 1.4947$

$\phi = [k/(k-1)]n_p = (1.495/0.495)(0.76) = 2.295$

$T_{out} = (460 + 430)r_p^{1/2.295} = 1436\ R$

If $n_p = 1.0$, then:

$\phi = 1.495/0.495 = 3.02$

$T_{out} = 890(3)^{1/3.02} = 1280\ R.$

Hence the exit temperature is lower for a more efficient compressor.

Heuristic: 22 *It may be less expensive to condense a vapor, compress the liquid, and then revaporize the liquid, than to compress the vapor directly.*

Heuristic: 23 *If a compression ratio greater than three is required in a preliminary design, then specify multiple stages in series to achieve the overall compression ratio.*

Heuristic: 24 *If high-pressure gases are available and gas pressures must be reduced to permit further processing, then consider using a turbine expander to recover shaft work, rather than throttling the gas. This technique is often very helpful in reducing production costs.*

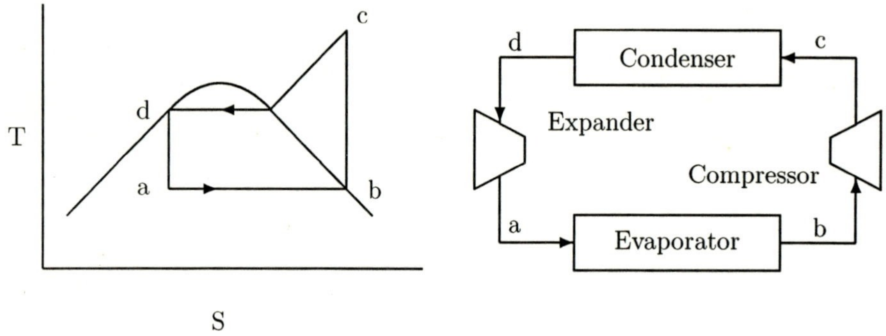

Figure 11.1: Vapor compression cycles may be represented either on pressure-enthalpy diagrams or temperature-entropy diagrams as shown here.

In general, it is desirable to avoid the two-phase region during compression since liquid droplets tend to cause erosion. Also, the condenser design should consider the fact that the temperature driving force is not linear. A discontinuity occurs at point c as indicated in Figure 11.1. This condition increases the transfer area requirements compared to the assumptions of a lognormal driving force.

11.1.2 Pumping Incompressible Fluids

During Preliminary Design, little or no information about piping layout, fitting requirements, etc. is available. Although pumps should be included in the PFD and listed as Major Equipment, their sizing and cost estimation at this point are approximate. Preliminary methods are therefore used to estimate pipe diameters and effective pipe lengths. The computed results should be reviewed during the subsequent Conceptual Design stage, if the project achieves that point. Unless pumps appear to be a major factor, economically speaking, time spent designing them should be minimized in favor of giving priority to the more expensive unit operations in the process.

The steps in Preliminary Design outlined below for liquid pumps include estimating: (1) pipe diameter, (2) resistance in terms of equivalent pipe lengths, and (3) hydraulic and brake horsepower requirements. Using Guthrie's Method (see Chapter 4), hydraulic horsepower is used to estimate the total installed cost of a pump, but brake horsepower should be used to estimate the energy requirements (e.g., electrical or steam).

Optimum Pipe Diameters

A preliminary estimate of pipe diameters is needed to compute pipe Reynolds numbers and friction factors. Nearly optimum pipe diameters for a given flow rate can be estimated from an approximate equation set provided by Peters and Timmerhaus [76]

$$D = \begin{cases} 3q^{0.36}\mu^{0.18}, & Re < 2100 \\ 3.9q^{0.45}\rho^{0.13}, & Re \geq 2100 \end{cases} \tag{11.6}$$

where D = the optimum inside diameter in inches
q = the flow rate in ft^3/s
ρ = the liquid density in lb/ft^3
μ = the liquid viscosity in centipoise

These results are preliminary and subject to change, depending upon the underlying economic environment. If cash flow analysis is used to determine the diameter, the apparent optimum can be reduced by a factor of two compared to the results predicted by Equation 11.6. More details on estimating optimum diameters are provided in Chapter 7.

Example: 11.2 *A liquid has a density of 1.1 g/cm^2 and a viscosity of 1.3 cp. A flowrate of 500 gpm is required. Estimate the optimal pipe diameter and the Reynolds number in the pipe.*

Solution: 11.2 $q = (300\ gal/min)(1\ min/60\ s)(1\ ft^3/7.48\ gal) = 0.668\ ft^3/s$
Assume laminar flow. Then
$D = 3q^{0.36}\mu^{0.18} = 3(0.668)^{0.36}(1.3)^{0.18} = 2.72\ in$
$A = \pi(2.72/24)^2 = 0.04035\ ft^2$
$V = q/A = 0.668/0.04035 = 16.56\ ft/s$
$Re = \rho VD/\mu = (1.1\ g/cm^3)(62.4\ lb_m - cm^3/ft^3 - g)(16.56\ ft/s)(2.72/12\ ft)$
$/(1.3\ cp)(6.72 \times 10^{-4}\ lb_m/ft - s - cp) = 294,926$
Since $Re > 2100$, *flow is turbulent. So*
$D = 3.9q^{0.45}\rho^{0.13} = 3.9(0.668)^{0.45}[1.1(62.4)]^{0.13} = 5.64\ in$
$A = \pi(5.64/24)^2 = 0.17349\ ft^2$
$V = q/A = 0.668/0.17349 = 3.85\ ft/s$
$Re_2 = Re_1(D_2/D_1)(V_2/V_1) = 294,926(5.64/2.72)(3.85/16.56) = 142,175$

Friction Factors

It is convenient to approximate the Fanning friction factor f using a spline defined in terms of the Reynolds number Re:

$$f = \begin{cases} 16/Re, & Re < 2100 \\ 0.04/Re^{0.16}, & Re \geq 2100 \end{cases} \tag{11.7}$$

However, the Colebrook [23] equation is a more accurate correlation for predicting the friction factor for fluid flow:

$$\frac{1}{\sqrt{f}} = -2.0 \ln\left[\frac{\epsilon/D}{3.7} + \frac{2.51}{Re\sqrt{f}}\right] \tag{11.8}$$

where ϵ is the absolute pipe roughness in feet, D is the internal pipe diameter in feet, and $Re = \rho vD/\mu$ is the Reynolds number for fluid flow. Since the friction factor appears on both sides of the Colebrook equation, trial-and-error is required to solve this equation. On the other hand, for transitional and turbulent flow ($Re > 2,100$) and at any relative roughness (ϵ/D), Serghides [88] suggests that a friction factor equation of the form below can be used:

$$f = \left[A - \frac{(B-A)^2}{C - 2B + A}\right]^{-2} \tag{11.9}$$

$$A = -2.0 \ln\left[\frac{\epsilon/D}{3.7} + \frac{12}{Re}\right] \tag{11.10}$$

$$B = -2.0 \ln\left[\frac{\epsilon/D}{3.7} + \frac{2.51A}{Re}\right] \tag{11.11}$$

$$C = -2.0 \ln\left[\frac{\epsilon/D}{3.7} + \frac{2.51B}{Re}\right] \tag{11.12}$$

Sizing Pumps

Liquid pumps are sized from the throughput (q) and total developed head (TDH) requirements. Head calculations depend primarily on the length and diameter of the connecting pipe, and the elevation of the suction and discharge lines.

Table 11.1: Loss coefficients for standard threaded pipe fittings [112].

Fitting	K_i	Fitting	K_i
Globe valve, wide open	10.0	Return bend	2.2
Angle valve, wide open	5.0	Tee	1.8
Gate valve, wide open	0.2	Elbow, 90°	0.9
Gate valve, half open	5.6	Elbow, 45°	0.4

If Equation 11.1 is applied to the immediate pump boundaries, then $\Delta u^2 = \Delta z = 0$ and for an incompressible fluid:

$$\frac{\dot{w}_f}{\dot{m}} - \frac{u^2}{2g_c}\left(\frac{4fL}{D} + \sum_i K_i\right) = \frac{\Delta p_p}{\rho} = \frac{TDH}{\rho} \tag{11.13}$$

The parenthetical term on the left hand side of Equation 11.13 represents the frictional losses due to the fluid motion within the pump casing. This term is complex and generally difficult to evaluate. For purposes of the preliminary sizing of pumps, it is often sufficient to define an intrinsic pump efficiency factor ϵ_i such that:

$$\frac{\dot{w}_p}{\dot{m}} = \frac{\epsilon_i \dot{w}_s}{\dot{m}} = \frac{\Delta p_p}{\rho} = \frac{TDH}{\rho} \tag{11.14}$$

The term on the right hand side of Equations 11.13 and 11.14 defines the total developed head (TDH), or pressure differential, Δp_p, that is developed by the pump and imparted to the fluid divided by the density of the incompressible fluid. The TDH times the volumetric flowrate yields the hydraulic horsepower of the pump, HP. So Δp_p in Equations 11.13 and 11.14 is the pressure rise across the pump whereas Δp in Equation 11.15 below is the pressure difference between the outlet and inlet of the connecting piping (i.e., the environmental Δp).

In order to estimate the TDH, the mechanical energy balance equation is applied to the environment surrounding the pump and

$$TDH = \Delta p + \frac{g\rho\Delta z}{g_c} + \frac{\rho u^2}{2g_c}\left(\frac{4fL}{D} + \sum_i K_i\right) \tag{11.15}$$

where Δp is the pressure difference between the outlet and inlet of the connecting piping, and Δz is the elevation difference through which the liquid must be pumped. The terms L and D are the length and diameter of the connecting pipe, f is the friction factor in the pipe and u is the average fluid velocity through the piping. The $\sum_i K_i$ represents the estimated frictional losses across any fittings and valves in the piping. Table 11.1 gives values for K_i for several types of fittings.

Equivalent Pipe Length

During a Preliminary Design, detailed counts of the actual fittings in the network are usually not available. In such cases, an average layout can be used. Hooper [46]

Table 11.2: Equivalent lengths per foot of pipe [13] (L_e/L) as a function of F_c and the pipe diameter.

Design		Nominal diameter (inches)								
Complexity	F_c	0.5	1	2	4	8	10	16	24	36
Complex Manifolds	4	2.9	3.2	3.8	4.6	5.8	6.2	7.3	8.5	10.1
Manifold piping	2	1.9	2.1	2.4	2.8	3.4	3.7	4.2	4.8	5.5
Normal piping	1	1.5	1.6	1.8	1.9	2.2	2.3	2.6	2.9	3.2
Straight run piping	1/2	1.2	1.3	1.4	1.4	1.6	1.7	1.8	1.9	2.1
Utility supply lines	1/4	1.1	1.2	1.3	1.3	1.4	1.4	1.4	1.4	1.5

indicates that the number of pipe fittings per foot of straight pipe is:

$$N_f = 0.35\left(\frac{F_c}{D^{1/2}}\right) \tag{11.16}$$

and the number of valves is:

$$N_v = 0.09\left(\frac{F_c}{D}\right) \tag{11.17}$$

where D is the nominal pipe diameter in inches and F_c is a complexity factor. Typical values for the complexity factor are summarized in Table 11.2. By using an average number and mix of valves and fittings, Brown [13] developed a simple expression for estimating the total equivalent feet of pipe per foot of straight pipe:

$$\frac{(L_e/D)}{L} = \frac{12}{D} + 11.9\left(\frac{0.35F_c}{\sqrt{D}}\right) + 28.8\left(\frac{0.09F_c}{D}\right) \tag{11.18}$$

where L is the length of straight pipe in feet and D is the nominal pipe diameter in inches. In this expression, the first term on the right hand side accounts for the straight pipe, the second accounts for the fittings, and the third for the valves. This expression can be further simplified to:

$$\frac{L_e}{L} = 1 + [0.34\sqrt{D} + 0.216]F_c \tag{11.19}$$

Rather than using $\sum_i K_i$ in Equation 11.15, fitting resistance may be expressed in terms of L_e/D, an equivalent length divided by the nominal pipe inner diameter. Table 11.3 gives several values for fittings.The L_e/D values in Table 11.3 are related to the K_i values in Table 11.1 using:

$$\left(\frac{4fL}{D} + \sum_i K_i\right) = \left(\frac{4fL_e}{D}\right) \tag{11.20}$$

and so,

$$\sum_i K_i = \frac{4f(L_e - L)}{D} \tag{11.21}$$

Table 11.3: Equivalent pipe lengths for frictional losses.

Fitting	L_e/D	Fitting	L_e/D
Flanges	0	90-deg elbows	16
Tees	40	Reducers	15
45-deg elbows	16	Plugs & caps	0
Block valves	20	Check valves	50
Globe valves	340		

Hooper thus provides a significant approximation for preliminary design. Initially, detailed piping diagrams are not available, so the number of elbows, fittings, etc. are not known. As a first approximation, it is often fairly simple to estimate the lengths of straight pipe required. Once the height of a distillation tower is known, for example, then the approximate length of straight pipe to handle the condenser is also known (i.e., two times the column height if the condenser is on the ground). Similarly, simple spacing approximations between unit operations can be used to approximate straight lengths of pipe at different points. The lengths can then be used to estimate L_e and to estimate pumping requirements without knowledge of the fitting and valve layouts.

Example: 11.3 *Compute an apparent friction factor for Globe valves using the information in Table 11.1 and 11.3.*
Solution: 11.3 *Starting with:*
$\sum_i K_i = 4f(L_e - L)/D$
If the valve only is considered, then $L = 0$. *So*
$f = K_i/4(D/Le) = 10.0/[4(340)] = 0.0074$

Liquid Pump Efficiency

In order to estimate the power requirements to drive a pump, one must also estimate the overall efficiency, ϵ_o where:

$$P = \frac{\dot{w}_s}{\epsilon_o} = \frac{\dot{m}\Delta p}{\epsilon_d \epsilon_i \rho} = \frac{\dot{m}\Delta p}{\epsilon_o \rho} \tag{11.22}$$

and the overall efficiency, ϵ_o, is defined as the product of the intrinsic efficiency, ϵ_i, and the drive efficiency, ϵ_dTypical values for ϵ_o are in the range of 40–85% for centrifugal pumps. Hendry [42] provides an approximate correlation for the overall efficiency:

$$\epsilon_o = \frac{\text{Hydraulic Horsepower}}{\text{Driver or Brake Horsepower}} = 1 - 0.8/HP^{0.2} \tag{11.23}$$

where HP (1 hp = 550 lb_f-ft/s) is the hydraulic horsepower delivered to the fluid:

$$HP = \frac{q(TDH)}{550} \tag{11.24}$$

q is the volumetric flowrate in ft^3/s and TDH is in lb_f/ft^2. This correlation has two limitations. First, it predicts negative efficiencies for very small pumps. Second, it predicts efficiencies approaching 100% for very large pumps. In reality, pumps can usually be designed to have overall efficiencies of at least 20%, but efficiencies seldom exceed 80%. So combining the Hendry model with the 80/20 rule gives a preferred estimation technique:

$$\epsilon_o = \begin{cases} 0.2 & \text{if } HP \leq 1.0 \\ 1 - 0.8/HP^{0.2} & \text{if } 1.0 < HP < 1024 \\ 0.8 & \text{if } HP \geq 1024 \end{cases} \tag{11.25}$$

Example: 11.4 *A pump application requires a hydraulic horsepower of 175. Estimate the brake (or driver) horsepower.*
Solution: 11.4 *Since* $1 < HP < 1024$, $\epsilon_o = 1 - 0.8/HP^{0.2} = 0.715$
The required brake horsepower is $BHP = HP/\epsilon_o = 175/0.715 = 245$ *hp.*

Example: 11.5 *What size motor is needed to pump oil to the top of an absorption tower? The oil must be pumped from an open tank at one atmosphere pressure with a liquid level in the tank of 10 ft above the floor, and forced through 150 ft of schedule number 40 steel pipe with a 3 inch nominal diameter. Estimate frictional losses using the concept of equivalent length,* L_e*, summarized in Table 11.2, and assuming normal piping complexity. The tower is 30 ft above the floor level, and its operating pressure is 64.7 psia with an oil flowrate of 50 gpm. The viscosity of oil is 15 cp and its density is 53.5 lb/ft*^{3}*. The pumping process is essentially adiabatic.*

Estimate the pump efficiency, the average linear velocity of the oil through the pipe, the required TDH*, the Reynolds number for the oil, the fanning friction factor and the brake horsepower of the pump.*

Solution: 11.5 *Interpolating from Table 11.2,* $L_e/L \approx (1.8 + 1.9)/2 = 1.85$,
$L_e = 1.85(150) = 278$ *ft, where* L_e *is the equivalent length.*

$D = 3\ in = 0.25\ ft$

$q = (50\ gal/min)(1\ ft^3/7.48\ gal)/(60\ s/min) = 0.1114\ ft^3/s$

$Linear\ Velocity = V = Q/A = (50\ gal/min)(1\ ft^3/7.48\ gal)(144\ in^2/ft^2)$
$/[\pi(3/2)^2\ in^2] = 136\ ft/min = 2.2667\ ft/s$

$N_{Re} = DV\rho/\mu = (0.25\ ft)(136\ ft/min)(53.5\ lb/ft^3)$
$/[(60\ s/min)(15\ cp)(0.0672 \times 10^{-2}\ lb_m/(ft - s - cp))] = 3008$

$f = 0.04/N_{Re}^{0.16} = 0.011$

Using equivalent lengths, $[4fL/D + \sum K_i] = 4fL_e/D$

$TDH = \Delta p + \rho \Delta z g/g_c + [\rho u^2/(2g_c)]\,[4fL_e/D]\ psi$

$\Delta p = 64.7 - 14.696 = 50\ psi$

$\rho \Delta z g/g_c = (53.5\ lb_m/ft^3)[(30 - 10)\ ft](1\ lb_f/lb_m)(1\ ft^2/144\ in^2) = 7.43\ psi$

$[\rho u^2]\,[2f(L_e/D)/g_c)] = [(53.5\ lb/ft^3)(2.2667\ ft/s)^2][2(0.011)(278/0.25)]$
$/[32.174\ lb_m - ft/(s^2 - lb_f)] = 209\ lb_f/ft^2 = 1.45\ psi$

$TDH = 50 + 7.43 + 1.45 = 58.9\ psi = 8481.6\ lb_f/ft^2$

$HP = q(TDH)/550 = (0.1114\ ft^3/s)(58.9\ psi)(144\ in^2/ft^2)/550 = 1.72\ hp$

$\epsilon = 1 - 0.8/HP^{0.2} = 0.28$

$BHP = HP/\epsilon = 6.14\ bhp$

Example: 11.6 *What size motor is needed to pump oil into the bottom of a* *solvent extraction column? The oil must be pumped from a surge tank at two atmospheres pressure. The liquid level in the tank cycles from 30 feet above the floor to five feet above the floor. The oil density is 0.8 g/cm^3. The oil is pumped from the surge tank to the bottom of the extraction colum that is 100 ft tall and full of a well-mixed water and oil mixture with an average density of 0.95 g/cm^3. The top of the extraction column is at one atmosphere pressure. The nominal distance between the surge tank and the extraction column is 600 ft. The base of the extraction column is at floor level. The required oil flowrate (steady state) is 800 gpm. Assume normal piping complexity. The viscosity of the oil is 8 cp.*

Estimate the pipe diameter, the average linear velocity of the oil through the pipe, the Reynolds number in the connecting pipe, the friction factor in the pipe, the equivalent length, the required TDH, and the brake horsepower of the pump. Design the pump so that it is always able to provide the required flowrate.

Solution: 11.6 *Use Equation 11.6 to estimate the pipe diameter. First assume* $Re < 2100$.

$q = 800\ gpm(1\ ft^3/7.48\ gal)(1\ min/60\ s) = 1.7813\ ft^3/s$

$D = 3q^{0.36}\mu^{0.18} = 3(1.7813^{0.36})(8^{0.18}) = 5.37\ in$

$A = \pi(D/2)^2 = \pi(5.37/24)^2 = 0.1573\ ft^2$

$V = 1.7813/0.1573 = 11.327\ ft/s$

$Re = 0.8(62.4\ lb_m/ft^3)(11.327\ ft/s)(5.369/12\ ft)/[(8\ cp)(6.72 \times 10^{-4}\ lb_m/ft - s)] = 47,066$

Since $Re > 2100$, *use* $D = 3.9q^{0.45}\rho^{0.13}$

$D = 3.9q^{0.45}\rho^{0.13} = 3.9(1.7813^{0.45})[0.8(62.4)]^{0.13} = 8.4076\ in$

$A = \pi(D/2)^2 = \pi(8.403/24)^2 = 0.38555\ ft^2$

$V = q/A = 1.7813/0.3855 = 4.6203\ ft/s$

$Re = 0.8(62.4\ lb_m/ft^3)(4.6203\ ft/s)(8.408/12\ ft)$
$/[(8\ cp)(6.72 \times 10^{-4}\ lb_m/ft - s)] = 30,059 > 2100$

Since $Re > 2100$, $f = 0.04/Re^{0.16} = 0.0.00768$

Since Normal piping, $F_c = 1.0$ *and*

$L_e/L = 1 + [0.34\sqrt{D} + 0.216]F_c = 1 + 0.34[\sqrt{8.4076} + 0.216](1) = 2.2019$

$L_e = 2.2019(600) = 1320$ *ft, where* L_e *is the equivalent length.*

Using equivalent lengths, $[4fL/D + \sum K_i] = 4fL_e/D$

$TDH = \Delta p + \rho\Delta zg/g_c + [\rho u^2/(2g_c)]\,[4fL_e/D]\ psi$

Calculate the pressure at the base of the extraction column, $\Delta P = \rho\Delta zg/g_c$

$\Delta P = \rho\Delta zg/g_c$
$= (0.95\ g/cc)(62.4\ lb_m/ft^3/(1\ g/cc))(100\ ft)(1\ lb_f/lb_m)(1\ ft^2/144\ in^2) = 41.17\ psi$

$\Delta p = 41.17 + 14.696 - 2(14.696) = 26.47\ psi$

Use the surge tank height at the lowest level to provide adequate capacity.

$\rho\Delta zg/g_c = (0.8(62.4)\ lb_m/ft^3)[(0 - 5)\ ft](1\ lb_f/lb_m)(1\ ft^2/144\ in^2) = -1.73\ psi$

$[\rho u^2]\,[2f(L_e/D)/g_c)] = [0.8(62.4\,lb/ft^3)(4.6203\ ft/s)^2][2(0.00768)(1320/(8.4076/12))]$
$/[32.174\ lb_m - ft/(s^2 - lb_f)] = 952.4\ lb_f/ft^2 = 6.614\ psi$

$TDH = 26.47 - 1.73 + 6.61 = 31.35\ psi = 4514.4\ lb_f/ft^2$

$HP = q(TDH)/550 = (1.7813\ ft^3/s)(4514.4\ lb_f/ft^2)/550 = 14.62\ hp$

$\epsilon = 1 - 0.8/HP^{0.2} = 0.53$

$BHP = HP/\epsilon = 27.6\ bhp$

11.2 Heat Transfer Equipment

11.2.1 Heat Exchangers

The Preliminary Design of heat exchangers usually requires that several steps be completed. First, the process conditions (stream flowrates, compositions, temperatures, and pressures) must be specified. Secondly, physical properties for the streams are required. After a heat exchanger type has been selected, it is then possible to complete the preliminary sizing calculations.

The basic design Equation [75] for heat exchanger sizing is:

$$dA = \frac{dQ}{U\Delta T} \tag{11.26}$$

where dA = differential element of heat transfer area
dQ = differential amount of heat transferred
U = the overall heat-transfer coefficient (local)
ΔT = the local temperature difference between two streams

The overall heat-transfer coefficient is related to the individual film heat-transfer coefficients, fouling, and wall resistances. Defining an overall heat-transfer coefficient, U_o, based upon the outside surface area, A_o, yields:

$$\frac{1}{U_o} = \frac{1}{h_o} + R_{do} + \frac{xA_o}{k_w A_{wm}} + \frac{A_o}{h_i A_i} + \frac{R_{di} A_o}{A_i} \tag{11.27}$$

where A_i = inside area of the heat transfer surface
A_o = outside area of the heat transfer surface
A_{wm} = mean wall area of heat transfer surface
h_i = inside wall film coefficient, Btu/h-ft^2-^{o}F
h_o = outside wall film coefficient, Btu/h-ft^2-^{o}F
k_w = thermal conductivity of heat exchanger wall, Btu-ft/h-ft^2-^{o}F
R_{di}, R_{do} = dirt or scale resistance on inner and outer surfaces, h-^{o}F/Btu
x = length of conduction path (heat exchanger wall thickness), ft.

The transfer area is then given by:

$$A_o = \int_0^Q \frac{dQ}{U_o \Delta T} \tag{11.28}$$

where U_o and ΔT must be known as functions of Q. In cases where U_o does not vary greatly in the exchanger, then a mean value, U_{om} may be used. Similarly, if a mean value for the temperature driving force, ΔT_m, can be used then:

$$A = \frac{Q}{U_{om}\Delta T_m} \tag{11.29}$$

and the design process is reduced to estimating an overall heat-transfer coefficient, U_{om}, and a mean temperature driving force. If the ΔT_m term is taken to be the log mean

Table 11.4: Overall heat-transfer coefficients for coils immersed in liquids [75]

Substance inside coil	Substance outside coil	Coil material	Agitation	U Btu/h-ft^2-^{o}F
Steam	Water	Lead	Agitated	70
Steam	Sugar & molasses	Copper	None	50–240
Steam	Boiling water solutions		600	
Cold water	Dilute organic dyes	Lead	Agitator, 95 rpm	300
Cold water	Warm water	Wrought iron	Air sparged around coil	150–300
Cold water	Hot water	Copper	None	105–180
Steam	Vegetable oil	Steel	Various	39–72
Cold water	Vegetable oil	Steel	Various	29–72

temperature difference based on a countercurrent operation, then:

$$\Delta T_m = \frac{\Delta T_1 - \Delta T_2}{\ln(\Delta T_1/\Delta T_2)} \tag{11.30}$$

where ΔT_1 and ΔT_2 are the temperature differences between the two process streams at either end of the exchanger. If the streams on both sides of an exchanger are pure fluids undergoing a phase change, then ΔT_m is simply a constant driving force throughout the exchanger and the heat load, Q, is the enthalpy difference between saturated liquid and vapor at the node from which the stream comes times the flowrate being condensed or vaporized.

Typical values for overall heat transfer coefficients are summarized in Tables 11.4 and 11.5.

11.3 Surge and Storage Capacity

Surge and storage capacity are significant cost components in most chemical processes. Surge capacity is needed whenever batch operations are interfaced with continuous steps, and to stabilize continuous unit operations from process upsets. Storage capacity is needed for raw materials and products. Approximate sizing of surge and storage vessels is appropriate for Preliminary Design.

Heuristic: 25 *Size surge and storage tanks so that they are never more than 80% full (i.e. with 20% freeboard), and never less than 20% full. Assume instantaneous transfers during Preliminary Design.*

Example: 11.7 *A surge tank is needed to interface a batch reactor with a continuous, liquid-liquid extraction column. The batch reactor produces 1000 gal/batch every 5 hours. The extraction column must be sized to accommodate the batchwise*

Table 11.5: Typical [75] overall heat-transfer coefficients in shell and tube heat exchangers.

Shell side	Tube side	U Btu/h-ft^2-^{o}F
Liquid/liquid media		
Fuel oil	Water	15–25
Fuel oil	oil	10–15
Gasoline	Water	60–100
Heavy oils	Heavy oils	10–40
Heavy oils	Water	15–50
Organic solvents	Water	50–150
Organic solvents	Brine	35–90
Organic solvents	Organic solvents	20–60
Water	Water	200-250
Wax distillate	water	15–25
Condensing vapor/liquid media		
Alcohol vapor	water	100–200
Asphalt (450°F)	Dowtherm vapor	40–60
Dowtherm vapor	Dowtherm liquid	80–120
Organic solvents	Water	100–200
Kerosene	Water	30–65
Kerosene	Oil	20–30
Steam	Water	400–1000
Gas/liquid media		
Compressed air	Water or brine	40–80
Air, N_2 (1 atm)	Water or brine	10–50
Water or brine	Compressed air	20–40
Water or brine	Air (1 atm)	5–20
Vaporizers		
NH_3	Condensing steam	150–300
Propane	Condensing steam	200–300
Water	Condensing steam	250–400

feed. Estimate the feed rate to the extraction column and the size of the surge tank for the extraction column feed.

Solution: 11.7 *The batch reactor produces (1000 gal/batch)(1 batch/ 5 hours) or 200 gal/h. This is the required feed rate to the extraction column. When a batch is pumped into the surge vessel, the vessel is initially 20% full and is 80% full after receiving the batch instantaneously. So*

$1000 + 0.2V = 0.8V, V = 1000/0.6 = 1667$ *gal*

Just before the next batch is added, the surge tank holds 333 gal, which is 20% of its capacity . At the end of instanteously transferring the batch to the surge tank, the tank hold 1333 gal, which is 80% of the tank capacity. Normally standard tank sizes are used, rather then requesting a special size, so a 1700 to 2000 gal tank would be appropriate.

Heuristic: 26 *Size vessels that normally have constant inventories by assuming that they are half full. Examples of such vessels include steady-state reactors (e.g., constant, stirred-tank reactors) and reflux accumulator drums for distillation columns.*

11.3.1 Reflux Accumulator Drums

Size reflux accumulator drums for distillation columns by assuming a nominal space time of 7.5 minutes and a length to diameter ratio of 2.0. The required volume in cubic feet is then given approximately as

$$Vol = \frac{7.5W}{60\rho_l} \tag{11.31}$$

where W is the liquid flow rate into the accumulator lb mol/hr and ρ_l is the liquid density in lb-mol/ft^3. The resulting drum diameter is:

$$D_d = \left(\frac{2Vol}{\pi}\right)^{1/3} \tag{11.32}$$

11.4 Reactor Vessels

Reactor design should be based on experimental kinetic measurements by the end of the Conceptual Engineeering Phase. During Preliminary Engineering, simplied approximations are frequently required due to a lack of experimental data. In some cases, reactor sizing must be based on a nominal residence time, τ. For fast reactions, τ can be quite short, say a few seconds or less. Such reactions often occur at high temperatures in the gas phase, and in the presence of a suitable catalyst. Because of the small residence time, the reactor costs can be relatively low, depending upon whether or not special materials of construction are required, and the costs of catalysts.

The residence time for liquid phase reactions may be much greater, sometimes on the order of hours or days. In such instances the reactor cost can be quite high because of the large anticipated reactor volume. When large residence times are suspected during Preliminary Design, additional experimental research should be immediately requested to help reduce the anticipated reactor costs.

Example: 11.8 *Fifty gallons per minute of a process stream must be reacted in a liquid phase reactor. Preliminary experimental studies suggest that reasonable conversion can be achieved in a constant, stirred-tank reactor if the residence time is about 1.2 hours. Estimate the reactor volume.*
Solution: 11.8 *The nominal residence time is $\tau = V/\dot{q}$, where V is the reactor volume and $\dot{q}$ is the volumetric flowrate through the reactor.*
So $V = \tau\dot{q} = (1.2\ h)(60\ min/h)(50\ gal/min) = 3600\ gal$.
Specify a 7200 to 7500 gal reactor vessel.

11.5 Mass Transfer Equipment

While the reactor is the heart of most chemical processes, the production costs are often dominated by the costs of purifying reagents and products.

Heuristic: 27 *Undersizing mass transfer equipment incurs the following costs: (1) cost of replacing the unit with a larger one that achieves the desired separation, (2) cost of removing the undersized unit, and (3) cost of being out of operation while the new unit is installed (downtime). These costs are sufficiently large as to justify 10% overdesign or more in mass transfer equipment.*

11.5.1 Distillation Towers

Minimum Theoretical Trays

A tower section is defined as a series of countercurrent trays that are located between two adjacent composition nodes. The minimum number of trays may be computed based on a split between pairs of species or using an averaging technique. The former method is best for sharp separations, but averaging helps when species are distributed or products are recovered as mixtures.

Sharp Separations In sharp separations, say between two species that are contiguous in volatility, then the minimum number of trays is well estimated using the classical Fenske [28] equation. The minimum tray requirements to effect a specified separation across section n at total reflux, N_n^* is calculated from:

$$\left(\frac{x_{it}}{x_{ib}}\right)\left(\frac{x_{jb}}{x_{jt}}\right) = \overline{\alpha}_{ij}^{N_n^*} \tag{11.33}$$

where subscripts t and b refer to composition nodes at the top and bottom of a section and the subscripts i and j refer to a component pair. The average relative volatility $\overline{\alpha}_{ij}$ is computed as the geometric mean[1] of two volatilities at the two

[1]The geometric mean of n values of x is $\bar{x} = (x_1 \cdot x_2 \cdots x_n)^{1/n}$.

composition nodes over the column section. The *maximum* value found for N_n^* by applying Equation 11.33 to all possible pairs of components i and j may be treated as the *minimum* number of trays for that section.

The controlling component pairs may vary from section-to-section[2]. A simple tower, has only two sections, but complex fractionators may have more. The minimum theoretical trays should be calculated for each section.

Example: 11.9 *Find the minimum number of trays between the feed and top of the column for the following four component system.*

	Feed Tray		Distillate	
Component	x_{fi}	K_{fi}	x_{di}	K_{di}
A	0.3	1.4	0.85	1.2
B	0.2	1.1	0.10	1.0
C	0.4	0.8	0.04	0.7
D	0.1	0.5	0.01	0.3

Solution: 11.9 *Use the geometric mean volatility:*

$$\overline{\alpha}_{AB} = \sqrt{\alpha_{AB_f}\alpha_{AB_t}} \qquad = \sqrt{\frac{K_{fA}}{K_{fB}}\frac{K_{dA}}{K_{dB}}}$$

$$\alpha_{AB_f} = 1.4/1.1 = 1.27 \quad \alpha_{AB_d} = 1.2/1.0 = 1.2$$

$$\overline{\alpha}_{AB} = \sqrt{(1.27)(1.2)} = 1.23$$

Similarly:

$$\overline{\alpha}_{BC} = 1.40 \quad \overline{\alpha}_{CD} = 1.93$$

$$N_n^*(AB) = 8.4$$

$$N_n^*(BC) = 5.6$$

$$N_n^*(CD) = 0.64$$

Hence the AB pair controls the minimum number of theoretical stages in this section and $N_n^* = 8.4$ *stages minimum.*

Distributed Components In cases where species distribute between the keys, then an averaging equation is better suited for estimated the minimum tray requirements. Shoaei and Tedder [92] found:

$$N = \frac{\sum_{i=1}^{NC} \ln\left[\left(\frac{x_{i,j+N}}{x_{r,j+N}}\right)\left(\frac{x_{r,j}}{x_{i,j}}\right)\right]}{\sum_{i=1}^{NC} \ln \alpha_i} \tag{11.34}$$

[2]A section is defined as a series of countercurrent trays. Either a column feed or a product draw point can define one end of a section. A simple tower has a distillate, a mid-point feed, and a bottoms product. The top section falls between the distillate and the mid-point feed. The bottom section falls between the mid-point feed and the bottoms product.

to be helpful. In this case, i is any distributed species in the tower section. Usually, $LK < i < HK$ when i refers to the species name list and HK is the heavy key, while LK is the light key.

Theoretical Trays

Sharp Separations The number of theoretical trays required in a given tower section, n, can be calculated from a correlation presented by Molokanov [67]:

$$Y = 1 - \exp\left[\left(\frac{1 + 54.4X}{11 + 117.2X}\right)\frac{X - 1}{\sqrt{X}}\right] \tag{11.35}$$

where

$$X = \frac{V_n - V_n^*}{V_n} = \frac{R - R_{min}}{R + 1}$$

$$Y = \frac{N_n - N_n^*}{N_n + 1}$$

where V_n^* = is the Underwood minimum vapor rate for section n
N_n^* = the minimum number of theoretical plates as found from the Fenske equation
V_n = the estimated actual vapor rate
N_n = estimated theoretical number of trays for section n

Equation 11.35 is basically a restatement of the Gilliland correlation which is an empirical relationship that is based on actual tower operation.

Distributed Components An extended form of the Gilliland equation was developed [90, 92] for refinery operations where the products are often mixtures of species rather than pure components. Based upon comparisons with equilibrium stage models, the most attractive correlation has the form:

$$Y = 1 - \exp[-0.1901 \cdot C \cdot D \cdot E \cdot O \cdot S] \tag{11.36}$$

where:

$$Y = \frac{N - N_{min}}{N + 1} \tag{11.37}$$

$$C = 0.6592 - 0.183X + 0.3615X^2 + X^3 \tag{11.38}$$

$$D = \frac{X - 1}{X^{0.1}} \tag{11.39}$$

$$E = 3.28 - 0.332f + 1.233f^2 - f^3 \tag{11.40}$$

$$O = \left|\frac{T_{top} - T_{bot}}{T_{sect} - T_{feed}}\right| \tag{11.41}$$

$$S = \left(\frac{x_{F,H}}{x_{F,L}}\right)^{0.01851} \tag{11.42}$$

$$X = \frac{V - V_{min}}{V} = \frac{R - R_{min}}{R + 1} \tag{11.43}$$

where f = the feed fractional vaporization
T_{top} = the temperature at the top of the column
T_{bot} = the temperature at the bottom of the column
T_{sect} = the temperature at the top of a column section if the section is above the feed or the temperature at the bottom of a column section if the section is below the feed
T_{feed} = the temperature of the column feed
$x_{F,L}$ = the mole fraction of the light key in the feed
$x_{F,H}$ = the mole fraction of the heavy key in the feed stream.

Overall Tray Efficiency

The overall efficiency of those trays in a given section can be estimated using a correlation from O'Connell [73]:

$$\log \epsilon_n = 1.6794 - 0.25 \log(\bar{\mu}\bar{\alpha}) + 0.3 \log \left(\frac{L}{V}\right) \tag{11.44}$$

where ϵ_n = the tray efficiency for section n, percent
L = molar liquid flowrate, lb-mole/h
V = molar vapor flowrate, lb-mole/h
$\bar{\alpha}$ = the average relative volatility between the key components for section n
$\bar{\mu}$ = the average liquid viscosity, cp

In this correlation, it is also assumed that sieve trays are being used with 0.5 inches static submergence and a slot height of 1.5 inches. Also, the key components are assumed to be those that result in the largest number of minimum theoretical plates as calculated from the Fenske equation.

Total Tower Height

The overall height of a distillation tower feet may be calculated from the number of trays and the tray efficiency in each section as:

$$HT = 10 + 1.1(1.5) \sum_n \left(\frac{100\% N_n}{\epsilon_n}\right) \tag{11.45}$$

where HT = the column height, ft.
N_n = the theoretical sieve trays required for section n
ϵ_n = the tray efficiency for section n, percent

The factor of 1.5 assumes a tray-to-tray spacing of 18 inches, and the factor 1.10 is included to provide a 10% overdesign [86] factor. The ten additional feet are an allowance for vapor–liquid separation space.

Column Diameter

Distillation columns are usually assumed to be a single diameter which may be taken as the maximum diameter required for any section as calculated from the following equation [75]:

$$D_t = \left(\frac{4V_n}{432\pi\rho_v}\right)^{1/2}\left(\frac{\rho_v}{\rho_L - \rho_v}\right)^{1/4} \tag{11.46}$$

where V_n = an average vapor rate for section n, lb-mole/hr
ρ_v = the mixture vapor density in section n, lb-mole/ft^3
ρ_L = the mixture liquid density in section n, lb-mole/ft^3
D_t = the column diameter in ft.

The vapor density can be estimated using an equation of state such as Chao-Seader or, perhaps, the ideal gas law. Mixture liquid densities are usually estimated using the average properties of the pure component liquids. In columns with high vapor and reflux, the largest diameter will normally be required at the top of the column, where the vapor density is the lowest and the volumetric flowrate is the greatest. In refinery towers where only a fraction of the feed is vaporized, the largest diameter requirements may occur in the lower sections of the tower due to the hydraulic requirements of the liquid. These latter requirements are not considered when using Equation 11.46.

11.5.2 Solvent Extraction Columns

The height equivalent to a theoretical stage ($HETS$) for reciprocating plate extraction columns has been correlated [99, 100] using:

$$\Gamma_L = 24.4 C_{dt}^{-0.075} N_{Re}^{-0.015} N_{We}^{-0.525} N_{Sl}^{-1.15}\left(\frac{U_d}{U_c}\right)^{0.210} \tag{11.47}$$

$$\Gamma_U = 28.8 C_{dt}^{-0.651} C_{dr}^{-0.997} N_{Bo}^{0.776}\left(\frac{U_c}{U_d}\right)^{0.582} \tag{11.48}$$

where $\Gamma_L = HETS/D$, $(Af) < (Af)_{opt}$, and $(Af)_{opt}$ is the reciprocating speed that minimizes the $HETS$. Γ_U predicts $HETS/D$ values for $(Af) > (Af)_{opt}$. These two expressions can be mathematically combined to yield an overall correlation:

$$\frac{HETS}{D} = \Gamma_L\left(1 + \left[\frac{\Gamma_U}{\Gamma_L}\right]^{20}\right)^{1/20} \tag{11.49}$$

where A = the amplitude of the reciprocating cycle, measured from the mean position to an extreme, or 1/2 of the stroke length, cm
(Af) = the reciprocating speed, cm/s or cm/min
$(Af)_{opt}$ = the reciprocating speed yielding the minimum HETS value, cm/s
D = the column diameter, cm
f = the reciprocating frequency s^{-1}
g = gravity acceleration, 980.66 cm/s^2
g_c = gravity constant, 1.0 g-cm/d-s^2
h = plate spacing in column mass transfer zone, cm
$HETS$ = height equivalent to a theoretical stage, cm
U_c = the continuous phase superficial velocity in cm/s
U_d = the dispersed phase superficial velocity in cm/s
U_T = net superficial velocity of phases = $U_d + U_c$, cm/s
$\Delta\rho$ = $\rho_d - \rho_c$, g/ml
Γ_L = suboptimal $HETS/D$ correlation for $(Af) < (Af)_{opt}$
Γ_U = the superoptimal $HETS/D$ correlation for $(Af) > (Af)_{opt}$
μ_c = viscosity of the continuous phase, cp
ρ_c = density of the continuous phase, g/ml
ρ_d = density of the dispersed phase, g/ml
σ = interfacial tension, d/cm
N_{Bo} = $D^2\Delta\rho g/g_c\sigma$, a Bond number
C_{dr} = $D\Delta\rho g/(Af)^2\rho_c$, a reciprocating drag coefficient
C_{dt} = $D\Delta\rho g/\rho_c U_T^2$, a column drag coefficient
N_{Re} = $\rho_c U_T D/\mu_c$, a Reynolds number
N_{Sl} = Af/U_T, a Strouhal number
N_{We} = $DU_T^2\rho_c/g_c\sigma$, a Weber number

Rearranging equation 11.47 gives the following dependencies:

$$HETS \propto D^{0.385}\sigma^{0.525}(Af)^{-1.15}U_T^{0.235}\rho_c^{-0.465}(\Delta\rho g)^{-0.146} \tag{11.50}$$

for $(Af) < (Af)_{opt}$. Similarly, rearrangement of equation 11.48 gives the dependencies:

$$HETS \propto D^{0.892}\sigma^{-0.776}(Af)^{1.99}U_T^{1.30}\rho_c^{1.65}(\Delta\rho g)^{-0.872} \tag{11.51}$$

for $(Af) > (Af)_{opt}$. Analysis of these dependencies can be helpful when comparing these correlations with those developed by other investigators.

The minimum $HETS$ for a column can be estimated from an estimate of $(Af)_{opt}$. The latter quantitity may be obtained by differentiating equation 11.49 and setting $\partial HETS/\partial(Af) = 0$.

$$(Af)_{opt} = \left\{\frac{\Gamma_U}{\Gamma_L}(Af)^{-3.15}\right\}^{-1/3.15} \tag{11.52}$$

$$(Af)_{opt} = \left\{1.24D^{0.515}\sigma^{-1.30}U_T^{1.06}\rho_c^{2.12}(\Delta\rho g)^{-0.8}\mu_c^{0.015}\left(\frac{U_c}{U_d}\right)^{0.794}\right\}^{-1/3.15}$$

which predicts $(Af)_{opt} \pm 6\%$.

Example: 11.10 *Estimate the minimum HETS for a two-inch diameter reciprocating plate column. The liquid/liquid interfacial tension is 30 dyne/cm. The solvent phase density and viscosity are 0.82 g/cm^3 and 1.3 cp, respectively. The aqueous phase density and viscosity are 1 g/cm^3 and 1.129 cp, respectively. Water is the continuous phase and its linear velocity down the column is $U_c = 0.40$ cm/s. The velocity of the dispersed solvent phase is $U_d = 0.12$ cm/s.*

Solution: 11.10 $U_T = U_c + U_d = 0.40 + 0.12 = 0.52$ *cm/s*

$\Delta\rho = \mid \rho_d - \rho_c \mid = \mid 0.82 - 1.0 \mid = 0.18$ *g/cm^3*

Substituting into equation 11.52 gives $(Af)_{opt} = 10$ *cm/s. Substitution of* $(Af)_{opt}$ *and the other properties into equations 11.47–11.49 gives*

$C_{dt} = D\Delta\rho g/\rho_c U_T^2 = 5.08(0.18)980.66/[1(0.52^2)] = 3316$

$C_{dr} = D\Delta\rho g/(Af)^2\rho_c = 5.08(0.18)980.66/[(10^2)1] = 8.967$

$N_{Re} = \rho_c U_T D/\mu_c = 1.0(0.52)(5.08)/[1.129(10^{-2})] = 234$

$N_{We} = DU_T^2\rho_c/g_c\sigma = 5.08(0.52^2)1.0/[1.0(30)] = 0.046$

$N_{Sl} = Af/U_T = 10/0.52 = 19.2$

$N_{Bo} = D^2\Delta\rho g/g_c\sigma = 5.08^2(0.12)980.66/30 = 151.8$

$\Gamma_L = 1.613$, $\Gamma_U = 1.636$, *and* $HETS = 8.5$ *cm.*

11.6 Summary

Simplified methods can often be used to complete preliminary sizing calculations. The methods that are summarized here are typical of those which are frequently used during the Preliminary Design Phase. Subsequent evaluations of a chemical process flowsheet usually involve more rigorous and detailed calculations which are more accurate, but more time consuming to complete. During a preliminary design study, shortcut methods are preferred to identify the best process structure and to establish near–optimal conditions whenever possible. As the design evolves, more detailed and accurate methods should be used to establish the final design conditions and to estimate equipment specifications.

11.7 Problems

11.1 A preliminary estimate is needed for a completely installed pumping system. A pipeline is to be used to deliver 510 gpm of water at 60^oF and steady state. The theoretical horsepower requirements (100% efficiency) are 75 hp. What is the optimal pipe diameter in inches? What is the Reynolds number in the optimum pipe?

11.2 A preliminary estimate is needed for a completely installed pumping system. A pipeline is to be used to deliver 250 gpm of water at 60^oF and steady state. The theoretical horsepower requirements (100% efficiency) are 10 hp. What is the optimal pipe diameter in inches? What is the Reynolds number in the optimum pipe? What is the total developed head for the pump? What is the brake horsepower using Equation 11.14 to estimate the pump efficiency?

11.3 Size a motor that is needed to pump oil to the top of an absorption tower. The oil must be pumped from an open tank (at one atmosphere pressure) with a liquid level in the tank of 10 ft above the floor and forced through 150 ft of schedule number 40 steel pipe with 3 inch nominal diameter. There are five 90° elbows in the line which have a standard radius and the top of the tower is 30 ft above the floor level. The tower operating pressure is 64.7 psia and the oil flowrate is 50 gpm. The viscosity of the oil is 15 cp and its density is 53.5 lb/ft^3. The pumping process is essentially adiabatic and changes in the fluid kinetic energy can be neglected. Estimate the pump efficiency, the average linear velocity of the oil through the pipe, the Reynolds number for the oil in the pipe, the Fanning friction factor for the pipe, the required TDH, and the brake horsepower.

11.4 If 100 minimum theoretical stages are required in a distillation tower and the V/V_{min} ratio at the pinch point is 1.1, how many theoretical stages are required?

11.5 Consider a distillation tower where the distillate composition is (0.821, 0.173, 0.00312, 0.00208) with corresponding K-values of (1.05, 0.71, 0.34, 0.27). Just above the feed tray, the liquid composition is (0.674, 0.131, 0.181, 0.014) and the K-values are (1.18, 0.91, 0.43, 0.38) for the four species in the mixture. How many minimum theoretical trays are required for this section of the tower?

11.6 How does an increase in a tower's internal vapor rates affect the tower dimensions that are required if the same product compositions are produced before and after the vapor increase? How are the column diameter and height affected?

11.7 A new pipeline and pumping system will be installed. The pump will produce a water flowrate of 600 gpm and deliver 25 theoretical hp to the pipeline. What is the optimal pipe diameter in inches? What is the approximate Reynolds number for the optimum pipe?

11.8 How many theoretical stages are required in the upper section of a distillation tower where the V/V_{min} ratio is 1.0192 and 5.4 minimum theoretical stages are required?

11.9 A new plant is to have a 6 inch diameter carbon steel pipe traversing a distance of 300 ft and carrying 500 gal/min of water ($\rho = 62.4$ lb$_m$/ft^3, $\mu = 6.72 \times 10^{-4}$ lb$_m$/ft-s). The piping is of normal complexity. What is the estimated frictional pressure drop if the absolute roughness, ϵ, is 0.0018 in? Ans 4.6 psi.

11.10 Estimate the horsepower requirements to compress an ideal gas from 1 atm to 3 atm if the volumetric flowrate is 300 ft^3/min at 1 atm and 32 ^{o}C and the gas is an equimolar mixture of methane and hydrogen.

11.11 A distillation tower requires 38 minimum theoretical stages. If $V/V_{min} = 1.15$ at the pinch, how many theoretical stages are needed?

11.12 What is the optimum pipe diameter for transporting a liquid phase with a viscosity of 2.5 cp and a density of 1.5 g/cc at a rate of 150 m^3/hr?

11.13 A vapor mixture must be compressed. It has an inlet temperature and pressure of 322°C and 20 psia. The compression ratio per stage is 2. The average compressibility factor is 0.85 between the inlet and outlet conditions. The mole average, constant pressure heat capacity of the mixture is 4.4 Btu/lb-mole-R. The compressor polytropic efficiency is about 0.76. What is the temperature of the vapor mixture exiting the compressor?

11.14 In Problem 13, what is required shaft work for the compressor if the vapor feed rate is 100 lb-mole/hr?

11.15 Estimate the equivalent length for frictional losses per foot pipe. The nominal pipe diameter is 4 inches and it is installed in a chemical process as part of the manifold piping.

11.16 A reflux drum is needed for a distillation column. The vapor leaves the column at a rate of 870 lb/min and is totally condensed. All of the condensate must pass through the reflux drum. The liquid density is 0.48 g/ml. What is the drum volume if the nominal space time is the drum is 10 min?

Chapter 12

Process Synthesis

The heuristic synthesis of chemical processes involves several distinct steps. These steps include the selection of the technology and the synthesis of networks. The selection of technology is extremely important and the introduction of new technology often results in significant changes in a processing industry. Alternatively, existing technology may be operated more economically through flowsheets, or process network modifications, that exploit the system's operation to the greatest extent. This latter activity represents a major focus in optimization since significant improvements are often possible in chemical plants simply by rearranging the manner in which different pieces of standard equipment (or technology) are hooked together.

Several different activities relating to the synthesis of networks have evolved. The activities include:

1. Product and Reaction Path Screening — the estimation of potential profits of a proposed reaction sequence,

2. Species Allocation — the assignment of approximate material balances,

3. Identification of Separation Problems, and

4. Task and Network Integration.

In chemical plant design, these steps are typically performed sequentially. The reactor and the product and reaction path selection determine the heart of any chemical process. They are usually examined first. Species allocation has to do with associating the gross features of the plant into some type of coherent system. Separation problems are not the heart of the plant, but they are often the most important economic aspects and should be identified early.

Task and network integration relate to methods of improving energy utilization, such as heat exchanger network synthesis, which can reduce the total demand for energy by the plant. In terms of synthesis, the Task and Network Integration steps usually are superimposed on the plant design after the previous three activities are largely completed.

Table 12.1: Effect of the number of final products and the design types on the number of different sequences that can be generated. The complex towers include both simple towers and five complex towers producing three products.

Final	Total Sequences Made From	
Products	Simple Towers	Complex Towers
2	1	1
3	2	7
4	5	30
5	14	194
6	42	1,162
7	132	7,782
8	429	52,239
9	1,430	364,705
10	4,862	2,579,577
11	16,796	18,592,431
12	58,786	135,564,936
13	208,012	999,905,592
14	742,900	7,441,209,400
15	2,674,440	55,825,343,240

12.1 Task Solution Heuristics

There are usually many alternative flowsheet configurations, or different ways in which pieces of equipment can be connected, to perform the same tasks. For example, Table 12.1 summarizes the number of different sequences of simple distillation towers, each receiving a single feed and producing two product streams, that can be generated as a function of the number of separation products. These numbers are generated using the expression:

$$S_N = \frac{[2(N-1)]!}{(N-1)!N!} \tag{12.1}$$

where: N = the number of products
S_N = is the number of simple sequences

Table 12.1 also enumerates the number of different distillation sequences that can be generated using both simple towers and complex towers (five different types), such as are described in Appendix D. Recursion formulas [101] can be used to generate column three in Table 12.1. As can be seen, the number of possible alternatives increases dramatically as soon as complex towers are permitted, but Table 12.1 does not enumerate all of the possibilities. For example, a single tower could receive a single feed and produce four products (e.g., a distillate, upper and lower sidedraws, and a bottoms), but this type of complex tower is not included as a possibility in Table 12.1.

Because of the large number of alternatives that exist, we need simple ways to quickly identify a few alternatives that may not be optimal, but that have good economic features. The Task Solution Heuristics represent one approach to this combinatorial problem. They have the advantage that they are very easy to apply, but they have the disadvantage of being ambiguous at times and of yielding more than one preferred flowsheet. However, the heuristics usually give an excellent starting flowsheet with which to begin an evolutionary search for improvements.

For purposes of discussion, these Task Solution Heuristics are divided into two categories — those based on species allocation rules, and those based on separation rules.

12.1.1 Species Allocation Heuristics

1. Remove corrosive and dangerous materials early.

2. Reduce total loads on a system by removing the plentiful materials early.

3. Favor using plentiful materials already in the process.

4. Use stream splitters and mixers where it is possible to reduce the total separation load.

5. Favor direct separations (*i.e.*, no separative agent used) where reasonable.

12.1.2 Separation Heuristics (Simple Distillation Only)

6. Do the easy separations first.

7. Save the difficult separations for last.

8. In distillation, favor the removal of products as overheads (to achieve a high degree of purity).

9. In distillation, favor removing species one at a time as overheads (since it is more expensive to distill several products as overheads rather than just one).

10. Favor equimolar splits.

11. The cost of separating components A and B is proportional to the molar feed rate to the tower divided by the normal boiling point temperature difference between A and B. Alternatively, the cost of separating components A and B is proportional to the molar feed rate to the tower divided by the term, $\alpha - 1$, where α is the relative volatility of species A to B.

12. Avoid excursions in temperature and pressure, but it is usually better to aim high in pressure rather than low. It is best to use cooling water in your overhead condensers, if possible.

It is possible to use the Task Solution Heuristics listed above to synthesize several separation systems with very few, if any, calculations. In addition, Task Solution Heuristic 11 can be used along with dynamic programming to synthesize a network. In this case, some calculations are required, but most of the information that is needed for the calculations is readily available.

12.2 Decane Chlorination Example

This example illustrates how the Task Solution Heuristics, especially Rules 1 through 10, can be used to synthesize one or more alternative separation flowsheets.

In this example chlorine is reacted with decane (DEC) to yield two products, monochlorodecane (MCD) and dichlorodecane (DCD). The MCD is the desired product, the DCD is an undesirable waste. The reactions are:

$$C_{10}H_{22} + Cl_2 \rightarrow C_{10}H_{21}Cl + HCl \tag{12.2}$$

$$\mathrm{DEC} + Cl_2 \rightarrow \mathrm{MCD} + HCl$$

$$C_{10}H_{21}Cl + Cl_2 \rightarrow C_{10}H_{20}Cl_2 + HCl \tag{12.3}$$

$$\mathrm{MCD} + Cl_2 \rightarrow \mathrm{DCD} + HCl$$

In order to sell the MCD, a minimum of three simple distillation towers are required to purify the reaction products. Specifically, the HCL must be separated from the chlorine which is recycled to the reactor, the MCD must be separated from the DEC which is also recycled to the reactor, and the MCD must be separated from the DCD which is discharged as a waste with the HCL. The relationships between the reactor, the separation system, and the required effluent and recycle streams are summarized in Figure 2.4.

It should be mentioned, however, that there may be some advantage in separating the chlorine from the decane in a fourth distillation tower. This option will be examined using both the Task Solution Heuristics in this chapter and dynamic programming in Section 12.4.

The analysis using the Task Solution Heuristics is straightforward. Each of the alternatives is simply enumerated, and the heuristics are examined to determine which, if any, favor the alternative and which are against it. The sequence which has the least heuristics against it and/or the most in favor of it is the preferred system.

In order to minimize the production of DCD in the reactor, a low chlorine to decane feed ratio of 1-to-5 is maintained. This feed results in the product mixture summarized in Table 12.2 together with some of the physical properties for the species which must be separated. As is apparent from the table, there are significant differences in the relative amounts of materials. Decane is the most plentiful species and chlorine and DCD are the least plentiful. The chlorine/DEC separation is the easiest, but it is not required. The MCD/DCD separation is the hardest and it is required.

Each of the five alternatives is examined in Figures 12.1–12.5.

Table 12.2: Decane chlorination reactor effluents and properties.

Species	Moles	NBP(^{o}C)	ΔT
HCl	0.95	-85	
			50
Chlorine	0.05	-35	
			209
Decane	4.05	174	
			48
Monochlorodecane	0.85	222	
			19
Dichlorodecane	0.05	241	

Figure 12.1: Heuristic evaluation of Sequence A — the direct sequence of simple towers.

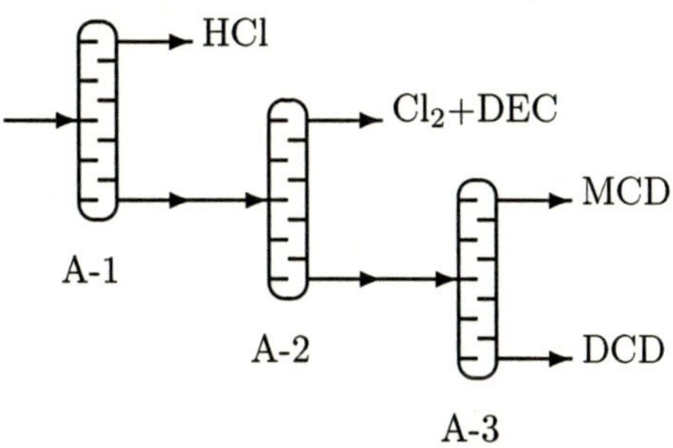

Heuristics	
For	Against
(1,HCl)	
(2,DEC)	
(6,HCl/Cl$_2$)	
(7,MCD/DCD)	
(8,MCD)	
(9,All splits)	

Figure 12.2: Heuristic evaluation of Sequence B.

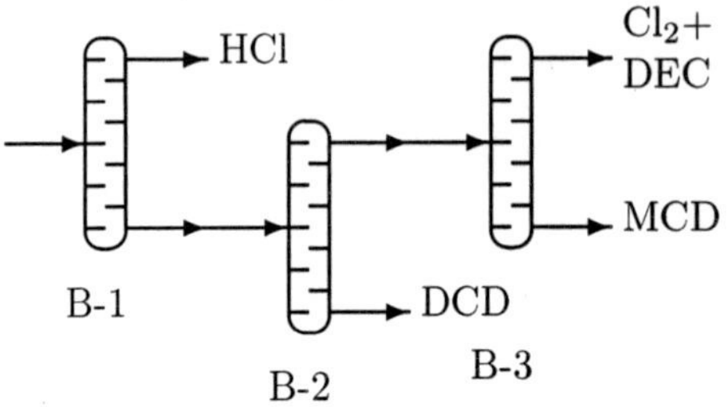

Heuristics	
For	Against
	(2,DEC)
(1,HCl)	(6,MCD/DCD)
(6,HCl/Cl_2)	(7,MCD/DCD)
	(8,MCD)
	(9,Cl_2/MCD)

Figure 12.3: Heuristic evaluation of Sequence C.

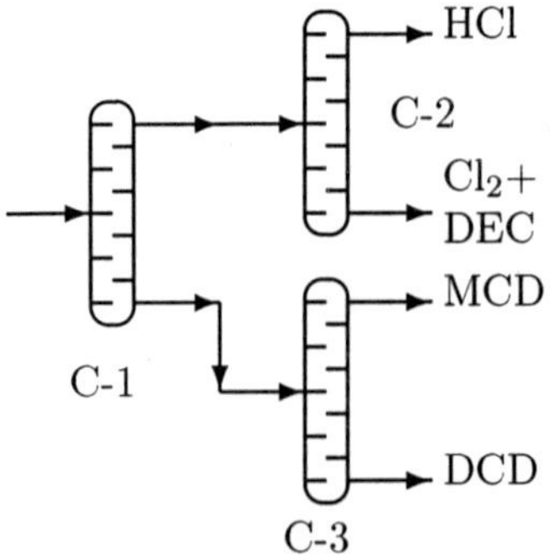

Heuristics	
For	Against
(2,DEC)	(1,HCl)
(7,MCD/DCD)	(6,DEC/MCD)
(8,MCD)	(9,C-1)

Figure 12.4: Heuristic evaluation of Sequence D — the inverted sequence of simple towers.

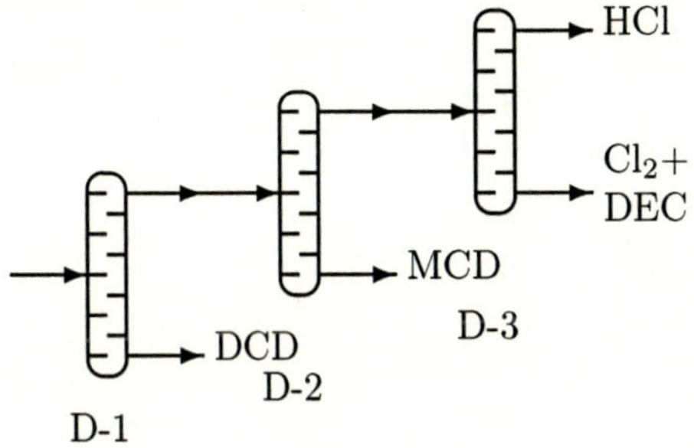

Heuristics	
For	Against
	(1,HCl)
	(2,DEC)
	(6,MCD/DCD)
	(7,MCD/DCD)
	(8,MCD)

Figure 12.5: Heuristic evaluation of Sequence E.

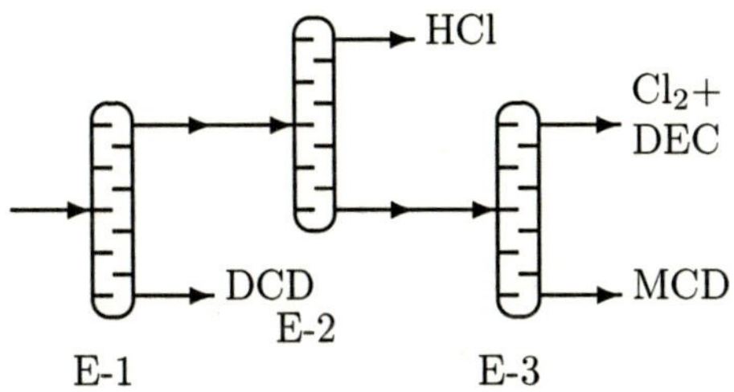

Heuristics	
For	Against
	(1,HCl)
	(2,DEC)
	(6,MCD/DCD)
	(7,MCD/DCD)
	(8,MCD)

Figure 12.6: Heuristic evaluation of Sequence F.

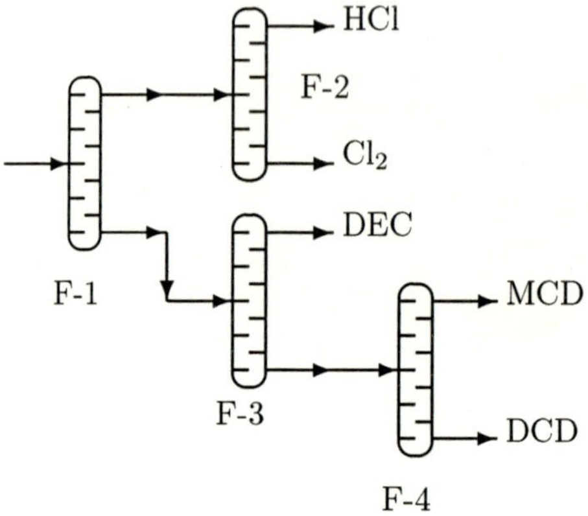

Heuristics	
For	Against
(2,DEC) (4,Cl_2/DEC) (6,Cl_2/DEC) (6,DEC/MCD) (7,MCD/DCD) (8,MCD)	(1,HCl)

In addition, a sixth alternative consisting of four distillation towers is examined as Figure 12.6.

If only the alternatives in Figures 12.1–12.5 are considered, then the Task Solution Heuristics seem to favor the sequence in Figure 12.1 most strongly. If one is willing to consider the separation of chlorine and decane, however, then the heuristics seem to indicate that there may be some economic incentives for it although it is not required. Analysis using dynamic programming and the task solution Heuristic 11, however, will suggest that the alternative in Figure 12.6 is actually less expensive. This analysis is described in Section 12.4.

12.3 Complex Tower Heuristics

All of the above studies are restricted in the sense that they only consider simple designs with structures analogous to those shown in Figure C.6. The economics of sloppy cuts and the possible advantages of taking sidestream products have not been examined as closely. Towers with multiple feeds have also been excluded, although King, Gantz, and Barnes [52] have shown that multiple feeds and sidestream draws can be useful in reducing the costs of demethanizing ethylene.

Parametric studies [101, 103] of ternary mixtures resulted in several heuristics for choosing complex towers which are summarized below and in Figures 12.7 and 12.8.

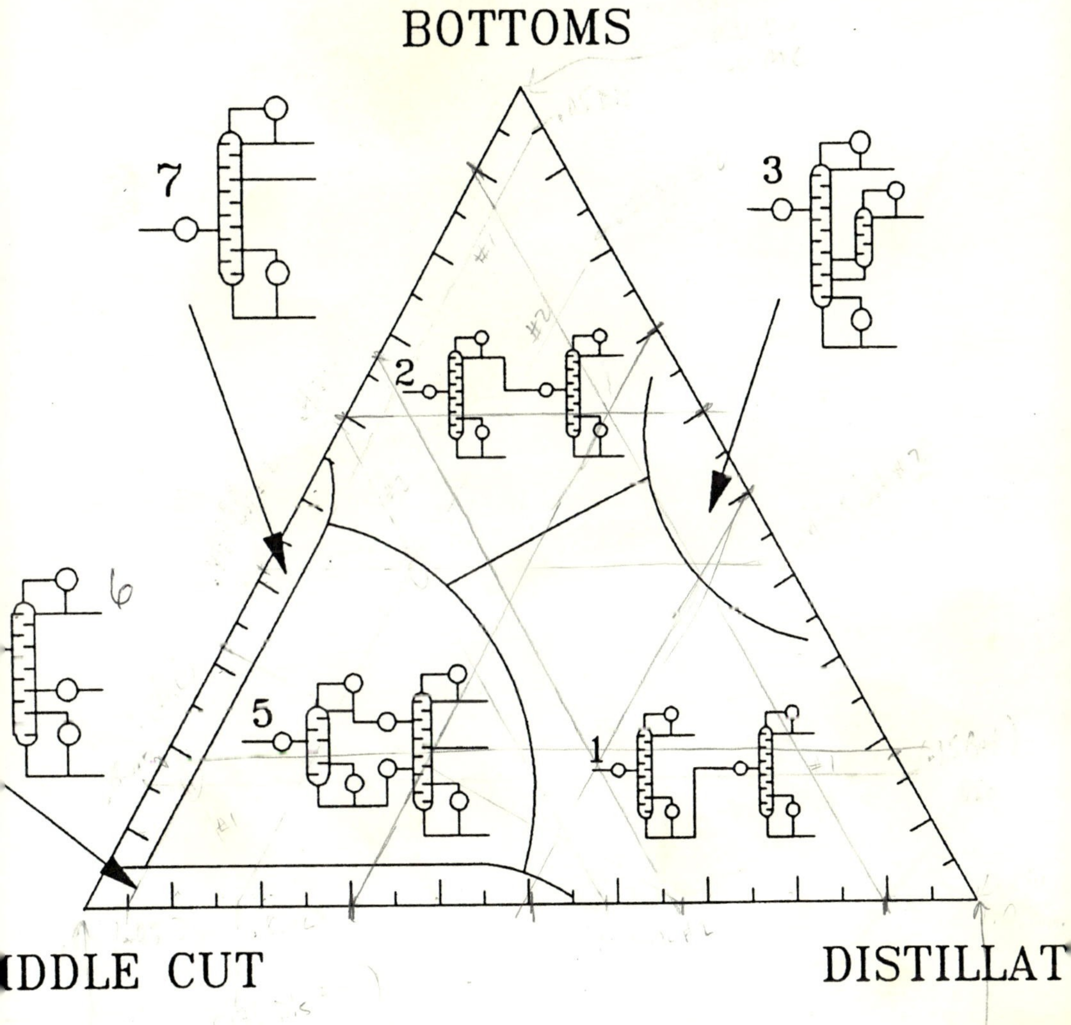

Figure 12.7: Regions of optimality for $ESI < 1.6$.

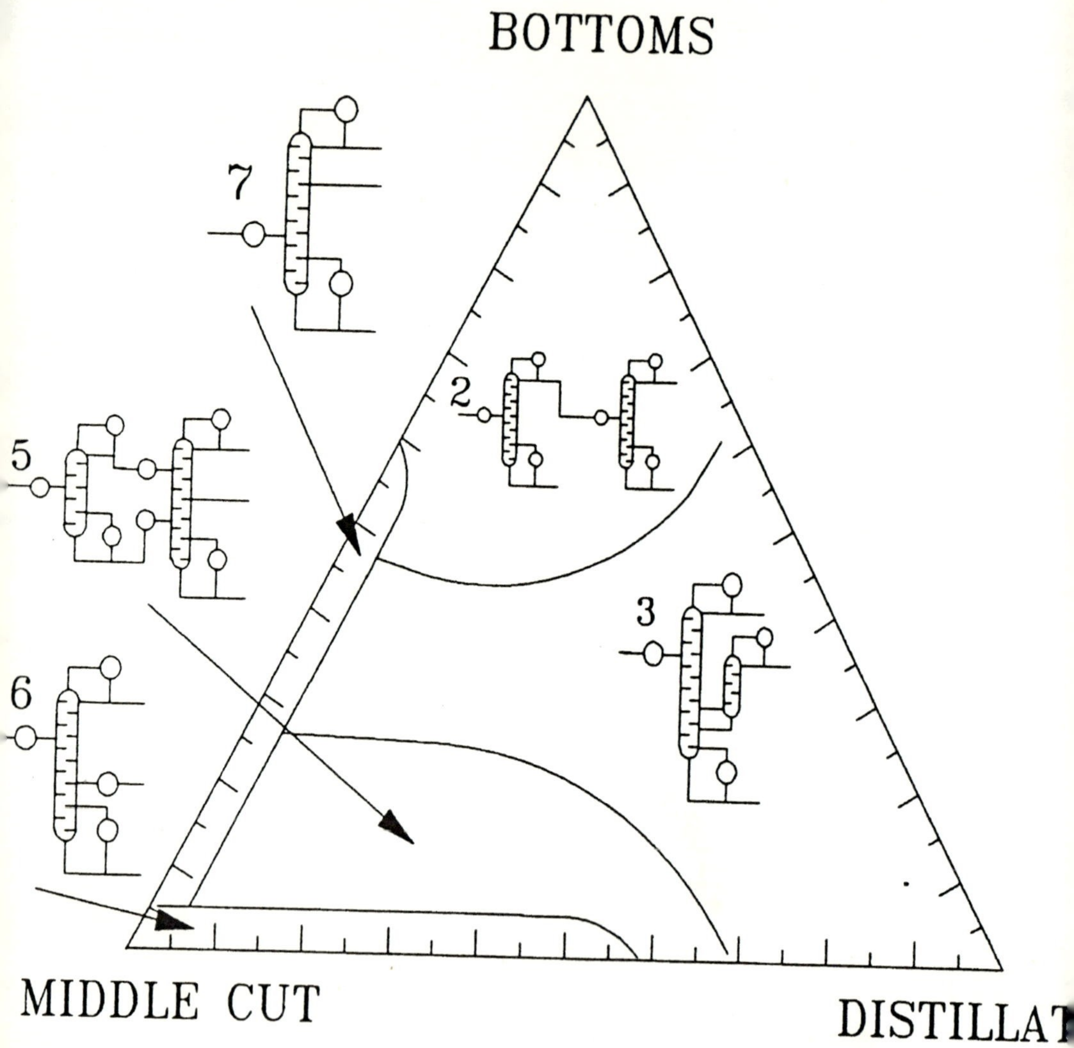

Figure 12.8: Regions of optimality for $ESI \geq 1.6$.

The following heuristics are based upon the Ease of Separation Index (ESI) which is defined as:

$$ESI(A/B/C) = \frac{K_A K_C}{K_B^2} \tag{12.4}$$

and

$$K_i = \frac{y_i}{x_i} \tag{12.5}$$

where the K values are the vapor/liquid equilibrium constants.

The ESI is simply a measure of the relative ease with which two different separations can be made. If the ESI is greater than unity, then the A/B split is easier than the B/C split. Otherwise, the B/C split is easier. These heuristics can also be applied to more than three component mixtures by defining pseudo ternary mixtures. For example, if the AB/CDEF/GH splits are being considered, then the ESI should be calculated from the ratio of the B/C relative volatility divided by the F/G relative volatility in order to generate a flowsheet performing those particular separations.

For a system of more than three components, an evaluation of the relative volatilities of neighboring components should be done. The first two splits should be performed at the positions with largest relative volatilities, consistent with the heuristic, "Do the easy separation first." If one separation is very much easier than all the rest, then it may be considered alone to generate an initial separation using a simple tower. Otherwise, the easiest separations may be considered two-at-a-time, together with Figures 12.7 and 12.8, to generate alternative separation networks. Since heuristics are being used to generate alternatives, subsequent more careful economic comparisons are needed to distinguish between those alternatives that are generated. The concept of using heuristics in this fashion is to generate a few likely alternatives from among the many possibilities, and then to use standard engineering principles to guide the selection among these alternatives.

Example: 12.1 *Given a 5 component mixture A to E, and K values of (3.2, 3.1, 1.8, 1.6, 1.0) respectively, synthesize a flowsheet using the* ESI *and Task Solution Heuristic 6, "Do the easy separation first." The feed composition for components A to E is (0.1, 0.1, 0.3, 0.4, 0.1).*

Solution: 12.1 *Compute the relative volatilities and choose the two easiest separations first.*

Species	x_i	K_i	$\alpha_{i,i+1}$
A	*0.1*	*3.2*	
			1.03
B	*0.1*	*3.1*	
			1.72
C	*0.3*	*1.8*	
			1.12
D	*0.4*	*1.6*	
			1.60
E	*0.1*	*1.0*	

Since the B/C and the D/E splits are the easiest, then $ESI = 1.72/1.6 = 1.08$. *The pseudo ternary composition of the feed is* $(x_{AB}, x_{CD}, x_E) = (0.2, 0.7, 0.1)$. *Since the* $ESI < 1.6$, *use Figure 12.7 to select the best design based on the amounts of bottoms, middle cut, and distillate produced. The pseudo feed composition,* $(0.2, 0.7, 0.1)$, *corresponds to the estimated amounts of distillate, middle cut, and bottoms product, respectively. In this case, the feed composition and the ESI favor Design 5 to complete the AB/CD/E separation first. The A/B and the C/D splits can then be completed using simple towers.*

Example: 12.2 *Consider the mixture of A-G with mole fractions and vapor-liquid K-values tabulated below. Use this information to synthesize two alternative methods for separating this mixture using distillation. Provide a brief justification of why these alternatives are likely to be attractive.*

Species	x_i	K_i	$\alpha_{i,i+1}$
A	*0.05*	*22.5*	
			1.957
B	*0.10*	*11.5*	
			1.237
C	*0.15*	*9.3*	
			1.134
D	*0.15*	*8.2*	
			1.108
E	*0.25*	*7.4*	
			2.387
F	*0.15*	*3.1*	
			1.240
G	*0.15*	*2.5*	

Solution: 12.2 *The A/B and E/F are the easiest two splits, based on relative volatilies. The ESI for this split is* $ESI = 1.957/2.39 = 0.82 < 1.6$, *so use Fig-*

ure 12.7 to select the most attractive design. The pseudo-ternary feed composition is $(x_d, x_m, x_b) = (0.05, 0.65, 0.30)$. *Using Figure 12.7, this feed composition falls close to the intersection line for Designs 5 and 7. So the economic differences between these two are small and either should be considered for the initial split.*

Since the BCDE mixture is generated as the middle cut in either case, the ESI heuristics should be used to generate the best way of separating that mixture. Based on relative volatilities, the B/C and C/D splits should be done next. So doing the (B/C/DE) split gives a pseudo-ternary feed composition of $(x_d, x_m, x_b) = (0.15, 0.23, 0.62)$ *and the ESI is,* $ESI = 1.237/1.134 = 1.09 < 1.6$, *so again use Figure 12.7 to select the best design. Based on the pseudo-ternary feed composition, Design 2 is favored. Since this second feed composition is not near any of the intersection lines, it is the only one that should be considered for this second separation.*

There are two sequences that should be considered further. First, consider using Design 5 to do the A/BCDE/FG split, followed by Design 2 to do the B/C/DE split, with simple towers to do the D/E and F/G splits. Second, consider using Design 7 to do the A/BCDE/FG splits, followed by Design 2 to do the B/C/DE split, with simple towers to do the D/E and F/G splits. This approach selects to alternative sequences out of 7,782 alternatives (see Table 12.1). Also, four of the towers in either sequence are the same which further reduces the amount of work in comparing these two alternatives.

Example: 12.3 *A hydrocarbon mixture consists of species A-H. It has the following flow rates in lb-mol/min for species A-H in this mixture: 5, 5, 9, 17, 16, 23, 10, 15. The Pure component vapor pressures for species A-H in kPa at 150 °C are: 140, 115, 110, 101, 100, 92, 88, 73. Use the Complex Tower (ESI) Heuristics to identify four likely candidate sequences for further consideration.*

Solution: 12.3 *Assume Raoult's law to calculate relative volatilities.* $K_i = P_i^{sat}/P$, $\alpha_{i,j} = P_i^{sat}/P_j^{sat}$. *Calculate the feed mole fractions and tabulate the results.*

Component	x_i	P_i^{sat}	$\alpha_{i,i+1}$
A	0.05	140	
			1.217
B	0.05	115	
			1.045
C	0.09	110	
			1.089
D	0.17	101	
			1.01
E	0.16	100	
			1.087
F	0.23	92	
			1.045
G	0.10	88	
			1.205
H	0.15	73	

The A/B and G/H splits are much easier than the others, based on relative volatilities, so they should be done first. The pseudo-ternary composition for the A/BCDEFG/H splits are $(x_d, x_m, x_b) = (0.05, 0.80, 0.15)$ *and the ESI is,* $ESI = 1.217/1.205 < 1.6$, *so use Figure 12.7 to identify the tower options. In this instance, the pseudo-ternary feed composition lies close to the line dividing the regions of optimality for Designs 5 and 7, so the cost differences between these two towers are small and either should be considered for this separation.*

The two next easiest splits are the C/D and E/F splits, both of which have relative volatives greater than 1.08 for the keys. So the BC/DE/FG splits should be considered next. In this instance the pseudo-ternary feed composition is $(x_d, x_m, x_b) = (14/80, 33/80, 33/80) = (0.175, 0.4125, 0.4125)$ *and the ESI is,* $ESI = 1.089/1.087 < 1.6$, *so use Figure 12.7 to identify the options. The pseudo-ternary feed composition in this case lies close to the line dividing the optimal regions for Designs 5 and 2, so either one of these designs may be preferred. The four alternatives sequences are therefore:*

(A/BCDEFG/H with 5) + (BC/DE/FG with 2) + (B/C + D/E + F/G with simple towers)

(A/BCDEFG/H with 7) + (BC/DE/FG with 2) + (B/C + D/E + F/G with simple towers)

(A/BCDEFG/H with 5) + (BC/DE/FG with 5) + (B/C + D/E + F/G with simple towers)

(A/BCDEFG/H with 7) + (BC/DE/FG with 5) + (B/C + D/E + F/G with simple towers)

So the ESI heuristics suggest that these four alternative sequences, out of 52,239 possibilities, should be considered for further evaluation and, perhaps, detailed simulation, optimization, and cost estimation.

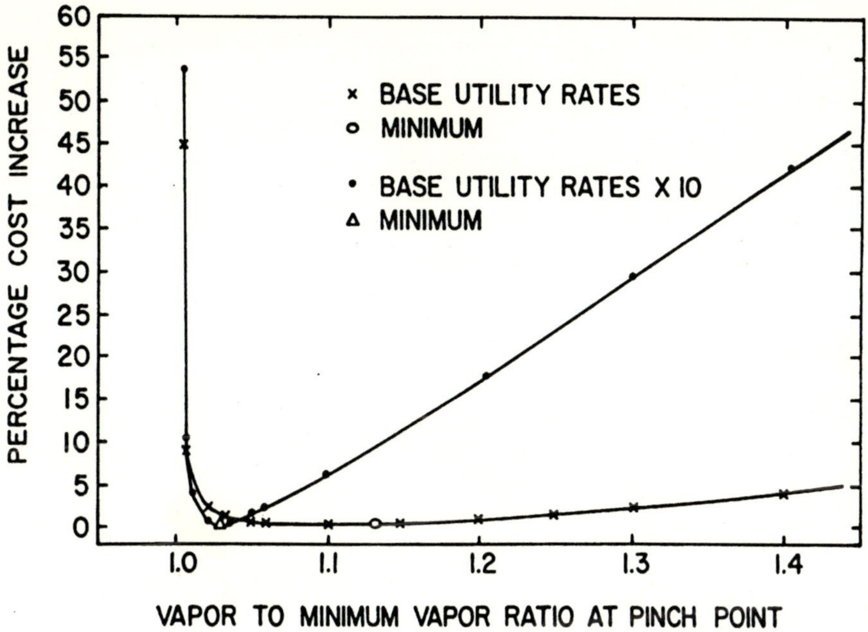

Figure 12.9: Effect of the vapor to minimum vapor ratio on the relative venture costs.

12.4 Optimal Distillation

A simple distillation tower receiving a single feed and producing two product streams was studied by calculating the effects of pressure, feed enthalpy, and vapor rate on its venture cost. In this example, the tower is separating a feed containing equal amounts of iso and normal butanes. Each of the three optimization variables was then varied parametrically while the others were set at their optimal values. In this manner, Figures 12.9-12.12 were generated. The estimated costs are presented as percentages of venture cost increase, relative to the minimum venture cost and arbitrarily scaled as a percentage of $250,000 in annual venture cost. For the low utility scale costs:

$$\% = (VC - \$140,377)/2500 \tag{12.6}$$

and for the higher utility costs:

$$\% = (VC - \$430,780)/2500 \tag{12.7}$$

so that the slopes of the curves in Figures 12.9- 12.12 are in comparable percentage points, although they have different references points. The range of utility costs that were assumed for the low and high cases is summarized in Table 12.3.

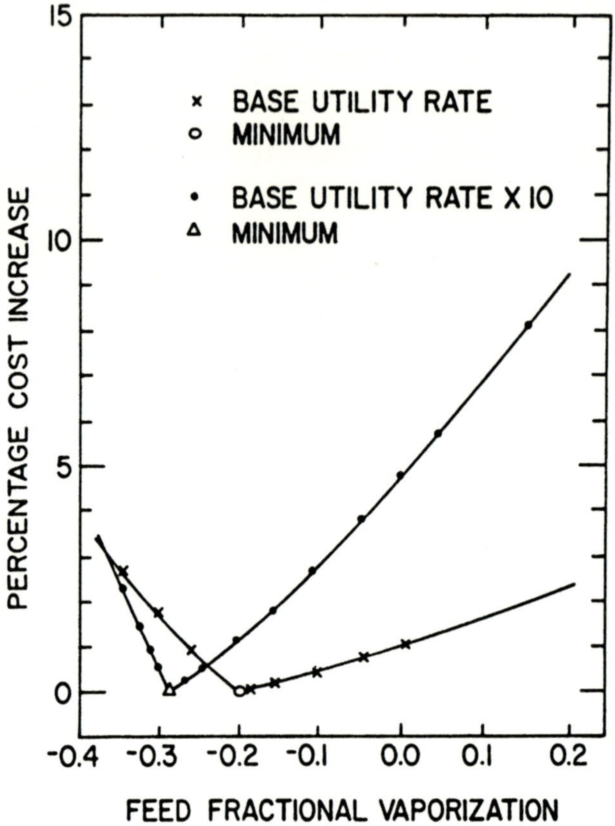

Figure 12.10: Effect of the feed fractional vaporization on the relative venture costs.

Table 12.3: Range of Utility Costs Used in the Parametric Studies

Utility	$T(^oK)$	Low	High
Steam($/GJ)	504	1.03	10.30
	442	0.77	7.70
	418	0.57	5.70
	389	0.42	4.20
Ammonia($/GJ)	274	1.65	16.50
	255	2.97	29.70
	246	3.96	39.60
Cooling Water(cents/m^3)	297	0.66	6.60

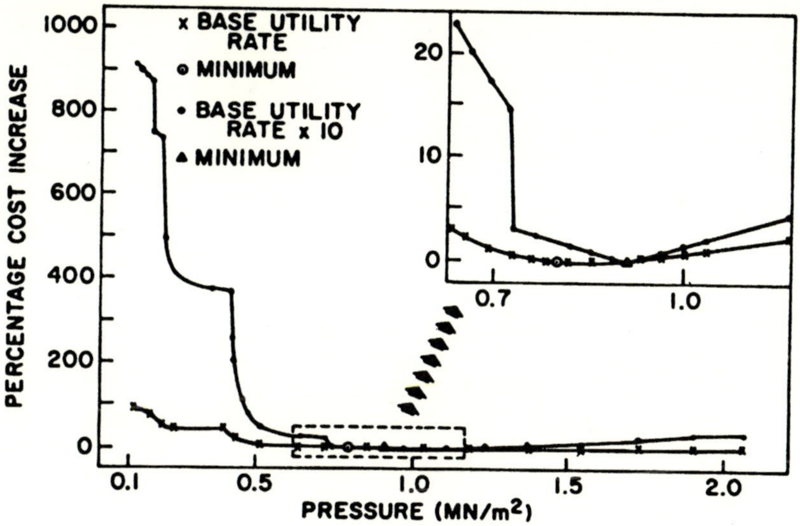

Figure 12.11: Effect of the operating pressure on the relative venture cost of a distillation tower.

Several effects can be seen from this study. First, the utility costs are seen to affect both the magnitude of the minimum venture cost and the location of the optimal point. Although the apparent shifts are small, they seem to be significant. Secondly, when the pressure variable is adjusted, then discontinuities appear in the venture cost surface. This effect results from changing the choice of utilities as the pressure changes. In particular, as the higher pressures are used, then warmer refrigerants can be used in the condenser and these changes result in large cost saving which appear as surface discontinuities in Figure 12.9. This latter effect gives rise to the existence of local minima on the response surface and illustrates the wisdom of the heuristic *aim at high pressures, rather than low.*

Heuristic: 28 *Avoid excursions in temperature and pressure, but aim high rather than low.*

The tower operating pressure does have a pronounced effect on venture cost. In fact, this effect is so strong that a reasonable correlation exists between the optimal overhead operating pressures and the normal feed bubble point as shown in Figure 12.12. For light hydrocarbons (*i.e.* $C_3 - C_8$), the optimal pressure is correlated using:

$$\ln P = \frac{3152}{T} - 1.799 \qquad (12.8)$$

where: P = optimal pressure in psia, $1 \leq P \leq 1000$ psia
T = tower feed normal boiling point in R, $362 < T \leq 1750$ R

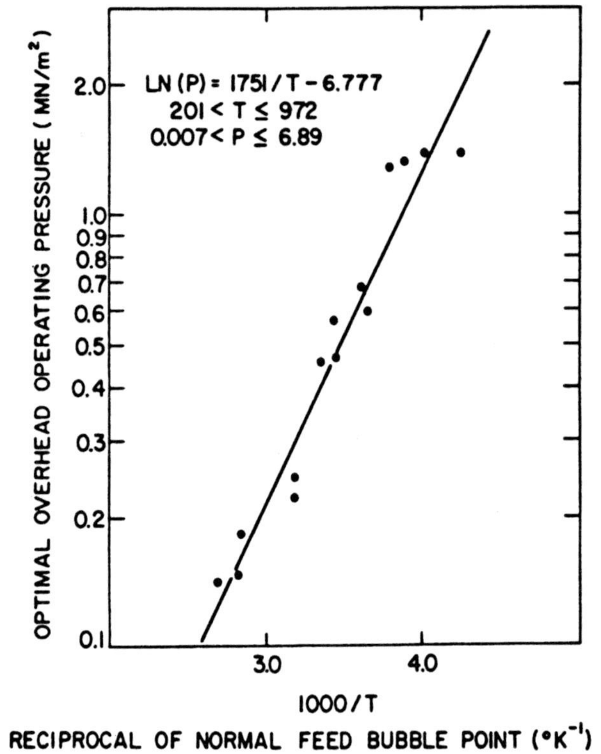

Figure 12.12: Effect of the normal bubble point pressure on the optimal tower operating pressure.

If Equation 12.8 is used for ethylene/ethane mixtures, it predicts optimal operating pressure above the critical pressures. These conditions are not feasible since the column must operate within the vapor/liquid two phase region on the pressure/temperture diagram. However, the optimal operating pressures in the case of ethylene and propylene purification are dominated by the refrigeration costs and they should be operated at conditions near the critical point. The optimal tower operating pressures are well predicted using a correlation [102] of the form:

$$\ln P^* = 2.677 - \frac{309.4}{T_{fb}} + \frac{1147.6}{T_{ob}} + \frac{1087.4}{T_{bd}} \tag{12.9}$$

with a standard deviation of± 15.3 psia for 10 observations.

where: P^* = the optimal overhead operating pressure for ethylene/propylene towers in psia

T_{bd} = the tower bottom product mixture dew point temperature in R evaluated at a total pressure of 485 psia

T_{fb} = the tower feed mixture bubble point temperature in R evaluated at a total pressure of 485 psia

T_{ob} = the tower overhead product mixture bubble point in R evaluated at a total pressure of 485 psia

These correlations are useful heuristics to guide other synthesis studies. In the case of grass roots design, they provide initial estimates of the best operating pressures for new towers.

Two algorithmic methods are discussed here — dynamic programming and ordered-branch search. Dynamic programming can also be used to search a tree of alternatives in their rank order (i.e., first finding the apparent best, then the apparent second best, etc). So the computed results are ranked, say by increasing cost. Ordered-branch search is not as selective. The ordered-branch search method is a tree searching algorithm that utilizes branch and bound to prune the alternatives, but alternatives are not generated in rank order.

First we describe how to generate an optimal base case using dynamic programming. As originally described by Bellman, dynamic programming (DP) was used to optimize a single structure with respect to its operating state. Later, investigators learned how to use DP to synthesize optimal structures.

12.5 Dynamic Programming

Dynamic programming is based upon the principle of optimality as stated by Bellman [8]:

> An acyclic system is optimized when its downstream components are suboptimized with respect to the feed they receive upstream.

For a chemical process consisting of serial unit operations, dynamic programming is applied as suggested in Figure 12.13. First the downstream subsystem 1 is optimized with respect to the feed it receives. Then subsystems 1 and 2 are suboptimized

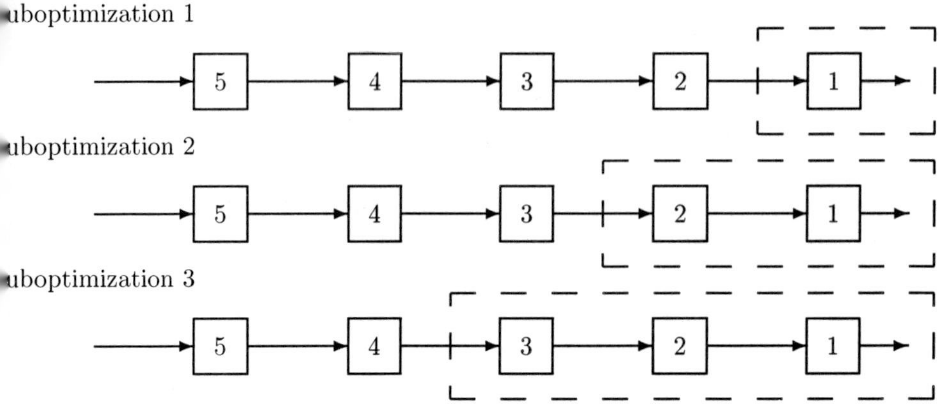

Figure 12.13: Use of dynamic programming in optimizing a serial process. The optimization direction is opposite to that of material flow. Downstream unit operations are suboptimized first.

together with respect to the feed that subsystem 2 receives. Then subsystems 1, 2, and 3 are suboptimized together with respect to the feed that subsystem 3 receives and so forth.

These suboptimization steps are repeated until the original feed to subsystem 5 in Figure 12.13 is reached. After subsystems 1–5 are optimized with respect to the feed received by subsystem 5, then the entire system is optimized. For each suboptimization, the feed conditions are varied over an anticipated range of values and the subsystem is optimized with respect to that feed.

When the feed to subsystem 1 is varied, the costs of subsystem 1 are computed and saved. This analysis yields a plot of the subsystem 1 costs as a function of the feed to subsystem 1. That is, Suboptimization 1 yields:

$$VC_1 = VC_1(F_1) \tag{12.10}$$

where: F_1 = is the subsystem 1 feed
VC_1 = the cost of subsystem 1 for a given value of F_1

The goal of Suboptimization 1 is:

$$VC_1' = \min_{F_1} \{VC_1(F_1)\} \tag{12.11}$$

During Suboptimization 2, the dynamic programming principle is used. This means that the results of Suboptimization 1, VC_1', are used in Suboptimization 2. Mathematically:

$$VC_2 = VC_2(F_2) \tag{12.12}$$

$$F_1 = g_2(F_2) \tag{12.13}$$

so that the value of F_1 is a dependent variable of F_2. When F_2 is specified, then F_1 is also fixed. For Suboptimization 2, the objective function is:

$$VC_2'(F_1, F_2) = \min_{F_1, F_2} \{VC_2(F_2) + VC_1(F_1)\} \tag{12.14}$$

subject to the feed constraint expressed as Equation 12.13. However, since the downstream subsystem 1 was optimized in Suboptimization 1, this result can be used during Suboptimization 2:

$$VC_2'(F_2) = \min_{F_2} \{VC_2(F_2) + VC_1'[g_2(F_2)]\} \tag{12.15}$$

During Suboptimization 3, the cost of subsystem 3 is:

$$VC_3 = VC_3(F_3) \tag{12.16}$$

and there are two feed constraints:

$$F_2 = g_3(F_3) \tag{12.17}$$

$$F_1 = g_2(F_2) \tag{12.18}$$

and the objective function is:

$$VC_3'(F_3) = \min_{F_3} \{VC_3(F_3) + VC_2'[g_3(F_3)]\} \tag{12.19}$$

where the results of Suboptimization 1 and 2 are incorporated into Suboptimization 3.

Because of the serial nature of the flowsheet, the feeds F_2 and F_3 are both specified when F_3 is specified since optimal decisions are always made downstream using the results generated during the previous suboptimizations. Hence, Bellman's principle of optimality is used during each step of the optimization.

Dynamic programming reduces the optimization in Figure 12.13 to a sequential optimization consisting of five, single dimensional suboptimizations. Alternatively, the entire process may be optimized as a single, five dimensional search subject to the constraints:

$$F_{i-1} = g_i(F_i) \tag{12.20}$$

$i = 2, 3, \ldots, 5$.

When subsystem 2 in Figure 12.13 is optimized with respect to its feed, the feed to subsystem 2 is varied over a range of values, but the objective function to be minimized in this case is the value obtained for subsystem 2 plus the value previously obtained for subsystem 1. Using the previously calculated optimal values for subsystem 1 as a function of its feed and generating them from a lookup table, a new lookup table is formed that consists of the sum of the costs for subsystems 1 and 2 as a function of the feed to subsystem 2.

This process is then repeated for subsystems 3, 4, *etc.* until the entire flowsheet is optimized with respect to the initial feed. Under these conditions, Bellman's principle guarantees that the optimal conditions for the entire flowsheet will result since this procedure accounts for any interactions that may exist between the subproblems in the sequence.

Figure 12.14: Example process consisting of a mixer followed by a heat exchanger

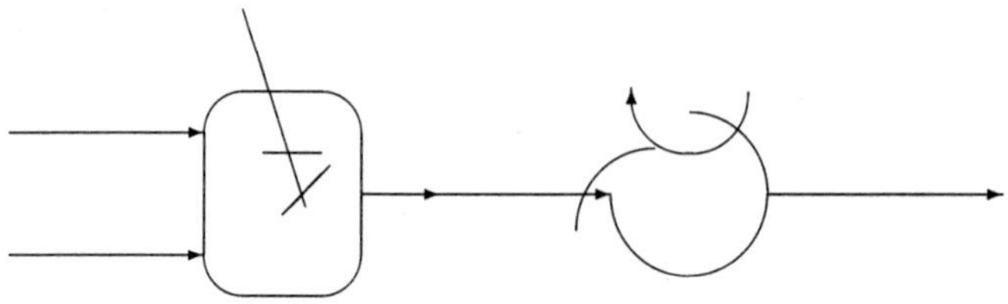

12.5.1 A Simple Example

Consider a chemical process battery limits operation that consists of only two unit operations in series: a mixer followed by a heat exchanger that are serially connected as shown in Figure 12.14. If two types of mixers are under consideration (say types a and b) and two types of heat exchangers (say types c and d), then one may define four subproblems. Two subproblems define mixers while the remaining two define heat exchangers. The design engineer can then combine or aggregate these two unit operations in $2(2) = 4$ different ways to synthesize four alternative processing structures.

In this example, each of the structures that may be synthesized by aggregation solves the original design problem (*i.e.* a serial configuration consisting of a mixer followed by a heat exchanger) and each structure is unique in that at least part of it is different from the other structures which may be generated. The alternatives can also be enumerated as the *ac, ad, bc* or *bd* structure.

Now the design engineer wants to select that alternative which maximizes the overall plant profitability. Therefore, the engineer is concerned not only with the costs of building and operating the battery limits, but also with the costs of building and operating the auxiliaries to support the battery limits process. Also suppose that a chemical reaction occurs in the heat exchanger and that the extent of this chemical reaction is a function of the mixing rate, m and the residence time τ_m in the mixer. In addition, suppose that the extent of reaction depends upon the residence time in the exchanger τ_e and the process stream exit temperature T from the heat exchanger. Then the selection problem may be represented as:

$$\max_{m,\tau_m,T,\tau_e} \{P_{ac}, P_{ad}, P_{bc}, P_{bd}\} \tag{12.21}$$

$$m_{min} < m < m_{max}, \quad \tau_{m,min} < \tau_m < \tau_{m,max}$$
$$T_{min} < T < T_{max}, \quad \tau_{e,min} < \tau_e < \tau_{e,max}$$

where the P_i is the estimated plant profitability of the *ith* structural alternative where $i = ac, ad, bc, bd$. Moreover, the profitability is a function of the four optimization variables which are constrained between minimum and maximum values for each, and the profitability depends upon the costs of the auxiliary systems supporting the battery limits structures.

Now this problem may be decomposed by fixing the material balances between the two unit operations and at the heat exchanger outlet and by fixing the utility costs using an internal pricing schedule (also see Section 6.2). For example, suppose that m and τ_m are constrained so that essentially perfect mixing is always achieved at the mixer outlet (*i.e.* the heat exchanger feed). Also assume that the heat exchanger is always operated so as to achieve the same degree of conversion. In this case, the system can be represented by four composition nodes as shown in Figure 12.15 and the optimization and synthesis problem is reduced to minimizing the venture costs sequentially for each the subproblems. For the mixers:

$$\min_{m,\tau_m,} \{C_i\} = \min_{m,\tau_m} \{\alpha I_i + \beta \lambda_{i,e} q_{i,e}\} \tag{12.22}$$

$$i = a, b, \quad m_{min} < m < m_{max}, \quad \tau_{m,min} < \tau_m < \tau_{m,max}$$

and I_i is the total capital investment for the mixer system only, $\lambda_{i,e}$ is the internal price of electricity, and $q_{i,e}$ is the demand for electricity by the mixer. The parameters α and β are constants that depend upon the type of economic criterion that is selected (*e.g.* venture cost, present cost, *etc.*).

For the heat exchangers, the problem reduces to:

$$\min_{T,\tau_e} \{C_j\} = \min_{T,\tau_e} \{\alpha I_j + \beta \lambda_{j,s} q_{j,s}\} \tag{12.23}$$

$$j = c, d, \quad T_{min} < T < T_{max}, \quad \tau_{e,min} < \tau_e < \tau_{e,max}$$

and I_j is the total capital investment for the heat exchangers only, $\lambda_{j,s}$ is the internal price of steam, and $q_{j,s}$ is the demand for steam by the heat exchanger.

Moreover, the synthesis problem reduces to:

$$\min_{i,j} \{C_i + C_j\} \tag{12.24}$$

$$i = a, b \quad j = c, d$$

Hence the decomposition procedure has decoupled the state optimization and the process synthesis steps in this example because the intermediate stream compositions and their characteristics (*i.e.* perfect mixing) were fixed.

Using dynamic programming [91], one may use the subproblem venture cost data base (C_a, C_b, C_c, C_d) to generate the four alternatives in their approximate rank order of increasing costs. Suppose, for example, that the minimum cost of

Figure 12.15: Composition node representation of the example process

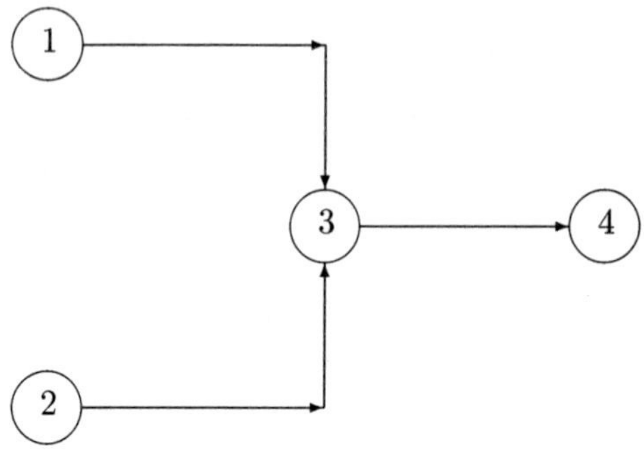

each after state optimization is given by ($C_a = 4, C_b = 7, C_c = 8, C_d = 10$), then the approximate rank order will be:

$$\begin{gathered} C_a + C_c < C_a + C_d < C_b + C_c < C_b + C_d \\ 4 + 8 < 4 + 10 < 7 + 8 < 7 + 10 \\ ac < ad < bc < bd \end{gathered}$$

Moreover, the engineer may choose to relax the initial assumptions at this point by considering imperfect mixing, the advantages of process integration with other unit operations, and the overall utility management costs for the plant. This rigorous optimization problem is more general than the one stated as the maximization objective, Equation 12.21, and the resulting model is more complex than the minimization functions, Equations 12.22 and 12.23.

However, because of the rank order that has already been established through the approximate analysis, the logical procedure will be to first optimize the *ac* combination, then the *ad* combination, *etc.* In each of these integrated optimizations, all process and utility interactions and integration possibilities will be completely evaluated. Moreover, if the economic interaction and integration effects are small, then the final rigorous rank order will be the same as above. Otherwise, one of the other alternatives may be the overall optimum structure while *ac* is only the approximate optimum structure.

In this simple example, the sequential approach described above probably represents much more work than the development of a superstructure [108, 89] and completing the synthesis and multivariate state optimization tasks simultaneously. However, the general approach is most useful for synthesizing structures involving thousands of subproblems. Moreover, each subproblem may involve as many as twenty or more degrees of freedom in its state state optimization vector and literally millions of alternative structures may be synthesized by aggregation. In the types of structures that are of interest, the dynamic programming method may be the only practical method that enables one to both screen millions of alternatives and rigorously account for subproblem interactions, process integration, and overall utility management in the plant.

12.5.2 Sequence Synthesis

Dynamic programming (DP) may be applied to the synthesis of distillation systems by treating the alternative subsystems as independent building blocks that are aggregated to obtain the optimal sequence. If the subsystems which have been defined are truly independent, then this approach guarantees an optimal flowsheet. On the other hand, if interactions exist between the subsystems or towers in the sequence, then the resulting network may not be optimal. However, even in this latter case it is possible to account rigorously for the interactions between the towers by starting with the downstream subsystem and optimizing the sequence in the direction opposite to that of material flow.

In summary, an exact application of dynamic programming to the synthesis of distillation sequences consists of the following steps:

1. Optimize the downstream towers as a function of their feed pressures.
2. Minimize $\sum_i^{N_{sub}} VC_i$ for the downstream towers in the sequence at that point.
3. Suboptimize and synthesize in the direction which is reverse to that of material flow.

On the other hand, an approximate DP solution to the distillation sequencing problem is given by:

1. Suboptimize each possible tower subproblem once using an arbitrarily chosen feed pressure. (This step ignores interactions between the towers.)
2. Synthesize the optimal network using the minimum costs for each of the subproblems that have been considered.

If the interactions between towers in a sequence are negligible, then each of these methods will give the same answer. Otherwise, the two methods may either yield the same sequence, but with slightly different optimal states, or else they may yield different sequences altogether. Since experience with the approximate method suggests that the interactions between distillation towers are often weak, the approximate method is recommended, but it is subject to failure.

The DP approach to synthesizing separation networks begins with the downstream towers. Hendry and Hughes [43] showed how a list splitting operation could be used together with dynamic programming [8] to synthesize separation networks. In this method the components in the original feed mixture are first ranked according to some physical property. In the case of straight distillation, decreasing volatility is a convenient ranking, but the method is applicable for any ranking of separation factors in general. Hendry and Hughes also showed that the method can easily be applied to extractive distillation. If the ranking changes, say with pressure, then both rank lists can be formed and the pressure optimization appropriately constrained. So the method is quite general.

For any one ranked list, L_N, so formed for an N component mixture, two contiguous rankings can be formed from it by deleting either the first or the last component from L_N. This process can be continued until $N-1$ lists, each containing only two contiguous components, are formed. In this manner all the subproblems derived from performing sharp cuts only are represented by the subgroup lists.

Once the subproblems for all lists have been identified, then each subproblem is state optimized. The structural optimization is performed by working backwards using the subproblems as building blocks to synthesize the structure having the minimum venture cost. Every three component mixture, for example, can be split in two ways. Each of these sharp splits between adjacent components will require one additional two component split.

Let VC_{ABC} represent the venture cost of generating product streams A, B, and C with two simple splits from a ternary feed. Then, comparing only the direct and inverted simple sequences:

$$VC_{ABC} = \min\{VC_{A/BC} + VC_{B/C}, VC_{AB/C} + VC_{A/B}\} \tag{12.25}$$

Similarly:

$$\begin{aligned} VC_{ABCD} &= \min\{VC_{A/BCD} + VC_{BCD}, VC_{ABC/D} + VC_{ABC}, \\ &\quad VC_{AB/CD} + VC_{A/B} + VC_{C/D}\} \end{aligned} \tag{12.26}$$

so that by working backwards in this fashion one can generate the optimal structure for separating N components using knowledge of the optimal structures for separating $N-1$, $N-2, \ldots$,3 and 2 components. The extension to consider all of designs in Appendix D by this method is straightforward.

12.5.3 DP and the $F/\Delta T$ Heuristic

Dynamic programming may be used to synthesize distillation sequences if the venture costs have been calculated for a collection of tower subsystems. It may also be used by applying Task Solution Heuristic 11 that was described in earlier in this chapter. That is, where the goal is:

$$\min_{State} \sum_{i=1}^{N_{sub}} \frac{F_i}{\Delta T_i} \tag{12.27}$$

where F_i = the feedrate to the ith subproblem
ΔT_i = the boiling point difference between the light and heavy keys in the ith tower (or subproblem)

Using the Task Solution Heuristic 11, the goal is to minimize $\sum F_i/\Delta T_i$'s over all possible sums. This can be easily accomplished by first minimizing the sums over all possible binary towers, then over all towers separating three components, then over all towers with four component feeds, etc. In other words, the dynamic programming principle is again applied, but starting with the downstream towers first.

Table 12.4: Two component subproblems.

Split	$F_i/\Delta T_i$	Tower Reference
A/B	0.101	C-2, D-3
B/C	0.103	B-3
C/D	0.047	A-3

Table 12.5: Three component subproblems.

Split	$F_i/\Delta T_i$	Tower Reference
A/BC	0.118	E-2
AB/C	0.123	D-3
B/CD	0.104	A-2
BC/D	0.263	B-2

12.5.4 DP Solution to the Decane Chlorination Example

Table 12.2 summarizes the species feedrates, boiling point temperatures, and ΔT_i values for this case. Using the data in Table 12.2, two cases will be examined.

Case 1: Four Components Separated

In this case, only the minimum four cuts will be considered. Here A = HCl, B = Cl_2 + decane, C = monochlorodecane (MCD), and D = dichlorodecane (DCD). In applying the method, the first step is to generate a set of subproblem lookup tables, starting with the binary mixtures. For this example, the required tables are shown as Tables 12.4-12.6.

Table 12.6: Four component subproblems.

Split	$F_i/\Delta T_i$	Tower Reference
A/BCD	0.119	A-1, B-1
AB/CD	0.124	C-1
ABC/D	0.313	D-1, E-1

Table 12.7: Subproblem costs.

Mixture	Split	Number	Cost k$/yr
ABCD	A/BCD	1	85
	AB/CD	2	254
	ABC/D	3	510
ABC	A/BC	8	59
	AB/C	9	197
BCD	B/CD	4	247
	BC/D	5	500
AB	A/B	6	15
BC	B/C	10	190
CD	C/D	7	420

The information in the lookup tables is then used as indicated in Figure 12.16.

Case 3: Five Splits Allowed

In this case, the separation of decane and chlorine is also permitted. Although it was not originally required from the problem statement, it is considered here as an alternative, and A = HCL, B = Cl_2, C = DEC, D = MCD, and E = DCD.

Using the same data as shown in Tables 12.4-12.6, and generating a similar table for the five component subproblems, a minimum value for $\phi(ABCDE) = 0.199$, the sum of the $F_i/\Delta T_i$, values is obtained. Since this value for ϕ is less than that value obtained in Case 1 (*i.e.* $\phi(ABCD) = 0.270$), the application of Task Solution Heuristic 11 and dynamic programming to this example suggests that savings will be realized by performing the additional chlorine/decane split even though it is not required. The optimal sequence for this case is shown as sequence F in Figure 12.6.

Example: 12.4 *Consider the separator costs which are presented in Table 12.7. Use dynamic programming to synthesize the optimal separation sequence. The alternatives for the ABC mixture are:*

Solution: 12.4

A/BC + B/C = 59 + 190 = 249
AB/C + A/B = 197 + 15 = 212
Hence the AB/C + A/B sequence is preferred.

The alternatives for the BCD mixture are:
B/CD + C/D = 247 + 420 = 667
BC/D + B/C = 500 + 190 = 690
Hence the B/CD + C/D sequence is preferred.

For the four component mixture, the principle of optimality is applied and the sequence choice is based on the optimal decisions for the downstream towers. The

Figure 12.16: Dynamic programming solution to the four component decane chlorination problem.

Note: $\phi_i = F_i/\Delta T_i$ Also see ϕ_i values in Tables 12.4–12.6

$\phi(ABC) = \min\{\phi(A/BC) + \phi(B/C),\ \phi(AB/C) + \phi(A/B)\}$
$\phi(A/BC) + \phi(B/C) = 0.118 + 0.103 = 0.221$
$\phi(AB/C) + \phi(A/B) = 0.123 + 0.101 = 0.224$
$\phi(ABC) = \min\{0.221,\ 0.224\} = 0.221$

$\phi(BCD) = \min\{\phi(B/CD) + \phi(C/D),\ \phi(BC/D) + \phi(B/C)\}$
$\phi(B/CD) + \phi(C/D) = 0.104 + 0.047 = 0.151$
$\phi(BC/D) + \phi(B/C) = 0.263 + 0.103 = 0.366$
$\phi(BCD) = \min\{0.151,\ 0.366\} = 0.151$

$$\begin{aligned}\phi(ABCD) &= \min\{\phi(A/BCD) + \phi(BCD), \\ &\quad \phi(AB/CD) + \phi(A/B) + \phi(C/D),\ \phi(ABC/D) + \phi(ABC)\}\end{aligned}$$

$\phi(A/BCD) + \phi(BCD) = 0.119 + 0.151 = 0.270$
$\phi(AB/CD) + \phi(A/B) + \phi(C/D) = 0.124 + 0.101 + 0.047 = 0.272$
$\phi(ABC/D) + \phi(ABC) = 0.313 + 0.221 = 0.534$
$\phi(ABCD) = \min\{0.270,\ 0.272,\ 0.534\} = 0.270$

Therefore, the A/BCD, B/CD, C/D (the direct sequence) is favored.

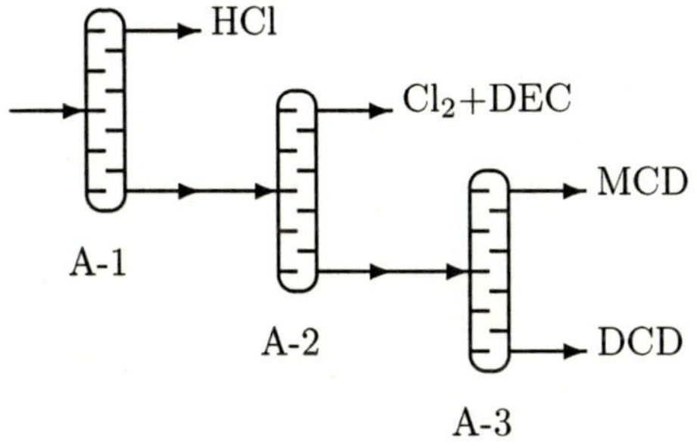

alternatives are:
A/BCD + B/CD + C/D = 85 + 667 = 752
AB/CD + A/B + C/D = 254 + 15 + 420 = 689
ABC/D + AB/C + A/B = 510 + 212 = 722
Hence the optimal sequence consists of towers AB/CD, A/B, and C/D (Separators numbers 2,6, and 7).

12.6 Rank Ordered Synthesis

Shoaei *et. al.* [91] developed a technique for using dynamic programming, together with list processing, to generate alternative separation sequences in the order of least expensive, second least expensive, third least expensive, etc. The alternatives thus generated are ranked by increasing costs, subject to assumptions that are made in the analysis.

This synthesis method is based upon a repeated DP analysis using subsets of the subproblem set resulting from the original process decomposition. At each point in the analysis, a list of sequences (or alternative process structures) and the subproblems involved in the sequences is available.

This generated list is used to generate a rank ordered list of alternative process structures without evaluating all possible combinations. Then starting with the apparently best, next best, *etc.* sequences, the engineer can systematically evaluate process interactions and integration options to find the best design overall. The structural accuracy of the initial rank order list depends upon the interactions between subproblems and energy integration options for the network.

Her methodology works as follows. First the original separation problem is decomposed into a set of separation subproblems. Each subproblem is then analyzed to estimate its venture cost, assuming that its cost is independent of other subproblems in the network. If the analysis is limited to simple towers only, then one could use Task Solution Heuristic 11 (i.e., the $F/\Delta T$ Heuristic) as a measure of relative cost. Analysis then proceeds as described above to synthesize the best network using dynamic programming.

Finding the second best network using DP follows from recognition of one simple fact. There must be something different about the second best network, compared to the best network already found. The best network already found consists of a set of subproblems. The second best network could include some or most of the subproblems as the best network, but at least one of the subproblems in the second best network is different from those in the best network. Dynamic programming can therefore be used to find the second best network by basing it on a subset of the original subproblems.

Suppose the best network consists of three subproblems. For simplicity, we call this set $A_1 = S_2, S_5, S_9$. Now the second best sequence, A_2, must have at least one subproblem that is different from the three already in A_1. So by repeating DP three times, where each time one of the three subproblems, either S_2, or S_5, or S_9 is not allowed, then the second best sequence, A_2, will be found.

Consider the trial where S_2 is disallowed. The resulting sequence may contain

S_5 and S_9, but not S_2. Now if S_2 is allowed, but S_5 is not allowed, then the same second sequence could be generated, or a third sequence that is different from the first two. Similarly, if S_2 and S_5 are allowed, but S_9 is not allowed to be part of the DP solution, then a fourth sequence may be generated, or else one of the earlier sequences (except not A_1) from the previous trials. In this manner, up to three different sequences, besides A_1, may be generated. One of the sequences, the least expensive, will be the second best sequence, A_2.

Generation of the third best sequence proceeds in a similar manner. Dynamic programming is carried out on a subset of subproblems that excludes the best and second best sequences, A_1 and A_2. This exclusion is carried out by not allowing at least one subproblem in A_1 and A_2 in each trial so that new sequences are generated. After all subproblem combinations in A_1 and A_2 have been tested, one of the new sequences thus generated will be the third best.

Example: 12.5 *Use the data base in Example 12.4 and the DP algorithm described above to synthesize the second best alternative for this problem.*
Solution: 12.5 *The best sequence consists of the towers AB/CD, A/B and C/D. This is* $A_1 = \{S_2, S_6, S_7\}$. *The second best sequence is generated by deleting the subproblems one at a time. First delete* S_7. *Then*
$\phi(ABC) = 59 + 190 = 212$, *the AB/C and A/B splits,* S_9 *and* S_6.
$\phi(BCD) = 690$, *BC/D + B/C,* $S_5 + S_{10}$, *since C/D is not allowed.*
$\phi(ABCD) = \min\{A/BCD + \phi(BCD), ABC/D + \phi(ABC)\}$, *since C/D is not allowed.*
$\phi(ABCD) = \min\{85 + 690, 510 + 212\} = \min\{775, 722\}$
Which is the sequence ABC/D, AB/C, A/B, So $T_1 = \{S_3, S_9, S_6\}$
Now allow S_7, *but disallow* $S_6 = A/B = 15$.
$\phi(ABC) = 59 + 190 = 249$, *A/BC + B/C,* $S_8 + S_{10}$
$\phi(BCD) = \min\{247 + 420, 500 + 190\} = \min\{667, 690\}$
$\phi(BCD) = 667$, *B/CD + C/D,* $S_4 + S_7$
$\phi(ABCD) = \min\{A/BCD + \phi(ABC), ABC/D + \phi(ABC)\}$
$\phi(ABCD) = \min\{85 + 667, 510 + 249\} = \min\{752, 759\}$
Which is the sequence A/BCD, B/CD, C/D, so $T_2 = \{S_1, S_4, S_7\}$
Now allow S_6, *but disallow* S_2, *AB/CD = 254.*
$\phi(ABC) = 212$, *AB/C + A/B,* $S_9 + S_6$
$\phi(BCD) = 667$, *B/CD + C/D,* $S_4 + S_7$
$\phi(ABCD) = \min\{A/BCD + \phi(BCD), ABC/D + \phi(ABC)\}$
$\phi(ABCD) = \min\{85 + 667, 510 + 212\} = \min\{752, 722\}$
Which is the sequence ABC/D + AB/C + A/B, So $T_3 = \{S_3, S_9, S_6\} = T_1$
So deleting the AB/CD split yields the same sequence as when C/D is disallowed.
And $A_2 = \min\{T_1, T_2\} = \min\{722, 752\} = 722$
Thus the second best sequence is $A_2 = \{S_3, S_9, S_6\} = 722$
Which is the sequence ABC/D + AB/C + A/B.

The generation of alternatives is continued until a sufficiently large number has been generated, but the characteristics of the rank order thus generated gives clues about how many alternatives need to be generated in order to be reasonably

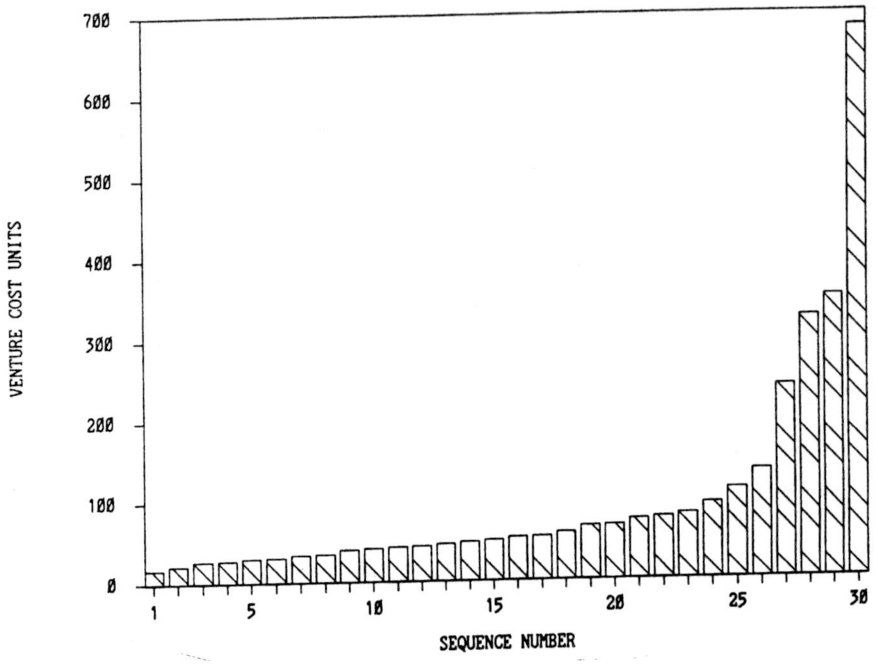

Figure 12.17: Rank order list for Example B from Shoaei *et al* [91].

certain that the best alternative has been found. Even if cost is used to evaluate subproblems, the economic interactions between them have to be neglected initially because we never know which subproblem preceeds another until the sequence is synthesized. Typically, however, the economic interactions between subproblems are on the order of ±5% of the venture cost of the best alternatives. Seldom are the interactions greater than ±10%. So when generating alternatives in rank order, there is little need to generate additional sequences when the cost of the next best one found is 10% greater than that of the best one found thus far.

Figures 12.17 and 12.18 illustrate the rank order sequence lists for two examples taken from Shoaei *et al.* [91]. In both examples, a single feed is separated into four essentially pure products. All subproblems were evaluated using venture cost, and subproblems were optimized with respect to their operating variables to minimize their venture cost prior to using them in dynamic programming synthesis. As can be seen, the costs rise quickly, and only a few alternatives have estimated venture costs within 10% of the apparent best sequence (i.e., Sequence Number 1 in Figures 12.17 and 12.18).

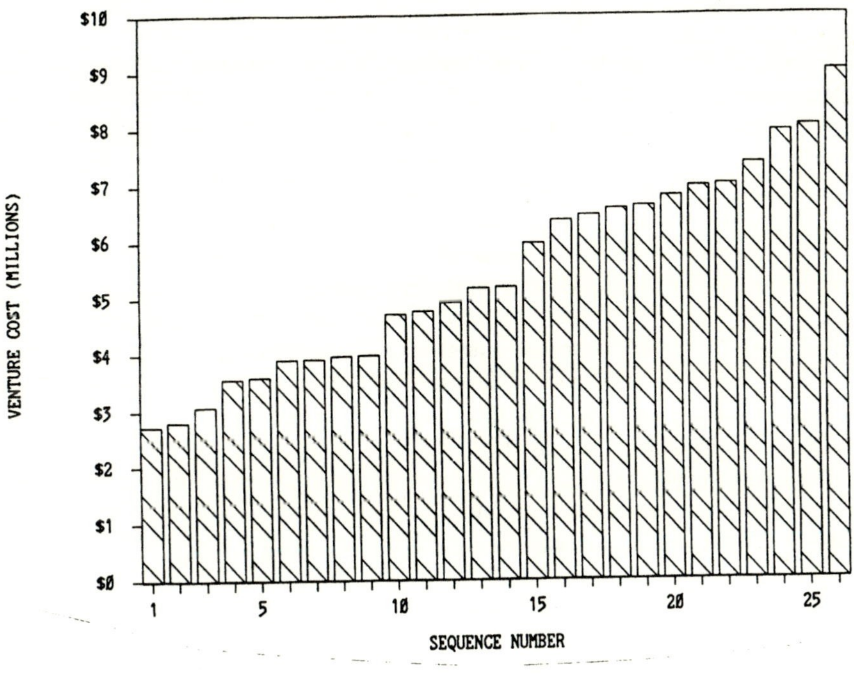

Figure 12.18: Partial rank order list for Example C from Shoaei *et al* [91].

12.7 Summary

Heuristic rules are available for reducing the design stage during the initial planning effort. These rules include the use of reaction screening to determine whether or not the process is worth evaluation and the use of Task Solution Heuristics to guide the development of the overall process flowsheet. The combinatorial problems associated with the sequencing of separators are also reduced by the use of heuristics. In addition, the Ease of Separation Index can be used to guide the choices among complex fractionators.

A limited number of heuristics are available for distillation optimization. In terms of cost, it is generally better to start at higher pressures in distillation towers, than at lower pressures. Avoid the use of refrigeration by providing sufficiently high temperatures to enable product condensation with cooling water.

Algorithmic synthesis techniques, such as dynamic programming and ordered-branch searching, have an advantage over purely heuristic synthesis techniques in that they tend to guarantee that the optimal sequence will be found. They are, however, more difficult to implement and they are also limited by the underlying assumptions upon which the modeling and evaluation effort is based. The most significant difficulty associated with both of these techniques is their limited flexibility for evaluating the economic interactions that may exist between the subproblems. In the examples which were presented, the synthesis steps assumed that the subproblems were economically independent. That is, they assumed that cost reductions will not occur if the aggregate is optimized at each point versus a suboptimization of each individual subproblem.

In some cases, chemical process flowsheets can be decomposed into economically independent subproblems. Distillation tower examples have been found where the cost of each tower is not a function of the operating conditions for the tower(s) which precedes it. On the other hand, other examples have been found where significant interactions do occur. Although these basic synthesis techniques can be extended to account for these interactions, the methodologies become even more complex. For industrial problems, a rigorous synthesis becomes impractical even when a design model is used to evaluate the alternatives. Consequently, the design engineer must recognize the shortcomings in these methods and should exert some caution in the elimination of processing alternatives.

12.8 Problems

12.1 Enumerate the 14 simple tower sequences for separating a five component mixture of species (A,B,C,D,E) into five essentially pure components. A is the most volatile species; E is the least volatile.

12.2 Enumerate the 30 sequences of analogues of Designs 1, 3, 4, 5, 6 or 7 that may be used to separate the four component mixture of (A, B, C, D) into four essentially pure products. Be sure to locate each reboiler and condenser in each sequence with a circle.

12.3 Use Task Solution Heuristics 1–10 to synthesize the best sequence of simple towers for separating the six component mixture shown below.

Species	Mole%	NBP(^{o}C)	K=y/x
Hydrogen	18	-253	200.00
Methane	15	-161	28.10
Ethylene	24	-104	11.20
Ethane	15	-88	7.20
Propylene	14	-48	3.30
Propane	14	-42	2.95

12.4 Work the previous separation problem using the K values that are given above and the ESI heuristics for complex fractionators.

12.5 Consider an equimolar mixture of species A–F with relative volatilities (3.5, 3.2, 2.9, 2.7, 2.2, 1.0), respectively. Based on this information, answer the following questions. Does the Task Solution Heuristic which says, *Do the easy split first* favor one of the splits? Does the TSH *do the hard split last* favor one of the splits? What sequence is favored by the ESI heuristics?

12.6 A six-component mixture has the following composition in terms of mole fractions (0.2,0.2,0.03,0.02,0.25,0.3) of species A through F, respectively. Their mixture K-values are (4.1, 3.7, 2.1, 1.7, 1.0, 0.92) at 2 atm total pressure. Use this information and the ESI heuristics to determine the optimal distillation sequence for producing final products A through F.

12.7 A seven-component mixture has the following composition in terms of mole fractions (0.1, 0.3, 0.1, 0.03, 0.03, 0.1, 0.34) of species A through F, respectively. Their mixture K-values are (4.8, 4.7, 4.2, 3.3, 3.2, 2.5, 2.4) at 4 atm total pressure. Use this information and the ESI heuristics to determine the optimal distillation sequence for producing final products A through G.

12.8 A feed mixture consists of six species which must be separated into essentially pure products. The feed composition and the relative volatilities of *adjacent* species are tabulated below. Species 1–6 are listed by decreasing volatilities.

i	1		2		3		4		5		6
x_i	0.05		0.2		0.05		0.4		0.25		0.05
α		1.1		1.2		1.5		1.1		1.2	

Use this information and the ESI heuristics to synthesize a separation network producing these six products. Draw the flowsheet.

12.9 Estimate the optimal operating pressure for a distillation tower if its feed has a normal boiling point temperature of 155°F.

12.10 Estimate the optimal operating pressure for an ethylene tower with feed mole fractions of (0.25, 0.20, 0.30, 0.25) in ethylene, ethane, propylene, and propane, respectively. Assume the ethane/propylene split will be performed.

12.11 Estimate the optimal overhead pressure (in psia) for a distillation tower whose feed has a normal boiling point of 108 °C.

12.12 Solve problem 3 using the $F/\Delta T$ heuristic and dynamic programming.

12.13 The tabulated data below summarize minimum venture costs for a collection of simple distillation towers separating A–F subsystems. Use the data and dynamic programming to synthesize the best network. What is the next best network and its cost?

Subproblem	Cost	Subproblem	Cost
A/B	12	BCD/E	18
B/C	5	C/DEF	10
C/D	3	CD/EF	19
D/E	7	CDE/F	12
E/F	6	A/BCDE	25
A/BC	15	AB/CDE	12
AB/C	6	ABC/DE	17
B/CD	8	ABCD/E	21
BC/D	5	B/CDEF	14
C/DE	4	BC/DEF	15
CD/E	13	BCD/EF	27
D/EF	9	BCDE/F	18
DE/F	11	A/BCDEF	32
A/BCD	18	AB/CDEF	16
AB/CD	10	ABC/DEF	19
ABC/D	19	ABCD/EF	23
B/CDE	9	ABCDE/F	36
BC/DE	7		

12.14 The data tabulated below summarize the minimum venture costs (in thousands of dollars) for a collection of simple distillation towers and solvent extraction columns separating species A–F. Use this data and dynamic programming to determine the optimal separation sequence. The process must separate species A-F.

Tower	Cost	Tower	Cost
A/B	10	C/ABD	3
B/C	1	A/BDF	33
B/D	7	AB/DF	16
C/D	1	ABD/F	9
C/E	2	B/CDE	34
D/E	1	BC/DE	19
D/F	5	BCD/E	13
E/F	1	CE/BD	3
A/BC	13	C/DEF	17
AB/C	1	CD/EF	17
A/BD	20	CDE/F	22
AB/D	10	CE/DF	4
B/CD	26	A/BCDE	29
BC/D	15	AB/CDE	40
B/DF	14	ABC/DE	23
BD/F	8	ABCD/E	16
C/DE	15	CE/ABD	4
CD/E	10	B/CDEF	52
CE/D	2	BC/DEF	29
C/BD	3	BCD/EF	20
D/EF	16	BCDE/F	26
DE/F	20	CE/BDF	5
E/DF	3	A/BCDEF	42
A/BCD	23	AB/CDEF	58
AB/CD	32	ABC/DEF	33
ABC/D	18	ABCD/EF	23
		ABCDE/F	29
		CE/ABDF	5

12.15 The tabulated data below summarize minimum venture costs for a collection of simple distillation towers. Using this data, synthesize the most economical distillation sequence using dynamic programming. Draw your best flowsheet after synthesizing it.

A/B	5	A/BCDE	17
B/C	7	AB/CDE	20
C/D	5	ABC/DE	18
D/E	11	ABCD/E	24
E/F	10	B/CDEF	23
F/G	9	BC/DEF	18
A/BC	9	BCD/EF	21
AB/C	10	BCDE/F	22
B/CD	10	C/DEFG	16
BC/D	12	CD/EFG	22
C/DE	8	CDE/FG	22
CD/E	13	CDEF/G	18
D/EF	15	A/BCDEF	21
DE/F	13	AB/CDEF	23
E/FG	14	ABC/DEF	24
EF/G	11	ABCD/EF	27
A/BCD	12	ABCDE/F	26
AB/CD	16	B/CDEFG	25
ABC/D	15	BC/DEFG	23
B/CDE	18	BCD/EFG	24
BC/DE	17	BCDE/FG	19
BCD/E	19	BCDEF/G	23
C/DEF	11	A/BCDEFG	24
CD/EF	18	AB/CDEFG	27
CDE/F	20	ABC/DEFG	26
D/EFG	18	ABCD/EFG	30
DE/FG	22	ABCDE/FG	27
DEF/G	16	ABCDEF/G	23

12.16 The tabulated data below summarize minimum venture costs for a collection of solvent extraction and simple distillation subproblems. The process must separate species A-F. Use this information and dynamic programming to synthesize the optimal separation network.

Unit	Cost	Unit	Cost
A/B	59	C/ABD	3
B/C	2	A/BDF	116
B/D	11	AB/DF	28
C/D	1	ABD/F	10
C/E	6	B/CDE	133
D/E	3	BC/DE	44
D/F	9	BCD/E	13
E/F	4	CE/BD	2
A/BC	106	C/DEF	14
AB/C	3	CD/EF	14
A/BD	78	CDE/F	144
AB/D	19	CE/DF	2
B/CD	91	A/BCDE	166
BC/D	30	AB/CDE	166
B/DF	20	ABC/DE	55
BD/F	7	ABCD/E	16
C/DE	35	CE/ABD	3
CD/E	10	B/CDEF	171
CE/D	2	BC/DEF	57
C/BD	2	BCD/EF	17
D/EF	9	BCDE/F	171
DE/F	98	CE/BDF	3
E/DF	5	A/BCDEF	204
A/BCD	124	AB/CDEF	204
AB/CD	124	ABC/DEF	68
ABC/D	41	ABCD/EF	20
		ABCDE/F	204
		CE/ABDF	4

12.17 Use dynamic programming to find the second best sequence in Problem 13.

12.18 Use dynamic programming to find the second best sequence in Problem 14.

12.19 Use dynamic programming to find the second best sequence in Problem 15.

Chapter 13

Process Integration

Flowsheet integration techniques are usually oriented toward improving the efficiency of energy utilization in a process. Although some techniques have been derived for flowsheet synthesis with simultaneous integration [79], it is more common to superimpose energy integration after the initial synthesis has been completed. In fact, state optimization of the unintegrated network can be more important than integration, so it is often appropriate to deal with process integration as a final step rather than attempting to incorporate it into the initial synthesis.

13.1 Heat Exchanger Matching

Integration through heat exchanger matching is one of the more extensively studied problems in this area. Both heuristic and algorithmic techniques have been devised for this task. Another important integration aspect is that of vapor reuse which often involves compressor design and analysis.

13.1.1 Enumeration of Matching Alternatives

In order to understand the synthesis problem, it is important to have a feeling for the number of possible matches which can be made between heat exchangers to form a network. Consider, for example, the four stream problem that is summarized in Table 13.1. In this case, streams B and D must be cooled by either matching them with streams A and/or C or with utilities. The *out* temperatures are the stream targets.

In order to enumerate the number of possible matches, a technique is adopted from Pho and Lapidus [78]. These results are summarized in Table 13.2, and although 47 alternatives are identified, additional designs are possible because no stream splitting or rematching between streams has been permitted. That is, combinations involving more than four exchangers have been eliminated.

It is easy to see that even with this relatively simple example an exhaustive search of all possible alternatives is not attractive. If the analysis in Table 13.2

Table 13.1: A four stream heat exchanger matching problem.

Stream		Temperature (^oF)		Flowrate-C_p
Label	Type	Supply	Target	(kBtu/h-^oF)
A	Cold	140	320	14.42
B	Hot	320	200	16.68
C	Cold	240	500	11.60
D	Hot	480	280	20.00

Table 13.2: Enumeration of heat exchanger matches with two hot and two cold streams.

		Number of Alternatives
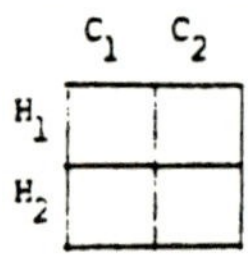	No matches at all	1
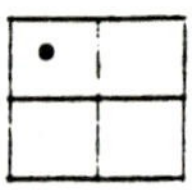	One match only	4
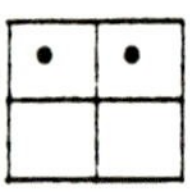	Two matches involving the same stream 4 x 2	8
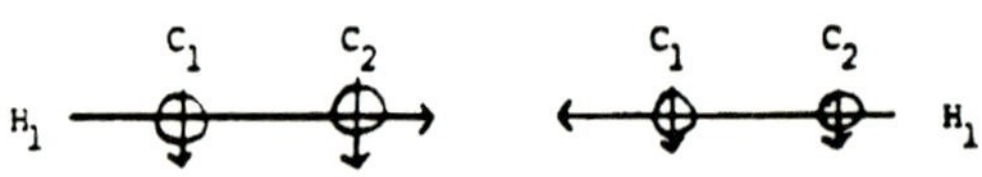		
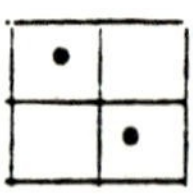	Two matches involving different streams	2
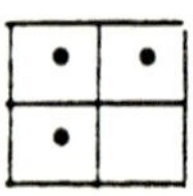	Three matches 4 x 2 x 2	16
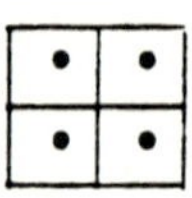	Four matches 2 x 2 x 2 x 2	16
	TOTAL	47

is extended to three cold and two hot streams, then the number of countable alternative networks increases to 847 flowsheets. Clearly, strategies are needed to approach this type of network problem.

13.1.2 Algorithmic Heat Exchanger Matching

A number of algorithms have been proposed for developing heat exchanger networks. The one presented here was first described by Linhoff and Flowers [57, 58, 60]. Variations [22] of this algorithm have also been devised which permit even greater detail in the analysis. It utilizes four simple steps:

1. Decompose the original problems into temperature intervals (TIs) (*e.g.* Table 13.3).

2. Using the temperature intervals, set up the TI Problem Table (*e.g.* Table 13.4) and determine the minimum utility requirements and the pinch temperature for the network.

3. From the pinch temperature and the process stream requirements, determine the minimum number of exchanger units (MNUs) via Equation 13.2.

4. Using the pinch temperature, the MNUs, and the minimum utility requirements as guidelines, synthesize networks starting at those temperature intervals closest to the pinch temperature (*e.g.* Figure 13.2).

Thus, the algorithm yields guidelines that tell the engineer which heat exchanger matches must be completed first. The key is to start in those temperature intervals that are closest to the pinch temperature. If at any point in the synthesis the utilities appear greater than the minimum requirements, then the engineer knows that an error has occurred, and some of the matches should be reconsidered. Similarly, if the final network requires more than the computed minimum number of exchanger units, then some exchangers can be consolidated.

An explanation of what is meant by temperature intervals and the pinch temperature is given below. In its simplest form, the TI method is based [60] on the following assumptions:

1. All process streams have constant flowrate heat capacities (*e.g.* Btu/h-F).

2. The minimum allowable approach temperature for heat transfer applies to all potential process/process and process/utility matches.

3. There are no processing, safety, or plant layout reasons which prevent matching pairs of streams.

4. There is only one hot utility level (at the extreme hot end of the problem) and one cold utility level (at the extreme cold end of the problem).

All of these assumptions can be relaxed so that the realism of the problem statement can be improved. However, the basic approach is not changed by these modifications. The analysis just becomes more complex.

Table 13.3: Temperature interval definition.

Stream		Inlet T (^{o}F)		Adjusted T (^{o}F)		Rank
Label	Type	Supply	Target	Supply	Target	Order
A	Cold	140		140		8
			320		320	3
B	Hot	320		(300)		4
			200		(180)	7
C	Cold	240		240		6
			500		500	1
D	Hot	480		(460)		2
			280		(260)	5

Notes: The hot stream temperatures have been adjusted by subtracting the minimum approach temperature (20 ^{o}F). These temperatures are in parentheses.

Use the Adjusted Temperatures to generate the rank order. The highest Adjusted temperature is rank 1, the lowest rank 8. These temperatures define the intervals (*i.e.* the TIs).

The Temperature Intervals

The initial step consists of decomposing the problem into temperature intervals. These intervals are then used to construct the TI Problem Table (and to explain what is meant by a pinch temperature). The temperature intervals are defined simply by the stream supply temperatures (the *Supply* values in Table 13.1) and the stream target temperatures (the *Target* values in Table 13.1).

The temperature intervals can be set up, for example, by reducing the hot stream temperatures by the minimum approach temperature (usually a given quantity) and using the cold stream temperatures without any adjustment. Any duplication of temperatures between intervals is ignored. The resulting temperatures are then ranked starting with the highest temperature.

Table 13.3, for example, was generated from the data in Table 13.1. Reducing the hot stream temperatures by a minimum approach temperature (20 ^{o}F as shown) ensures that the minimum approach temperature will automatically be satisfied during the final network synthesis step. The adjusted temperatures are in parentheses.

The TI Problem Table

The next step is to develop the TI Problem Table which uses the information in Table 13.3 and the energy balances around the resulting TIs to analyze the problem. Table 13.4, for example, shows the TI Problem Table from the data in Tables 13.1 and 13.3. There are seven TIs and, therefore, seven exchanger subnetworks that must be analyzed. The stream temperature ranges are also shown with the arrows pointing up for streams that are to be heated and down for streams that are to be

Table 13.4: Temperature interval problem table for heat exchanger problem in Table13.1.

i	T_i	Ranges A	B	C	D	(1) $\sum F_i C_{pi}$	(2) ΔH_i	(3) $\sum \Delta H_i$	(4) $Q_h + \sum \Delta H_i$	
	500			↑				0	**464**	← $Q_{h,min}$
1				|		-11.6	-464			
	(460)			|	⊤			**-464**	0	← T_{pinch}
2				|	|	8.4	1176			
	320	↑		|	|			712	1176	
3		|		|	|	-6.02	-120.4			
	(300)	|	⊤	|	|			591.6	1055.6	
4		|	|	|	|	10.66	426.4			
	260	|	|	|	↓			1018	1482	
5		|	|	|		-9.34	-186.8			
	(240)	|	|	⊥				831.2	1295.2	
6		|	|			2.26	135.6			
	(180)	|	↓					966.8	1430.8	
7		|				-14.42	-576.8			
	140	⊥						390	**854**	← $Q_{c,min}$

Q_a	=	-14.42(320-140)	=	-2595.6	Q_b	=	16.68(320-200)	=	2001.6
Q_c	=	-11.6(500-240)	=	-3016.0	Q_d	=	20(480-280)	=	4000.0
		Total sink	=	-5611.6			Total source	=	6001.6

$$\text{Net} = \text{Source} + \text{Sink} = 390 \text{ Btu/h}$$

$$\text{MNU} = 1 + 4 = 5$$

cooled. This layout is convenient for the development of the subsequent columns in the table. It is also helpful to show the stream flowrate heat capacities immediately below the stream designations (A, B, *etc.*).

The net heat transferred within each temperature interval (column (2) in Table 13.4) is calculated using the information already in the TI Problem Table and the relationship:

$$\Delta H_i = \left[\sum_{\text{Hot Streams}} FC_p - \sum_{\text{Cold Streams}} FC_p \right] (T_i - T_{i+1}) \tag{13.1}$$

where the summation only includes those streams that fall within the *ith* temperature interval as is indicated by the arrows in Table 13.4. Note that the flowrate-heat capacities are expressed in kBtu/h-°F in column (1) of Table 13.4. Also, the enthalpy balance for each TI is negative for a net heat input and positive when net heat is removed over that TI as defined by Equation 13.1 and column (2) in Table 13.4.

Column (3) in Table 13.4 shows the net energy balance for the system when it is constrained to satisfy the minimum approach temperature over each TI. It is computed simply by placing a zero in the top row of column (3) and adding the energy increments as computed from column (2). The largest negative number in column (3) indicates the minimum heating requirements, Q_h, for the network. In Table 13.4, this number is inside a box in column (3).

Column (4) is computed by adding the absolute value of the largest negative number in column (3) (-464 kBtu/h) to each number in column (3). The numbers in column (4) then indicate the minimum utility heat requirements at the top of the column, the minimum utility cooling requirements at the bottom of the column, and the pinch temperature for the network (*i.e.* that temperature in column (4) where the sum equals zero). These quantities are all identified in Table 13.4.

In the event that no negative numbers occur in column (3), then the number at the bottom of column (3) represents the minimum cooling utility requirement. In the event that the largest negative number occurs at the bottom of column (3), then the problem requires no cooling utility, and that largest negative number represents the minimum heating utility requirement. In these latter two cases, the pinch temperature is either at highest or lowest temperature in the network respectively.

The Composite Enthalpy-Temperature Diagram

A graphical representation of where the pinch temperature occurs in this example is shown as Figure 13.1 — a composite enthalpy-temperature diagram. At the pinch temperature, the composite curve for the hot streams and the composite curve for the cold streams are seen to intersect. The difference in energy transfer rates between the two curves at the highest temperature equals the minimum heating utility and the difference in energy transfer rates at the lowest temperature equals the minimum cooling utility. The two curves intersect only at the pinch temperature and hence the name.

Equation 13.1 uses a difference between the hot streams sum and the cold streams sum in each TI. In Equation 13.1, the difference between these sums is multiplied by the TI. On the other hand, if the hot and cold stream sums in Equation 13.1 are multiplied separately by the TI, and these sums are then added separately over all the TIs (similar to column (3) in Table 13.4), then the composite hot stream and the composite cold stream curves can be generated.

The only additional differences are that the summations in Figure 13.1 began with a zero at the lowest temperature for the hot streams while the cold streams sum was increased (offset) by the minimum cooling utility. This offset is required to force the two curves to intersect only at the pinch temperature. In Table 13.4, column (3) was initially zeroed at the top.

As constructed, the composite hot stream curve is always hotter than the composite cold stream curve by the minimum approach temperature. This minimum difference always occurs at the pinch temperature and this type of behavior is generally observable for all types of energy transfer networks.

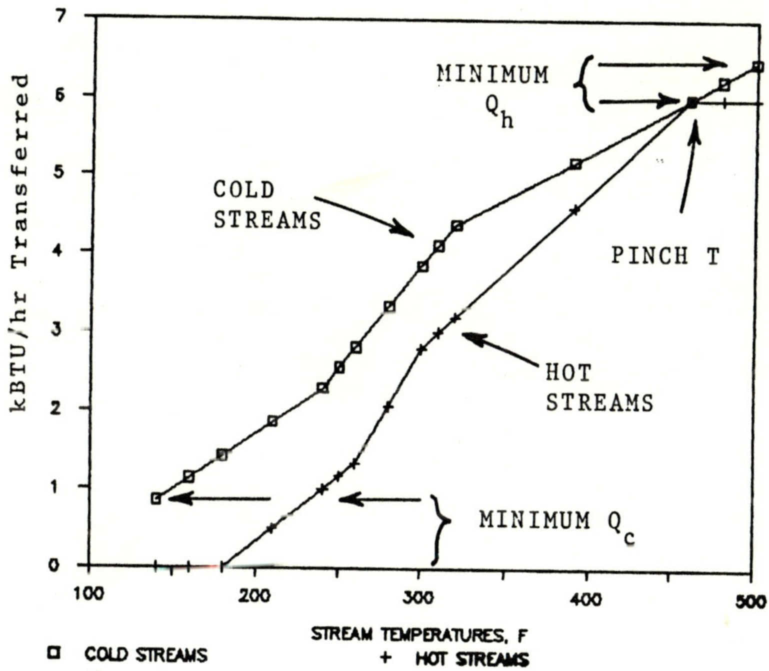

Figure 13.1: Composite enthalpy-temperature curves.

The Minimum Number of Exchanger Units

The minimum number of heat exchanger units (MNUs) required for a particular network problem is given simply by:

$$MNU = \begin{bmatrix} \text{Number of Process} \\ \text{Streams Transferring} \\ \text{Energy at } T > T_{pinch} \end{bmatrix} + \begin{bmatrix} \text{Number of Process} \\ \text{Streams Transferring} \\ \text{Energy at } T < T_{pinch} \end{bmatrix} \tag{13.2}$$

These principles also suggest a general approach for integration. The problem should always be divided at the pinch temperature and heat exchanger matching above and below the pinch temperature should be carried out separately. For the problem in Table 13.1, these principles indicate that steam must be used to heat stream C from 460 to 500°F, but that only cooling water is required to meet the targets for the remaining streams.

Additional rules are needed to guide the synthesis of the network:

3. The initial matches should always be those which are closest to the pinch. In this region, the fewest options exist. As one moves further from the pinch, there are more alternatives, and the choice may be made based on other considerations such as cost, safety, layout or controllability.

4. Each match should satisfy the requirements of one stream. If a match fails to satisfy the requirements of at least one stream, then the final design will usually require more than the minimum number of exchangers (Linhoff calls this heuristic the *ticking off* rule).

5. For initial matches above the pinch, first choose matches where the flowrate-heat capacity of the cold stream is greater than or equal to the flowrate-heat capacity of the hot stream.

6. For initial matches below the pinch, first choose matches where the flowrate-heat capacity of the hot stream is greater than or equal to the flowrate-heat capacity of the cold stream.

7. Insofar as utility requirements do not exceed the minimum requirements and options exist, place heaters and coolers to minimize the total transfer area of network.

At this point a graphical representation is helpful in developing the network. Figure 13.2, for example, consists of a series of diagrams illustrating a matching technique which obeys these rules. The hot streams are depicted as arrows going from left to right. The cold streams are depicted as arrows going from right to left. The vertical lines indicate the temperature intervals and the pinch temperature. As is shown, a heater is used to elevate Stream C to its target temperature. There are no options for this assignment, and it requires the minimum heating utility as indicated in Table 13.4. To the right of the pinch, the first match must be between streams D and C. Using the ticking off rule, this match satisfies the requirements for stream C (indicated by the check mark) and leaves a Stream D residual at 332.4°F.

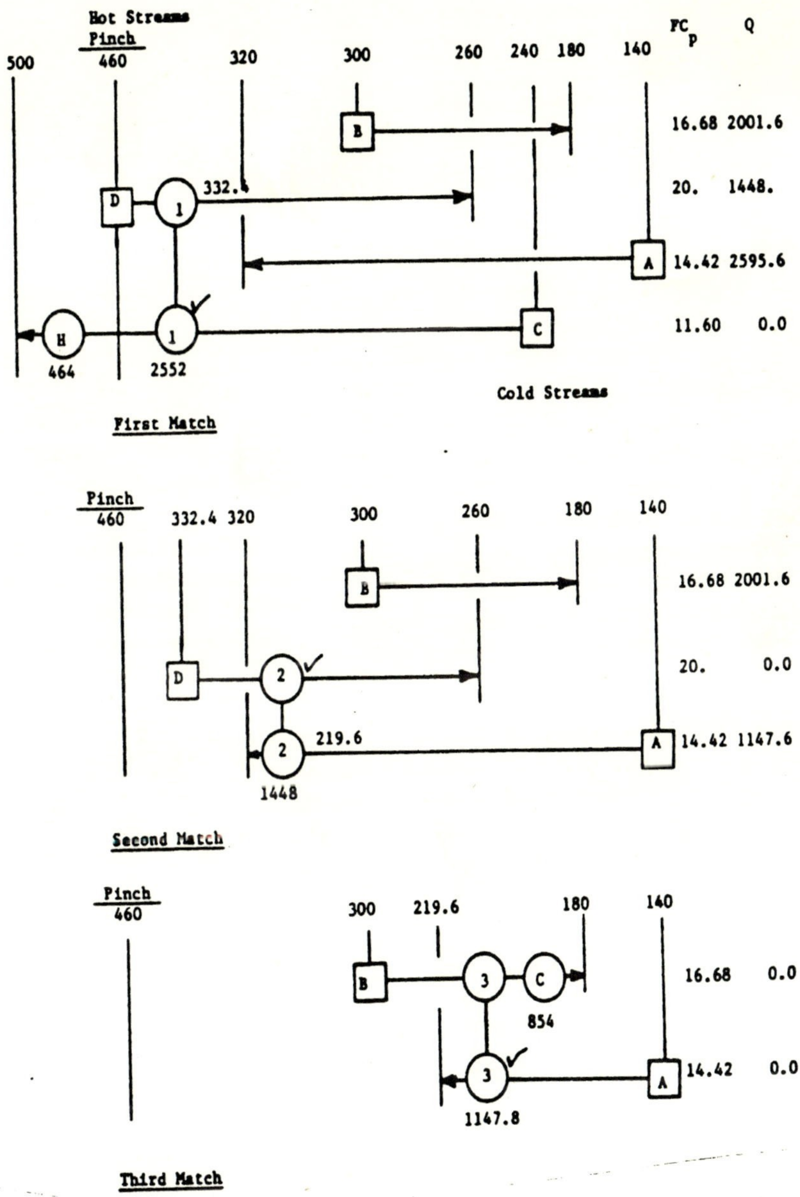

Figure 13.2: Algorithmic solution to heat exchanger matching problem in Table 13.1.

Table 13.5: Example problem with stream splitting. The minimum approach temperature is 20 °F.

Stream		Temperature (°F)		Flowrate-C_p
Label	Type	Supply	Target	(Btu/h-°F)
1	Cold	50	580	100
2	Cold	70	520	400
3	Hot	490	70	200
4	Hot	540	70	650

Options exist for the second match. It could be between streams B and A or between streams D and A. Hence, the second heat exchanger match is somewhat arbitrary. In actual practice, both options should be evaluated, but this choice ticks off stream D and leaves a Stream A residual at 219.6°F.

The third match is between streams B and A. In this case, a cooler is required to meet the requirements for stream B. However, it should be noted that the cooler could precede the third match. This situation would increase the temperature driving force between the cooler and Stream B, and it may reduce costs.

As shown, the network in Figure 13.2 meets the specified minimum utility requirements and requires only the minimum number of heat exchangers. However, this network is not unique, and others may be more cost effective. In general, this algorithm does not completely avoid the need to evaluate alternatives, but it does reduce the number of alternatives that need evaluation, and it enables the identification of those matches that are essential.

Example: 13.1 *Consider the heat exchanger matching problem defined in Table 13.5. Construction of the TI Problem Table gives the following results:* $Q_C^{min} = 162,500$ *Btu/h,* $Q_H^{min} = 6,000$ *Btu/h,* $T_p = 520$ *°F,* $MNU = 1 + 4 = 5$.

Solution: 13.1 *Plotting the streams and synthesizing a network gives the results in Figure 13.3. Stream 1 is the only one appearing above the pinch temperature, so a heater is required to reach its target of 580 °F. The duty is* Q_H^{min}*, as expected.*

Three streams appear in the temperature interval from 470 to 520 °F. One stream is hot, and two are cold. So the hot stream must be split in order to heat both cold streams to the pinch temperature. These matches satisfy cold Streams 1 and 2, and cools Stream 4 to a residual temperature of 170.77 °F. After matches 1 and 2, there are no more cold streams, so coolers must be used to cool Stream 3 and Residual 4 to their target temperatures. The duty of the two coolers equals Q_C^{min}.

13.2 Vapor Reuse Schemes

Because of rising energy costs, vapor reuse schemes have became more popular in chemical process designs. Related issues include boiler waste heat recovery [118, 50, 31, 7] and overall plant energy management [59, 24]. Although many important vapor reuse applications exist, such as in multiple effect evaporators [10, 17], the

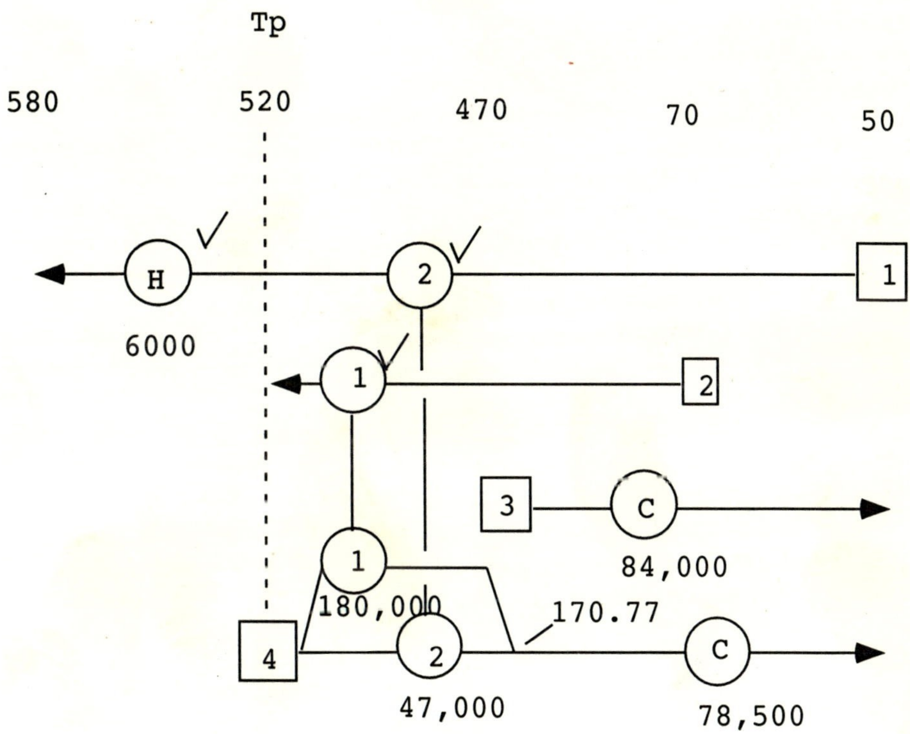

Figure 13.3: Heat exchanger network with stream splitting.

primary emphasis here is on process applications as they relate to distillation. One common feature many of these issues have in common is the need to design vapor compressors. For that reason, a brief summary is included.

13.2.1 Vapor Compressor Design

Since compressors are relatively complex pieces of mechanical equipment, their maintenance costs tend to be high when compared to other unit operations. For that reason, a detailed design analysis usually requires the evaluation of surge control techniques [32, 5], intercoolers, and aftercoolers [15] to maintain the compressor operation near steady state conditions. On the other hand, a vapor reuse scheme may be rejected as uneconomical simply by evaluating the anticipated costs of the compressor and the estimated energy savings. If a preliminary analysis on this basis suggests that the application is economical, then a more detailed study may be justified. In addition, the compressor itself tends to be the most expensive item in the network so this simplification is frequently a good approximation.

Guidance on anticipated compressor performance for various applications is available from a number of sources [56, 69, 66]. Ultimately, the design engineer must depend upon vendor estimates, but reasonably sophisticated evaluation techniques can often be easily implemented. For example, Bogart [11] provides an algorithm for a hand calculator that includes compressibility factor estimation and efficiency estimation techniques which consider details like the rotational speed of the compressor.

For preliminary design it is often sufficient to compute the pressure-enthalpy diagram for the fluid being processed and to estimate the polytropic head requirements. For pure fluids, the P-H diagram can often be found tabulated in a suitable reference. Estimation of the polytropic head is described in Section 11.1.1.

Although the enthalpy change from point a to b in Figure 13.4 is computed from the polytropic head, the remaining points are determined largely by the characteristics of the fluid which is being compressed.

In addition to the polytropic head, an initial design requires a description of the proposed compressor cycle on a pressure-enthalpy diagram. Figure 13.4 summarizes a typical cycle together with a flowsheet indicating the reference conditions in the cycle. This information is used to compute the recycle rates through the compressor network.

In general, it is desirable to avoid the two-phase region during compression since liquid droplets tend to cause erosion. Also, the condenser design should consider the fact that the temperature driving force is not linear. A discontinuity occurs at point C as indicated in Figure 13.4. This condition increases the transfer area requirements compared to the assumptions of a lognormal temperature driving force.

13.2.2 Vapor Reuse in Distillation

A number of variations in vapor reuse are possible. For example, towers can be cascaded as shown in Figure 13.6. In this case, the feed is split between two towers

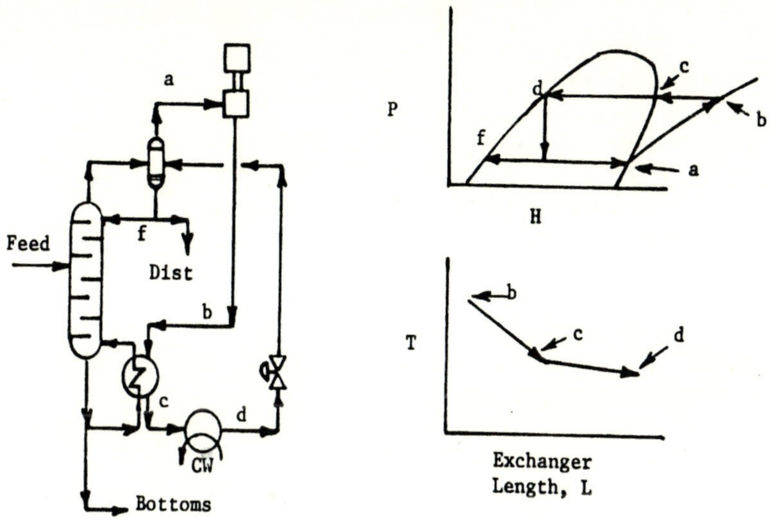

Figure 13.4: Vapor reuse scheme. Heat pump with overhead vapor compression.

and the condenser of the higher pressure tower is matched with the reboiler of the lower pressure tower. Thus, the pressure optimization of the towers is constrained by the requirements for the minimum temperature driving force. Nonetheless, this configuration can be optimized.

The cascade arrangement in Figure 13.6 results in energy degradation over a wider temperature interval than is the case with a single tower which processes the entire feed. Because it involves two separate distillation towers, rather than just one, the capital investment requirements are often higher than for the single tower case. On the other hand, the cascade system can reduce energy usage by nearly a factor of two although the thermodynamic efficiency is less than the single tower case due to the wider differences in the reboiler and condenser temperatures.

Heuristics for this type of design are limited. Tyreus and Luyben [106] suggest that a cascade may be preferred whenever more than 30,000 lb/hr of steam may be replaced with the system. Two reasonable heuristics are:

1. Favor cascades for very large diameter towers whenever the temperature drop across a single tower is approximately equal to or less than half the temperature difference between the coolest available steam source and cooling water.

2. Avoid cascades whenever the temperature drop across a single tower performing the required separation is such that the cascade will require either refrigeration or high pressure steam.

Another variation on vapor reuse is shown in Figure 13.7. This system consists of a high pressure beer still followed by a low pressure ethanol water fractionator. The

Table 13.6: Heuristics for heat pump applications

1.	Heat pumps have extensive applicability to distillation whenever direct refrigeration or chilled water is required to condense products.
2.	Heat pumps have limited ranges of applicability whenever reboiler temperatures are above 300 ^{o}F because cooling water can often be used in the condenser. If the condenser can produce low pressure steam, then the range of applicability is further reduced.
3.	Heat pumps are seldom justified whenever the condenser temperatures are between 100 and 250 ^{o}F and if the plant steam system is used both for power (either direct turbine drives or electricity generation) and heating. In such cases, low grade steam and/or hot water can be integrated with the power system more economically than using a heat pump. When steam is used for heating only, heat pumps are more attractive.
4.	Heat pumps become less attractive as the temperature drop across a tower and the required compression ratio in the heat pump increase. Most practical uses are for isomer towers where the temperature drop is relatively small.

dilute beer feed is introduced to the top of the beer still which is usually operated without reflux due to the dissolved solids in the beer which tend to cause plugging. Live steam is introduced at the bottom of the beer still and the vapor leaving this column is used to provide reflux to the low pressure fractionator. Although the feed rate to the fractionator is significantly less than that to the beer still, the vapor traffic in the fractionator is comparable to that in the beer still due to the increasing reflux requirements as the ethanol/water azeotrope is approached.

In refinery operations, it is common to have either large cooling or heating requirements for a process feed. In the former case, say when a high temperature reactor effluent must be cooled before it can be purified, this effluent may be matched with the reboiler in its purification tower to provide vapor reuse. In the latter case, say where cold crude oil must be heated for fractionation, it is common to match this feed with process condensers.

There are a number of possible ways in which heat pumps (compressors) can be used to provide vapor reuse. Some of these possibilities are described in more detail by Null [72] and Freshwater [30] and are diagrammed in Figure 13.5. Each of these systems exhibits different operating characteristics and may be desirable under somewhat different circumstances. The heat pump option to provide auxiliary reflux, for example, is only of value when the pinch point occurs at the feed tray. In cases where the pinch point occurs near a product tray (*e.g.* for ethanol/water mixtures), this technique would be less valuable.

A few heuristics are available for the use of heat pumps. These are summarized in Table 13.6.

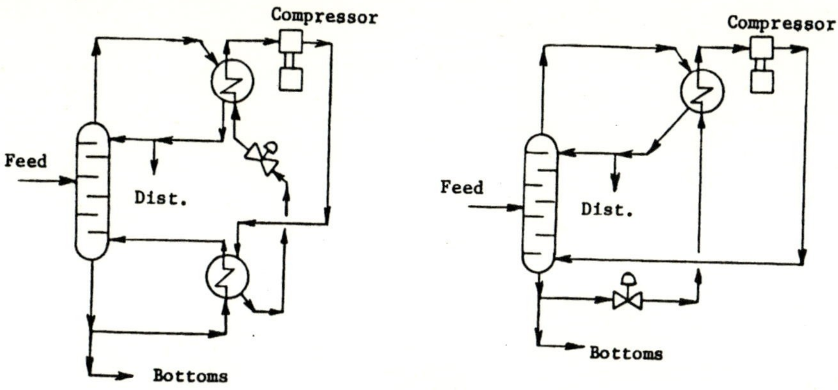

Figure 13.5: Several integrated uses for heat pumps.

13.3 Summary

Process integration can lead to important savings. The algorithm for heat exchanger matching is well developed and can be carried out manually. Vapor reuse schemes can also be important, but are currently more difficult to evaluate. In many instances, a high-speed computer is still required.

13.4 Problems

13.1 Process streams A and B are used to cool and heat each other in a countercurrent heat exchanger. Stream A has a flowrate heat capacity of 5000 Btu/h-^{o}F and stream B has a flowrate heat capacity of 10,000 Btu/h-^{o}F. The two streams are to be matched as much as possible, but the temperature approach at either end should not be less than 20°F. Both streams are below their bubble points and an overall heat transfer coefficient of 100 Btu/h-ft^2-^{o}F can be realized. Stream A is available at 425°F. Stream B is available at 530°F. What is the exit temperature of streams A and B if the match is carried as far as possible? How much heat is transferred? What are the transfer area requirements?

13.2 Consider the heat exchanger matching problem shown below. The minimum approach temperature is 20°F. Use the algorithmic methods described in this chapter to synthesize several heat exchanger networks.

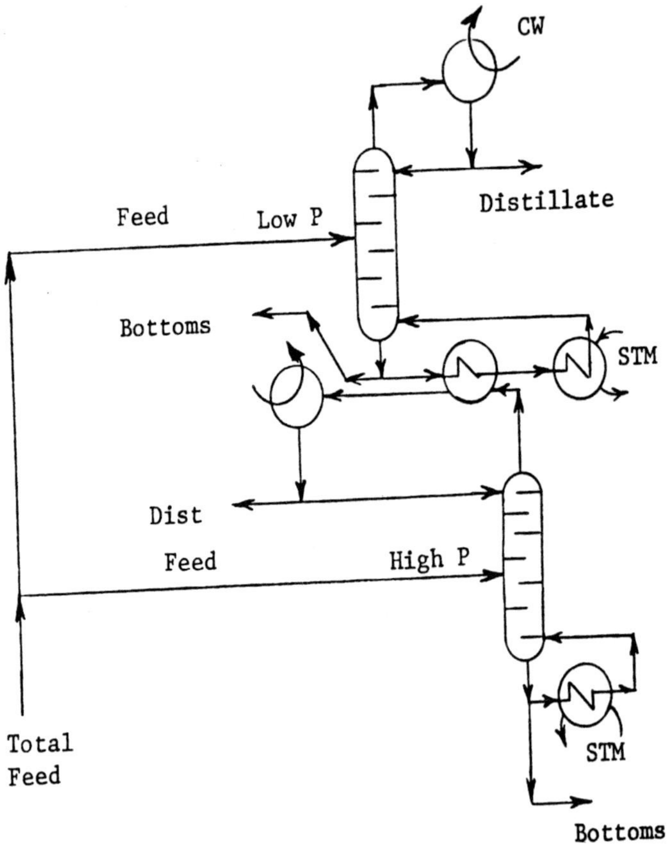

Figure 13.6: Two tower cascade. The high pressure tower dumps heat to the low pressure unit.

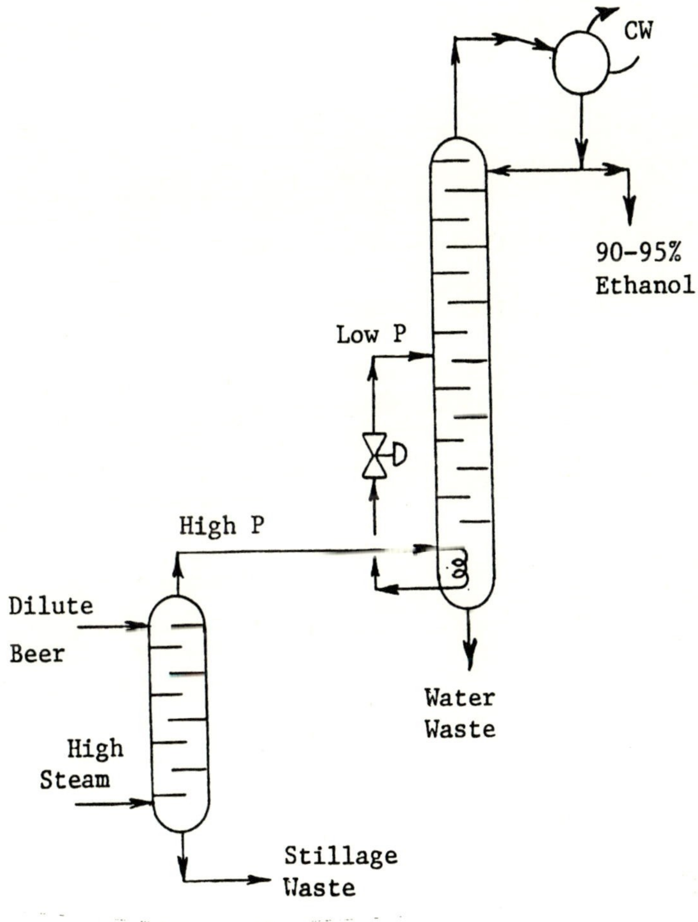

Figure 13.7: High pressure beer still with low pressure fractionator.

Stream Number	Type	Temperature (°F) Supply	Target	kBtu/h-°F
1	Cold	120	245	20
2	Hot	255	160	30
3	Cold	175	240	15
4	Hot	250	120	40
5	Cold	100	190	10

13.3 Consider the heat exchanger matching problem shown below. The minimum approach temperature is 10°F. Use the algorithmic method to synthesize several networks.

Stream Number	Type	Temperature (°F) Supply	Target	kBtu/h-°F
1	Cold	200	400	16
2	Cold	100	430	16
3	Hot	590	400	25
4	Cold	300	400	42
5	Hot	471	200	14
6	Cold	150	280	27
7	Hot	533	150	11

13.4 A vapor reuse scheme is under consideration. It will use a heat pump with overhead vapor compression to optimize a distillation tower as illustrated in Figure 13.4. The vapor is essentially a pure component with a normal boiling point of 250 K and a critical temperature of 410 K. The vapor enthalpy is given by the correlation:

$$H_v = 6000 + 10T + 0.01T^2$$

where H_v is in Btu/lb-mole of vapor and T is in °F.

As shown in Figure 13.4, the heat pump will compress Stream a to produce Stream b. The compression ratio is 2.0. The vapor entering the compressor at point is saturated and has a temperature of 140°F. The compressor has a polytropic efficiency of 0.76.

Based on this information, answer the following questions. What is the fraction of the recycle Stream d in Figure 13.4 that vaporizes as it passes through the throttle? What are the stream enthalpies for the vapor at point a, the liquid at point f, and the saturated liquid at point d? What is the total pressure at points a and b? What is the saturation temperature for the stream pressure at b?

HINTS: Estimate the latent heat of vaporization at the normal boiling point. Use the Watson correlation (see Equation B.29) to estimate the latent heats at the various stream pressures. Assume that $\partial H_v/\partial P = 0$. Use the compressor design equations to estimate the temperature at point b.

13.5 Consider the set of hot and cold streams tabulated below. Synthesize a heat exchanger matching network that enables the stream targets to be met using the minimum heating and cooling utilities and with the minimum number of heat exchangers. The minimum approach temperature is 10°F.

What are the minimum heating and cooling requirements? How many heat exchangers are required? What is the network pinch temperature?

		Stream Temperatures, °F	
Stream	$\dot{m}C_p$, kBtu/h-°F	Source	Target
1	30	120	250
2	20	250	160
3	20	180	200
4	30	300	110
5	10	100	150

13.6 Consider the set of hot and cold streams tabulated below. Synthesize a heat exchanger matching network that enables the stream targets to be met using the minimum heating and cooling utilities and with the minimum number of heat exchangers. The minimum approach temperature is 10°F.

What are the minimum heating and cooling requirements? How many heat exchangers are required? What is the network pinch temperature?

		Stream Temperatures, °F	
Stream	$\dot{m}C_p$, kBtu/h-°F	Source	Target
1	40	280	140
2	30	350	160
3	50	180	300
4	30	130	450
5	40	100	150

13.7 Consider the heat exchanger matching problem summarized below. There are seven streams which must either be heated or cooled. The minimum approach temperature is 10°F.

Stream	Flowrate-	Temperatures (F)	
Number	C_p, MBtu/h-°F	Supply	Target
1	1.0	110	460
2	4.5	140	510
3	2.5	380	85
4	1.1	250	125
5	0.8	370	85
6	6.9	275	75
7	0.179	380	105

What is the pinch temperature in °F? Synthesize a network that achieves minimum utilities and only uses the minimum number of heat exchangers.

13.8 (35 points) Consider the set of hot and cold streams tabulated below. Synthesize a heat exchanger matching network that enables the stream targets to be met using the minimum heating and cooling utilities and with the minimum number of heat exchangers. The minimum approach temperature is 20°F.

What are the minimum heating and cooling requirements? How many heat exchangers are required? What is the network pinch temperature?

		Stream Temperatures, °F	
Stream	$\dot{m}C_p$, 10^7Btu/h-°F	Source	Target
1	0.5	100	480
2	0.4	150	520
3	1.0	140	350
4	2.0	430	120
5	0.5	450	320

13.9 A heat exchanger network must be synthesized. This network should use the fewest possible heat exchangers and minimize the required utilities. Six process streams must be matched as shown below. The minimum approach temperature is 10 °F.

Stream		$\dot{m}C_p$	Temperatures (°F)	
Number	Type	kBtu/h-°F	Source	Target
1	Cold	500	120	480
2	Cold	20	150	525
3	Cold	20	150	420
4	Hot	120	495	205
5	Hot	900	300	210
6	Hot	92	570	420

Use this information and the temperature interval algorithm to synthesize a network meeting the design goals.

Part III

Appendices

Appendix A

Constants and Conversion Factors

Table A.1: Temperature conversion

Degrees	Symbol	Value
Kelvin	$T^oK =$	$T^oC + 273.16$
Rankine	$T^oR =$	$T^oF + 459.69$
Fahrenheit	$T^oF =$	$1.8(T^oC) + 32$

Table A.2: Ideal gas volumes

State	Moles	Volume
0^oC and 760 mm Hg	$1\ g - mole =$	$22.4140\ liters$
	$1\ kg - mole =$	$22.4140\ m^3$
	$1\ lb - mole =$	$359.05\ ft^3$
60^oF and 760 mm Hg	$1\ lb - mole =$	$378.7\ std\ ft^3\ (SCF)$

Table A.3: SI and other Prefixes

Factor	Prefix	Symbol	Factor	Prefix	Symbol
10^{18}	exa	E	10^{-1}	deci	d
10^{15}	peta	P	10^{-2}	centi	c
10^{12}	tera	T	10^{-3}	milli	m
10^{9}	giga	G	10^{-6}	micro	μ
10^{6}	mega	M[a]	10^{-9}	nano	n
10^{3}	kilo	k	10^{-12}	pico	p
10^{2}	hecto	h	10^{-15}	femto	f
10^{1}	deka	da	10^{-18}	atto	a

[a] Historically, M has been used to indicate 10^3; MM has been used to indicate 10^6.

Table A.4: Physical Constants

Constant	Symbol	Value
Gas law	$R =$	$1.987\ Btu/(lb-mole)(^oR)$
		$1.987\ cal/(g-mole)(^oK)$
		$0.08206\ (atm)(m^3)/(kg-mole)(^oK)$
		$0.08320\ (bara)(m^3)/(kg-mole)(^oK)$
		$0.73\ (atm)(ft^3)/(lb-mole)(^oR)$
		$1.314\ (atm)(ft^3)/(lb-mole)(^oR)$
		$10.73\ (psia)(ft^3)/(lb-mole)(^oR)$
		$82.06\ (atm)(cm^3)/(g-mole)(^oK)$
		$1545.0\ (lb_f/ft^2)(ft^3)/(lb-mole)(^oR)$
		$8320.0\ (J)/(kg-mole)(^0K)$
Gravity acceleration	$g_0 =$	$32.174\ ft/s^2$
		$980.665\ cm/s^2$
Joule's	$J_c =$	$778.16\ (lb_f-ft)/Btu$
		$41.84\ Merg/cal$
Avogadro's	$N_0 =$	$6.023 \times 10^{23}\ molecules/(g-mole)$
Boltzmann's	$\kappa = R/N_0 =$	$1.3805 \times 10^{-16}\ erg/(molecule)(^oK)$
Faraday's	$\mathcal{F} =$	$9.652 \times 10^4\ abs-couloumbs/(g-equivalent)$
Planck's	$h =$	$6.624 \times 10^{-27}\ erg-sec$
Stafan-Boltzmann	$\sigma =$	$1.355 \times 10^{-12}\ cal/(sec)(cm^2)(^oK^4)$
Electronic charge	$e =$	$1.602 \times 10^{-19}\ abs-coulomb$
Speed of light	$c =$	$2.99793 \times 10^{10}\ cm/s$

Table A.5: Conversion factors

Quantity	Symbol	Basis	Factor
Density	ρ	$1\ g/cm^3 =$	$62.428\ lb/ft^3$
Heat	Q	$1\ Btu =$	$251.996\ cal$
Length	L	$1\ Angstrom =$	$10^{-10}\ m$
		$1\ inch =$	$2.540\ cm$
		$1\ foot =$	$30.48\ cm$
Mass	m	$1\ lb =$	$453.5924277\ g$
		$1\ kg =$	$2.2046\ lb$
		$1\ ton =$	$2,000\ lb$
Mechanical energy	E_m	$1\ J =$	$1\ watt - s$
		$1\ J =$	$10^7\ erg$
Mechanical heat	J	$1\ cal =$	$4.1873 joule$
		$1\ kw - h =$	$3412.75\ Btu$
		$1\ Btu =$	$778.26\ lb_f - ft$
Newton's-law	g_c	$1\ g_f =$	$980.665\ g_m - cm/s^2$
		$1\ lb_f =$	$32.174\ lb_m - ft/s^2$
Pressure	p	$1\ atm =$	$14.696\ lb_f/in^2\ (psia)$
		$psig =$	$psia - 14.696$
		$1\ atm =$	$760\ mm\ Hg$
		$1\ atm =$	$29.92\ in\ Hg$
		$1\ atm =$	$1.01325 \times 10^5\ N/m^2\ (Pa)$
		$1\ atm =$	$101.325\ kPa$
Power	P	$1\ hp =$	$550\ lb_f - ft/s$
		$1\ hp =$	$0.74548\ kw$
		$1\ kw =$	$56.87\ Btu/min$
Specific energy		$1.8 Btu/lb - mole =$	$1\ g - cal/g - mole$
Specific heat	c	$1\ Btu/(lb_m)(^oF) =$	$1\ cal/(g)(^oC)$
Viscosity	μ	$1\ centipoise =$	$6.72 \times 10^{-4}\ lb_m/(ft - s)$
		$1\ centipoise =$	$2.089 \times 10^{-5}\ lb_f - s/ft^2$
		$1\ centipoise =$	$2.42\ lb_m/ft - h$
Volume	V	$1\ ft^3 =$	$28.316\ liter$
		$1\ m^3 =$	$1\ kiloliter$
		$1\ ft^3 =$	$7.48\ U.S.gal$
		$42\ gal =$	$1\ barrel(bbl)$

Appendix B

Physical Properties

A variety of physical properties are needed to model and design chemical processes. The accuracy of physical property estimates strongly affect the accuracy of the cost estimate. On the one hand, the design engineer wants to use the most accurate estimates that are available. On the other hand, approximations must often be made in order to control computational difficulties and to minimize the time requirements to complete the job.

The equations that are summarized below will enable one to compute a number of physical and thermodynamic properties that are commonly required for design. This summary is simply illustrative of physical property estimation techniques. It is intended to point out the general importance of physical property estimation to any design study and to emphasize some of the more useful approximations. A comprehensive discussion can be found elsewhere [81].

The procedures outlined here assume that a set of compositions for the appropriate phase (or phases) are available. Except for the bubble and dew point temperature calculations, these estimation methods don't require trial–and–error, at least whenever a composition is available. Of course, in certain instances one must estimate a composition simultaneously with the physical property calculations and then trial–and–error calculations are still required.

Physical property estimation begins with a selection of those design equations that will be used and the preparation of a suitable physical property data base. While there are many correlations that may be used, in each case a set of parameter values must be collected and organized into a suitable library. In such data bases, the name of a species and its physical properties usually comprise one *record*. The data base is simply a collection of such records.

For example, Tables B.1-B.3 summarize those parameter values that are required to use the equations described in this chapter for a few chemical species. If such information is organized into a data base, then a library name as indicated in the Species column of Tables B.1 through B.3 may be used to identify the component. Computer programs utilizing the data base simply reference the species name as it appears in the library and then all associated parameter values are copied from the data record into the appropriate storage locations of the computer program during

Table B.1: Physical properties for selected species from Cavett [18].

Species	T_{ci} (R)	P_{ci} (psia)	$\bar{\omega}_i$	δ_i $(\sqrt{cal/cc})$	$V_{\omega i}$ cc/g-mole
I-BUTANE	734.65	529.1	0.1825	6.240	13.370
N-BUTANE	756.31	550.7	0.1953	6.730	13.000
I-PENTANE	829.80	483.0	0.2104	6.710	15.360
N-PENTANE	845.60	489.50	0.2518	7.021	15.270
N-HEXANE	914.10	439.70	0.2972	7.266	17.640
ETHYLENE	509.51	742.1	0.0949	5.800	6.880
ETHANE	549.77	708.3	0.1064	5.880	7.880
PROPADIENE	707.00	674.9	0.1193	6.854	9.650
PROPYLENE	657.20	667.0	0.1451	6.200	9.690
PROPANE	665.95	617.4	0.1538	6.000	10.350
1,2-BUTADIENE	824.50	589.7	0.0987	7.950	10.940
1-BUTENE	755.30	583.0	0.1917	6.760	12.170
BENZENE	1012.7	714.0	0.2130	9.158	12.260
ETHYLBENZENE	1115.5	540.0	0.2936	8.787	17.230
M-XYLENE	1114.6	510.0	0.3045	8.818	1.280
O-XYLENE	1138.0	530.0	0.2904	8.987	17.030
TOLUENE	1069.20	590.0	0.2591	8.915	14.830

initialization. This technique enables one to accumulate a large library of species data, but to operate computer programs that minimize storage requirements.

The data required for each component in order to use these equations include the pure component critical temperature, T_{ci}, and pressure, P_{ci}, its modified acentric factor, $\bar{\omega}_i$, solubility parameter, δ_i, normal boiling point, T_{bi}, and λ_{bi}, its normal heat of vaporization. Several other constants that are specific for a particular correlation are also provided and their use is defined below.

B.1 Vapor Molar Enthalpy

The enthalpy of an ideal gas mixture of composition y_i at a temperature T, may be computed as the mole fraction weighted sum of the pure component vapor enthalpies:

$$H^{ig} = \sum_{i=1}^{NC} y_i H_i^{ig} \tag{B.1}$$

where $H_i^{ig} = H_i^{ig}(T)$ only since for an ideal gas $\left(\partial H_i^{ig} / \partial P\right)_T = 0$.

Furthermore, the pure component vapor enthalpies may be correlated [18] as a simple polynomial in temperature where:

$$H_i^{ig} = AH_i + BH_iT + CH_iT^2 + DH_iT^3 \tag{B.2}$$

Table B.2: Vapor phase enthalpy parameters from Cavett [18] yielding estimates in Btu/lb-mole.

Species	AH_i	BH_i	CH_i	DH_i
I-BUTANE	0.60231+04	0.19998+02	0.20441-01	-0.41461-05
N-BUTANE	0.66547+04	0.20545+02	0.18652-01	-0.30168-05
I-PENTANE	0.74756+04	0.24654+02	0.24465-01	-0.44373-05
N-PENTANE	0.80398+04	0.25189+02	0.23486-01	-0.41745-05
N-HEXANE	0.94750+04	0.29523+02	0.29141-01	-0.59119-05
ETHYLENE	0.37919+04	0.91448+01	0.78194-02	-0.15227-05
ETHANE	0.42201+04	0.11202+02	0.89381-02	-0.80662-06
PROPADIENE	0.44006+04	0.12635+02	0.96435-02	-0.22616-05
PROPYLENE	0.47148+04	0.13502+02	0.11469-01	-0.17972-05
PROPANE	0.50863+04	0.15108+02	0.15406-01	-0.29387-05
1,2-BUTADIENE	0.55404+04	0.17029+02	0.13548-01	-0.27142-05
1-BUTENE	0.59347+04	0.17689+02	0.17636-01	-0.39011-05
BENZENE	0.47399+04	0.16575+02	0.22876-01	-0.53555-05
ETHYLBENZENE	0.74046+04	0.25995+02	0.31878-01	-0.70263-05
M-XYLENE	0.73948+04	0.26078+02	0.29221-01	-0.59774-05
O-XYLENE	0.77375+04	0.27563+02	0.28048-01	-0.55709-05
TOLUENE	0.59813+04	0.20865+02	0.26105-01	-0.58166-05

Table B.3: Physical properties for selected species from Cavett [18].

Species	T_{bi} (R)	λ_{bi} (Btu/lb-mole)	AV_i (cp)	BV_i (K)	Molecular weight
I-BUTANE	470.07	9160.2	0.016946	698.03	58.12
N-BUTANE	490.68	9633.6	0.022860	608.59	58.12
I-PENTANE	541.80	10620.0	0.018717	733.66	72.151
N-PENTANE	556.74	11088.0	0.018717	733.66	72.15
N-HEXANE	615.42	12412.8	0.018879	820.48	86.18
ETHYLENE	305.10	5826.6	0.015872	391.37	28.05
ETHANE	332.28	6330.5	0.02292	368.69	30.07
PROPADIENE	429.66	8178.8	0.0220	368.6	40.0
PROPYLENE	405.90	7923.6	0.0175	600.0	42.08
PROPANE	415.98	8076.5	0.019045	542.32	44.09
1,2-BUTADIENE	511.08	9408.9	0.016	650.	54.0
1-BUTENE	480.42	9426.6	0.016	650.	56.1
BENZENE	635.94	13235.4	0.00724	1325.27	78.11
ETHYLBENZENE	736.92	15478.2	0.016165	1097.605	106.16
M-XYLENE	742.14	15660.0	0.0183	1033.91	106.16
O-XYLENE	751.68	15841.8	0.0131	1211.72	106.16
TOLUENE	690.84	14279.40	0.0169	1044.02	92.0

Table B.4: Constants for ideal enthalpy prediction using equations B.3- B.6.

i	c_i	i	c_i
0	3.1130335+02	10	-1.4886832-08
1	3.5442502	11	8.8395563-02
2	-5.2716944-02	12	4.2934703-03
3	3.5353777-04	13	-3.0836492-04
4	4.3401618+01	14	3.1073165-04
5	-8.9530784	15	-5.8323815-06
6	3.5118581-01	16	-5.6334748-08
7	-8.1862040-01	17	9.1745019-11
8	-2.8568658-05	18	-5.9563889-03
9	1.7024654-06		

where AH_i, BH_i, CH_i, and DH_i are constant coefficients for a given species. If the parameter values in Table B.2 are used, then the temperature must be expressed in ^{o}F and H_i^* is in Btu/lb-mole of pure gas.

The coefficients, AH_i, BH_i, CH_i, and DH_i may be [18] correlated from the relationships:

$$AH_i = M_i(c_0 + c_1A + c_2A^2 + c_3A^3 + c_4K + c_5K^2 + c_6K^3 + c_{18}AK) \tag{B.3}$$

$$BH_i = M_i(c_7 + c_8A + c_9A^2 + c_{10}A^3 + c_{11}K + c_{12}K^2 + c_{13}K^3) \tag{B.4}$$

$$CH_i = M_i(c_{14} + c_{15}A) \tag{B.5}$$

$$DH_i = M_i(c_{16} + c_{17}A) \tag{B.6}$$

where M_i = the molecular weight of the ith species
A = the degrees API, defined in equation B.7
K = the UOP characterization factor, defined in equation B.8
c_i = empirical parameters that are summarized in Table B.4.

The degrees API[1] is a density scale defined as:

$$A = \frac{141.5}{G} - 131.5 \tag{B.7}$$

where G is the species liquid density measured at 60°F divided by the density of water[2] at the same temperature. The UOP[3] characterization factor relates the normal boiling point in R, T_b, to G, the specific gravity at 60°F:

$$K = \frac{T_b^{1/3}}{G} \tag{B.8}$$

While equations B.3-B.6 were developed for use with true boiling point curves in refinery fractionation, Cavett [18] also used them for pure components to predict

[1] API stands for the American Petroleum Institute.
[2] The density of water at 60°F is 0.99905 g/cc
[3] UOP stands for the Universal Oil Products Company

ideal enthalpies. More generally, ideal gas phase enthalpies may be obtained by integrating from a reference temperature:

$$H_i^{ig}(T_o) = H_i^{ig}(T) + \int_{T_o}^{T} C_p^{ig}\, dT \tag{B.9}$$

In order to correct enthalpy for pressure effects, an equation of state may be used to integrate at constant T:

$$H(T,P) = H^{ig}(T,0) + \int_0^P \left[V - T\left(\frac{\partial V}{\partial T}\right)_P \right] dP \tag{B.10}$$

or more conveniently to integrate at constant T:

$$\frac{H - H^{ig}}{RT} = \frac{H^R}{RT} = -T \int_0^P \left(\frac{\partial Z}{\partial T}\right)_P \frac{dP}{P} \tag{B.11}$$

where $H^R = H - H^{ig}$ is the residual enthalpy.

B.2 Phase Changes

Whenever two or more phases are in thermodynamic equilibrium, the enthalpies of the phases are related to each other by one or more latent heats. For example, the latent heat of vaporization, λ_{vi}, relates the enthalpies of the *ith* species in saturated vapor and liquid phases:

$$H_{vi} = H_{Li} + \lambda_{vi} \tag{B.12}$$

for each species in an ideal mixture. However, since the latent heat of vaporization is zero at the critical point, we also have the restriction:

$$\lambda_{vi} = \psi_i = \begin{cases} \lambda_{vi} & T < T_{ci} \\ 0 & T > T_{ci} \end{cases} \tag{B.13}$$

Also, for the saturated liquid mixture:

$$H_L = \sum_{i=1}^{NC} x_i H_{Li} \tag{B.14}$$

if the heat of mixing can safely be ignored.

Similarly, the enthalpies of solid and liquid phases at equilibrium are related by the heat of fusion:

$$H_{Li} = H_{si} + \lambda_{fi} \tag{B.15}$$

and the enthalpy change of sublimation is the sum of the two:

$$\lambda_{si} = \lambda_{fi} + \lambda_{vi} \tag{B.16}$$

B.3 Vapor Pressures

The vapor pressure change with temperature of a substance and phase is related to the latent enthalpy through the Clapeyron equation. The change in pure component vapor pressure that results from heating a liquid is:

$$\frac{dP}{dT} = \frac{\lambda_v}{T(V_v - V_L)} \tag{B.17}$$

where V_v and V_L are the vapor and liquid phase molar volumes, respectively.

Similarly, the Clapeyron equation relates the vapor pressure change from heating a solid to the latent heat of fusion:

$$\frac{dP}{dT} = \frac{\lambda_f}{T(V_L - V_s)} \tag{B.18}$$

where V_s is the molar volume of the solid.

B.3.1 Cox-Antoine

Cox-Antoine relationship is often useful for estimating pure component vapor pressures over a limited temperature range. In this case, the vapor pressure is estimated as:

$$\log_{10} P_i^{sat} = A - \frac{B}{T - C} \tag{B.19}$$

If experimental data are available, it may be convenient to estimate the coefficients of the Cox–Antoine equation simply by plotting data over the range of interest. It is also possible to estimate these parameters from the boiling properties of the mixture. Thompson [104] has suggested some simple rules for estimating the Cox-Antoine coefficients. For monatomic elements and all substances with T_b less than 125 K:

$$C = -0.3 + 0.034T_{bi} \tag{B.20}$$

For all other substances:

$$C = -18 + 0.19T_{bi} \tag{B.21}$$

where T_{bi} is the normal boiling point in K. The remaining coefficients may then be calculated from:

$$B = \frac{0.651\lambda_{vbi}}{R}\left(\frac{T_{bi} - C}{T_{bi}}\right)^2 \tag{B.22}$$

where λ_{vbi} is the latent heat of vaporization at the normal boiling point in cal/g-mole, P_i^{sat} is the pure component vapor pressure in atm, and:

$$A = \frac{B}{T_{bi} - C} \tag{B.23}$$

B.3.2 Relative Volatilities

The average relative volatility of two species is related to the normal boiling points of the two species and their latent heats of vaporization by an expression developed by Wagle [114]:

$$\alpha = \exp\left[0.25164 \times \left(\frac{1}{T_{b1}} - \frac{1}{T_{b2}}\right) \times (\lambda_{b1} + \lambda_{b2})\right] \tag{B.24}$$

where α = is the relative volatility between the two species in the temperature range T_{b1} to T_{b2}

T_{bi} = the ith species normal boiling points in K

λ_{bi} = the species latent heats of vaporization at their respective boiling points in cal/g-mole.

B.4 Vaporization Enthalpies

Perhaps the simplest method of estimating a latent heat of vaporization at the normal boiling point is given by Trouton's rule:

$$\lambda_{vbi} = 21 T_{bi} \tag{B.25}$$

where T_{bi} is in K and λ_{vbi} is in cal/g-mole.

Vetere's modification of the Kistiakowsky equation [81, 114] gives improved more accurate estimates than Trouton's rule. For simple hydrocarbons:

$$\lambda_{vbi} = T_{bi}\left\{13.91 + 3.27\log_{10} M_i + 1.55\frac{\left[T_{bi} - (263M_i)^{0.581}\right]^{1.037}}{M_i}\right\} \tag{B.26}$$

For alcohols and acids:

$$\begin{aligned}\lambda_{vbi} &= T_{bi}\left[19.388 + 3.1269\log_{10} T_{bi} - 6.1589\frac{T_{bi}}{M_i} + 0.035021\frac{T_{bi}^2}{M_i}\right. \\ &\qquad \left. -5.1056 \times 10^{-5}\frac{T_{bi}^3}{M_i}\right]\end{aligned} \tag{B.27}$$

For other polar compounds:

$$\begin{aligned}\lambda_{vbi} &= T_{bi}\left[10.604 + 3.664\log_{10} T_{bi} + 0.09354\frac{T_{bi}}{M_i}\right. \\ &\qquad \left. +1.035 \times 10^{-3}\frac{T_{bi}^2}{M_i} - 1.345 \times 10^{-6}\frac{T_{bi}^3}{M_i}\right]\end{aligned} \tag{B.28}$$

where M_i = the ith species molecular weight

T_{bi} = the normal boiling point in K

λ_{vbi} = the latent heat at the normal boiling point in cal/g-mole.

If a latent heat of vaporization at one boiling point temperature is known, then the latent heat may be estimated at other temperatures using the Watson correlation [82]:

$$\lambda_{vi}(T_r) = \lambda_{vbi}\left(\frac{1-T_r}{1-T_{rbi}}\right)^{0.38} = \lambda_{vbi}\left(\frac{T_c-T}{T_c-T_{bi}}\right)^{0.38} \tag{B.29}$$

where T_r and T_{rbi} are the reduced temperatures at the conditions of interest and the normal boiling point, respectively.

Example: B.1 *Estimate the latent heat of vaporization of n-butane at its normal boiling point using (1) Trouton's rule, (2) Vetere's modification of the Kistiakowsky equation. What is the percent error compared with the experimental value in each case?*

Solution: B.1 *From Perry's Handbook, the experimental value is 5352.27 cal/g-mol. The normal boiling point is 490.6 R or 272.6 K. The molecular weight is 58.12. Using Trouton's Rule:*

$$\lambda_{vap_i} = 21T_{bi} = 21(272.6) = 5724.6$$

with $(5725.6 - 5352.27)/5352.7 \times 100\% = 6.96\%$ *error.*

Vetere's modification of the Kistiakowsky equation states:

$$\lambda_{vap_i} = T_{bi}\left\{13.91 + 3.27\log_{10} M_i + 1.55\frac{[T_{bi} - (263M_i)^{0.581}]^{1.037}}{M_i}\right\}$$

which in this case gives $\lambda_{vap_i} = 5352.27$ *cal/g-mol or 0.62% error. Both estimates are higher than the experimental value.*

B.5 Solubility Parameter

The solubility parameter is related to the cohesive energy density, c_{ii}, and to the latent heat of vaporization. For a pure liquid i, the cohesive energy density is defined as:

$$c_{ii} = \frac{\Delta U_i}{V_{Li}} \tag{B.30}$$

where ΔU_i is the energy required to isothermally evaporate liquid i from the saturated liquid to the ideal gas. At temperatures well below the critical point:

$$\Delta U_i \approx \lambda_{vi} - RT \tag{B.31}$$

where λ_{vi} is the molar enthalpy of vaporization of the pure liquid at temperature T and V_{Li} is the molar liquid volume. The solubility parameter, δ_i, is defined as:

$$\delta_i = (c_{ii})^{1/2} = \left(\frac{\Delta U_i}{V_{Li}}\right)^{1/2} \tag{B.32}$$

B.6 Molar Volumes

In computing vapor/liquid equilibria, it may be necessary to estimate both the vapor and liquid phase molar volumes. The former quantity may be estimated using an appropriate equation of state while the latter may be estimated using experimental liquid densities and empirical relationships that estimate the temperature dependence of the liquid phase.

B.6.1 Vapor Phases

The molar volume of a vapor mixture may be calculated from the Redlich–Kwong equation of state as given in Reid and Sherwood [82] and rearranged into compressibility factor form:

$$\frac{PV}{RT} = Z = \frac{1}{1-h} - \left(\frac{A^2}{B}\right)\left(\frac{h}{1+h}\right) \tag{B.33}$$

where

$$h = \frac{BP}{Z} \tag{B.34}$$

$$A = \sum_{i=1}^{NC} y_i a_i \tag{B.35}$$

$$B = \sum_{i=1}^{NC} y_i b_i \tag{B.36}$$

$$a_i^2 = \frac{0.4278}{T_{ri}^{2.5} P_{ci}} \tag{B.37}$$

$$b_i = \frac{0.0867}{T_{ri} P_{ci}} \tag{B.38}$$

where Z = the compressibility of the vapor mixture
y_i = vapor phase mole fraction of ith species

Equation B.33 is a cubic in Z and so it can have as many as three real roots. As discussed more fully by Edmister [27], the vapor phase compressibility factor is always the largest root.

B.6.2 Liquid Phases

Molar volumes for liquid mixtures may be estimated as the mole fraction weighted sum of the pure component volumes given in Reid and Sherwood [82]:

$$V_{Li} = V_{\omega i}(5.7 + 3.0 T_{ri}) \tag{B.39}$$

where the liquid expansion factor, $V_{\omega i}$, is a constant for each species.

Table B.5: Empirical coefficients for predicting pseudo critical temperatures and pressures in refinery mixtures.

i	a_i	b_i
0	768.07121	2.8290406
1	0.17133693+01	0.94120109-03
2	-0.10834003-02	-0.30474747-05
3	-0.89212579-02	-0.20876110-04
4	0.38890584-06	0.15184103-08
5	0.53094920-05	0.11047899-07
6	0.32711600-07	-0.48271599-07
7		0.13949619-09

For true boiling point[4] mixtures, the liquid expansion factor may be correlated using [18]:

$$V_{\omega i} = \frac{V_{Li}}{5.7 + 1610.07/T_{pc}} \tag{B.40}$$

where V_{Li} is the molar volume of the liquid mixture measured at 25°C (ml/g-mole) and:

$$V_{Li} = \frac{< M >}{\rho_{25^oC}} \tag{B.41}$$

while $< M >$ is the average molecular weight of the mixture and T_{pc} is a pseudo critical temperature estimated by:

$$T_{pc} = a_0 + a_1 T + a_2 T^2 + a_3 AT + a_4 T^3 + a_5 AT^2 + a_6 A^2 T^2 \tag{B.42}$$

where A is the density in degrees API and T is the molar average boiling point temperature in °F. The empirical coefficients, a_i, are summarized in Table B.5 together with coefficients for estimating a pseudo critical pressure via:

$$\log_{10} P_{pc} = b_0 + b_1 T + b_2 T^2 + b_3 AT + b_4 T^3 + b_5 AT^2 + b_6 A^2 T + b_7 A^2 T^2 \tag{B.43}$$

B.7 Acentric Factor

The acentric factor is defined as:

$$\omega \equiv - \log_{10} P_r|_{T_r=0.7} - 1.000 \tag{B.44}$$

for a pure component. For a mixture, using the pseudo critical pressure, P_r' (evaluated at $T_r = 0.7$) is substituted for P_r.

[4]For refinery oils, the composition is approximated by a *true boiling point* distillation, commonly called a *TBP* distillation. TBP procedures are described elsewhere [70], but the method involves a batch distillation using a large number of stages and high reflux ratios so that the temperature at any point on the temperature–volumetric yield curve represents the true boiling point at that volume percentage point.

In the Chao–Seader method described below, a modified acentric factor, $\bar{\omega}$, is used. For many species, the modified acentric factor is approximately equal to ω.

B.8 Phase Equilibrium Constants

At equilibrium, the component fugacity in a liquid mixture is equal to the fugacity of that species in the coexisting vapor phase, ($i.e.$ $\hat{f}_{Li} = \hat{f}_{vi}$). For the liquid phase:

$$\hat{f}_{Li} = x_i \gamma_{Li} f^o_{Li} \tag{B.45}$$

where γ_{Li} is the liquid phase activity coefficient at T and P and with the composition $x_i \cdots$. The f^o_{Li} term is the fugacity of the pure component i under the pressure and temperature of the system *and as a liquid.* If the system pressure, P, is greater than the pure component vapor pressure, P_i^{sat}, then the system is stable and the liquid is subcooled. In this case:

$$f^o_{Li} = \phi_i^{sat} P_i^{sat} exp\left[(1/RT)\int_{P_i^{sat}}^{P} V^o_{Li} dP\right] \tag{B.46}$$

$$f^o_{Li} \approx \phi_i^{sat} P_i^{sat} exp\left[\frac{V^o_{Li}(P - P_i^{sat})}{RT}\right] \tag{B.47}$$

$$\tag{B.48}$$

where the exponential is known as the Poynting[5] factor.

For the vapor phase:

$$\hat{f}_{vi} = y_i \hat{\phi}_{vi} P \tag{B.49}$$

where $\hat{\phi}_{vi}$ is the fugacity coefficient of species i in the vapor mixture, y_i is the mole fraction of i in the vapor phase, and P is the total system pressure.

In order to compute $\hat{\phi}_{vi}$, one must assume an equation of state and integrate:

$$\ln \hat{\phi}_{vi} = \int_0^P \left[\frac{\partial(nZ - n)}{\partial n_i}\right]_{P,T,n_j} \frac{dP}{P} \tag{B.50}$$

and since $\partial(nZ)/\partial n_i = \bar{Z}_i$ and $\partial n/\partial n_i = 1$, we may write:

$$\ln \hat{\phi}_{vi} = \int_0^P \left(\bar{Z}_i - 1\right) \frac{dP}{P} \tag{B.51}$$

where $\bar{Z}_i$ is the partial molar compressibility for species i. Using the Redlich–Kwong equation of state and integrating, then:

$$\begin{aligned} \ln \hat{\phi}_{vi} &= (Z-1)\frac{B_i}{B} - \ln|Z - BP| \\ &\quad - \frac{A^2}{B}\left[\frac{2A_i}{A} - \frac{B_i}{B}\right] \ln\left|1 + \frac{BP}{Z}\right| \end{aligned} \tag{B.52}$$

[5] John Henry Poynting (1852-1914), British physicist.

In the case of an ideal vapor solution, then $\hat{\phi}_{vi}^{id} = \phi_{vi}$ and integration of the Redlich–Kwong equation of state yields:

$$\ln \phi_{vi} = (Z-1) - \ln|Z(1-h)| - \psi \ln|1+h| \tag{B.53}$$

where:

$$\psi = \frac{A^2}{B} = \frac{A_i^2}{B_i} = \frac{4.9343}{T_{ri}^{1.5}} \tag{B.54}$$

$$BP = \frac{0.0867 P_{ri}}{T_{ri}} \tag{B.55}$$

$$Z = \frac{1}{1-h} - \psi\left(\frac{h}{1+h}\right) \tag{B.56}$$

$$h = \frac{BP}{Z} \tag{B.57}$$

The vapor/liquid equilibrium constant, K_i, is defined as

$$K_i = \frac{y_i}{x_i} = \left(\frac{\gamma_{Li}}{\hat{\phi}_{vi}}\right)\left(\frac{f_{Li}^o}{P}\right) \tag{B.58}$$

The various factors in the right hand side of the equation above are functions of temperature, pressure, and composition and may be determined using the Chao–Seader [82] method which is valid for non–polar mixtures up to about 800°F and 1500 psia. With this method, correlating equations are required for three different terms. That is:

$$K_i = f\left[\gamma_{Li}, \hat{\phi}_{vi}, f_{Li}^o/P\right] \tag{B.59}$$

where γ_{Li} is the liquid phase activity coefficient, f_{Li}^o/P is a pure component liquid fugacity coefficient, and $\hat{\phi}_{vi}$ is the partial molar vapor fugacity coefficient.

On the other hand, when a system obeys Raoult's law, then:

$$K_i = \frac{y_i}{x_i} = \frac{f_{Li}^o}{P} = \frac{P_i^{sat}}{P} \tag{B.60}$$

and at low to moderate pressures, non–ideal liquids may obey a *modified* Raoult's law:

$$K_i = \frac{y_i}{x_i} = \frac{\gamma_{Li} P_i^{sat}}{P} \tag{B.61}$$

However, with the Chao-Seader method, the liquid phase activity coefficients, γ_{Li}, are estimated from the Scatchard-Hildebrand regular solution theory:

$$\ln \gamma_{Li} = \frac{V_{Li}(\delta_i - \bar{\delta})^2}{RT} \tag{B.62}$$

and for the mixture:

$$\bar{\delta} = \frac{\sum_{i=1}^{NC} x_i V_{Li} \delta_i}{\sum_{i=1}^{NC} x_i V_{Li}} \tag{B.63}$$

Table B.6: Empirical coefficients for the Chao–Seader correlation for K-values.

	Simple fluid	Methane	Hydrogen
d_0	2.05135	1.36822	1.50709
d_1	-2.10899	-1.54831	2.74283
d_2			-0.02110
d_3	-0.19396	0.02889	0.00011
d_4	0.02282	-0.01076	0
d_5	0.08852	0.10486	0.008585
d_6		-0.02529	0
d_7	-0.00872		
d_8	-0.00353		
d_9	0.00203		

where δ_i = the Hildebrand solubility parameter
V_{Li} = the molar liquid volume for pure species i
x_i = the mole fraction in the liquid phase.
Values for the solubility parameter are included in Table B.1 for several pure components. They are generally computed using equations B.30-B.32.

The standard state liquid fugacity term, f^o_{Li}/P, may be obtained from a relation proposed by Chao and Seader [82] which involves reduced temperature and pressure, a modified acentric factor, and the other empirical constants that depend on the species.

$$\log_{10}\left(\frac{f^o_{Li}}{P}\right) = \log\nu^o + \bar{\omega}\log\nu' \qquad \text{(B.64)}$$

$$\begin{aligned}\log\nu^o &= d_o + \frac{d_1}{T_r} + d_2T_r + d_3T_r^2 + d_4T_r^3 + (d_5 + d_6T_r + d_7T_r^2)P_r \\ &\quad +(d_8 + d_9T_r)P_r^2 - \log_{10}P_r \end{aligned} \qquad \text{(B.65)}$$

and where the empirical coefficients, d_i, are as shown in Table B.6.

$$\log\nu' = -4.23893 + 8.65808T_r - \frac{1.2206}{T_r} - 3.15224T_r^3 - 0.025(P_r - 0.6) \qquad \text{(B.66)}$$

where d_i = are empirical coefficients in the Chao–Seader correlation
T_r = the reduced temperature for ith species
P_r = the reduced pressure for ith species
$\bar{\omega}$ = the modified acentric factor for the ith species

B.9 Bubble and Dew Point Temperatures

At a fixed composition and pressure, the bubble point of a liquid mixture may be found by a trial–and–error solution of the relationship:

$$\sum_{i=1}^{NC} y_i = \sum_{i=1}^{NC} K_i x_i = 1.0 \tag{B.67}$$

Similarly, the dew point temperature of a vapor mixture of composition $y_i, i = 1, \ldots, NC$ at a total vapor pressure, P, may be taken to be the temperature at which the equation:

$$\sum_{i=1}^{NC} x_i = \sum_{i=1}^{NC} \frac{y_i}{K_i} = 1.0 \tag{B.68}$$

is satisfied. Trial-and-error calculations are required in these cases since the pure component vapor pressure always has a temperature dependence.

B.10 Bubble and Dew Point Pressures

For a given composition and temperature, a bubble point pressure is computed from equation B.67. If the Chao–Seader method is used to compute K-values, then trial–and–error calculations are still required since equations B.65-B.66 involve the reduced pressures for the species in the mixture. However, if the VLE system obeys Raoult's law, then trial–and–error calculations are not required and:

$$P_{bub} = \sum_{i=1}^{NC} P_i^{sat} x_i \tag{B.69}$$

Similarly, trial–and–error calculations are not required for systems obeying a modified Raoult's law and:

$$P_{bub} = \sum_{i=1}^{NC} P_i^{sat} \gamma_{Li} x_i \tag{B.70}$$

Trial–and–error calculations are required for Chao–Seader mixtures to compute dew point pressures, but not for Raoult's law mixtures where:

$$\frac{1}{P} = \sum_{i=1}^{NC} \frac{y_i}{P_i^{sat}} \tag{B.71}$$

is obeyed. A modified Raoult's law is not applicable in this case since γ_{Li} always refers to non–ideality in the liquid phase.

B.11 Liquid Viscosities

The Andrade [82] equation may be used to estimate pure component liquid viscosity.

$$\mu_i = AV_i e^{BV_i/T} \tag{B.72}$$

where AV_i and BV_i are constant for a given component, and temperature is in K. Mixture viscosity may be approximated using the Kendall–Monroe equation:

$$\mu^{1/3} = \sum_{i=1}^{NC} x_i \mu_i^{1/3} \tag{B.73}$$

which is applicable to non–electrolytic, non–associating liquid pairs whose component molecular weight and viscosity differences are small.

B.12 Summary

Chemical process work can only be as accurate as the physical properties models that are employed in the evaluation. The models which are presented here are generally accurate for nonpolar hydrocarbon mixtures.

B.13 Problems

2.1 A dew point vapor consists of three species, A, B, and C, with mole fractions (0.8,0.1,0.1) respectively. The vapor is entering a heat exchanger where it will be totally condensed, but not subcooled. The vapor/liquid equilibrium constants for these species can be approximated as:

$$\begin{aligned} K_A &= 3 + 0.15T \\ K_B &= 2 + 0.10T \\ K_C &= 1 + 0.05T \end{aligned}$$

where the temperature, T, is in ^oF. What is the temperature of the vapor stream entering the heat exchanger? What is the temperature of the liquid stream exiting the heat exchanger?

2.2 An equimolar mixture of species 1, 2, and 3 is flowing through an isothermal pipeline. The mixture is ideal and each species obeys the Cox-Antoine pure component vapor pressure correlation shown below with the corresponding parameter values for each of the three species.

$$\log P_s = A - \frac{B}{T - C}$$

where T is in ^oC and P_s is in mm Hg.

Species	Parameter Values		
	A	B	C
1	7.3	1550	-210
2	6.9	1345	-219
3	7.2	1417	-211

If the pipeline is at 200 psia and 440.6 ^oF, is the mixture a liquid or a vapor? If it is a liquid, then at what pressure will it begin to boil? If it is a vapor, then at what pressure will it begin to condense? Answer these same three questions if the isothermal pipeline is at 100 psia and 440.6 ^oF. What are the K-values for these three species at 150 psia and 510 ^oF? Which species is the most volatile? Answer these two questions for all conditions which may be mentioned.

2.3 Consider the three species mentioned in problem 2. Which is the most volatile? Which is the least volatile? What are the K values of the species at 150 psia at the mixture bubble and dew point temperatures if the mixture is 25% A and 50% B on a molar basis? What are the relative volatilities of A and B to C at these conditions? Which is easier, the A/B or the B/C split?

2.4 Suppose that a bubble point liquid, consisting of species A, B, and C with mole fractions (0.1,0.1,0.8) and K values as expressed in Problem 1 is entering a reboiler where the liquid will be totally vaporized, but not superheated. What is the temperature of the vapor stream exiting the reboiler? What is the temperature of the liquid stream entering the reboiler?

2.5 Estimate the latent heat of vaporization at 300 ^oF for a species whose normal boiling point is 105 ^oF with a latent heat of vaporization equal to 565 BTU/lb-mole at the normal boiling point, and whose critical temperature is 935 K.

2.6 A pipeline carries a species whose vapor pressure over small temperature intervals (the range of interest) can be approximated using the equation:

$$P_s = 100 + 2.5T$$

where P is in psia and T ^oF. At reference point 1 in the pipeline, the temperature of the vapor is 100 ^oF and the temperature profile across a differential section of the pipe at that point is essentially flat. However, as the vapor moves downstream, the average temperature of the vapor drops 5 $^oF/ft$ of pipe length. At what length from the reference point will the vapor pressure equal 180 psia?

2.7 Species 1, 2, and 3 exhibit vapor liquid equilibrium K values which can be correlated using the expression:

$$\begin{aligned} K_1 &= 7.5 + 0.4T \\ K_2 &= 4.5 + 0.24T \\ K_3 &= 1.5 + 0.08T \end{aligned}$$

where T is in ^oF.

A saturated vapor stream with composition (0.75,0.1,0.15) must be totally condensed, but not subcooled. Based upon the vapor properties of the mixture, estimate the temperature of the condensate. Ans: -15.8

2.8 A bubble point liquid stream with composition (0.05,0.15,0.8) must be totally vaporized, but not superheated. Based upon the mixture properties in problem 7, estimate the temperature of the vapor exiting the reboiler. Ans: -8.0

2.9 A process stream has a latent heat of vaporization of 970.33 Btu/lb at 212 K, the stream's normal boiling point. Its critical temperature is 647 K. Between the temperatures of 32 - 212 ^oF, its liquid heat capacity is essentially constant at 1 Btu/lb-F. Between the temperatures 212 - 235 ^oF, the liquid heat capacity for the stream is 1.2 Btu/lb-F.

If the specific enthalpy of the saturated liquid is 0 Btu/lb at 32 ^oF, then estimate the enthalpy of the stream if it is a saturated vapor at 235 ^oF. Ans: 1161 Btu/lb

2.10 Estimate the specific heat of the stream in problem 9 as a saturated vapor between the temperatures 212 - 235 ^oF. Ans: 0.44 Btu/lb-^{o}F

2.11 A pure species exhibits a normal boiling point temperature of 85^oC and it has a critical temperature of 320 ^oC. Estimate the latent heat of vaporization for this species at a temperature of 225 ^oC. Ans (in cal/g-mole): 5331

2.12 Estimate the compressibility of an equimolar mixture of n-pentane and n-hexane at 500°F and 400 psia.

2.13 Estimate the latent heat of vaporization of n-pentane at its normal boiling point using (1) Trouton's rule, (2) Vetere's modification of the Kistiakowsky equation. What is the percent error compared with the experimental value in each case?

2.14 Repeat the calculations for problem 13 at 550°F using the Watson correlation to adjust the latent heat estimates for the new conditions.

2.15 Estimate the viscosity of an equimolar mixture of m-xylene, o-xylene, and toluene at 200°F.

2.16 Compute the K–values for an equimolar mixture of ethane, ethylene and propylene at 100°F and 300 psia assuming: (1) Raoult's law with the Thompson rules to compute pure component vapor pressures and (2) using the Chao–Seader correlation assuming a non–ideal vapor phase mixture and (3) using the Chao–Seader correlation assuming an ideal vapor phase mixture.

2.17 Estimate the molar volume of liquid ethylbenzene at 500°F.

2.18 Estimate the molar volume of ethane at 300°F and 400 psia using the Redlich–Kwong equation of state.

2.19 Estimate the compressibility factor for an equimolar mixture of n-pentane and n-butane vapors at 200°F and 100 psia.

2.20 Estimate the relative volatility of benzene to that of toluene in the temperature range of their normal boiling points. Ans: 2.375

2.21 A pure species exhibits a normal boiling point temperature of 311 °C and it has a critical temperature of 980 °C. Estimate the latent heat of vaporization for this species at a temperature of 510 °C.

2.22 A pure species exhibits a normal boiling point temperature of 350°C and it has a critical temperature of 780°C. Estimate the latent heat of vaporization for this species at a temperature of 510°C.

2.23 A mixture of species 1 and 2 is pumped through an isothermal pipeline. The mixture is 40 mol% species 1 and 60 mol% species 2. Each species obeys Raoult's law and the mixture is ideal. The pure component vapor pressures are 16 and 12 psia for species 1 and 2, respectively. At what pressure in the pipeline is exactly half of the mixture vapor and half liquid? What is the composition of the liquid phase at that pressure?

2.24 An air-free tank contains a vapor/liquid mixture of a single chemical species. Assuming that the liquid forms an ideal gas when vaporized, estimate the tank pressure at 230°C. The species normal boiling point is 85°C.

2.25 A sensitive liquid thermometer is needed for an online controller. One engineer suggests immersing a flask with an open-ended glass tube attached to it in the process stream. The glass tubing will serve as a sight glass above the flask, and it has an internal diameter of exactly 1 mm. The flask and tube are initially filled with 100 ml of n-pentane at 25°C to an initial temperature mark that is exactly 5 cm above the top of the flask. How high above the initial temperature mark will the liquid meniscus be at 100°C? Assume that the tube above the flask is also maintained at 100°C.

2.26 A sealed container holds a mixture that is at equilibrium in a constant temperature bath. At the equilibrium conditions, the K values of species A, B, and C are 1.9, 1.1, and 0.8, respectively. What is the condition of this mixture? How many and what phases are present?

Appendix C

Material and Energy Balances

Approximations in material and energy balances are helpful during the Preliminary Design Phase. Steady-state stream conditions are usually needed to define the design basis. In the case of batch operations, it is often helpful to define equivalent steady-state flows. In any case, the goal is to obtain reasonably accurate estimates of the project cost, based on approximate material and energy balances.

Material balances may be approximated as necessary to achieve design goals as long as several points are remembered. As approximations are made, the intent is to use them to complete preliminary equipment sizing calculations and cost estimates. Therefore, approximations should be made in such a way as to overdesign the equipment by a reasonable amount (e.g., 10%). If a distillation tower is being designed then the material balances used in its prelminary sizing should slightly overestimate the theoretical tray requirements and the tower diameter, because these are the parameters upon which the cost estimate is based.

Second, it should be recognized that all approximations must subsequently be reviewed and verified by experiment. If the design engineer has done a good job, then the resulting estimate of equipment sizing and costs will be less than computed during the Preliminary Design. In order to realize this goal, the engineer must understand how material balance assumptions impact sizing calculations and cost estimates.

The structure of many chemical processes is such that they can be modelled (or simulated) rigorously only by completing trial-and-error calculations. Countercurrent, equilibrium stage models of unit operations are one important class of models that require trial-and-error calculations. Seader and Henley [87] describe equilibrium-based methods for modeling multicomponent absorption, stripping, distillation and extraction processes. They summarize general strategies of numerical solution for such models which involve the solution of coupled sets (tridiagonal) of equations.

C.1 Material and Energy Balance Heuristics

Whenever information is lacking in a design project, engineers resort to assumptions in order to complete a Preliminary Design. These assumptions must eventually be confirmed by experiment, but reasonable assumptions, even when based upon very crude approximations or heuristics, can often be helpful in getting things started. Several suggestions are listed below. They should be used with caution.

Heuristic: 29 *(80/20 Rule) In the absence of better information, assume that stream splits are no better than 80/20. In waste management cleanup operations, for example, it is often approximately true that about 80% of the contamination, say on a molar basis, is placed in roughly 20% of the waste containers. So about 80% of the waste containers will only have low levels of contamination, but about 20% will contain much higher contamination levels.*

Heuristic: 30 *Focus on the mass and molar flowrates for the most plentiful components. They usually determine the size requirements for equipment. Try to make slightly pessimistic assumptions about trace components. for example, apply the 80/20 rule or some reasonable modification to fractional conversions or recovery factors to estimate trace components.*

Heuristic: 31 *Assume worst case scenarios for trace components if they may be a problem (e.g., excessive* NO_x *emissions, requiring additional cleanup equipment). Present your initial design with the assumed additional equipment added as line items in your budget (overdesign if it simplifies your analysis). No one is likely to complain later if you are able to take this equipment out because subsequent analyses proves that it is not needed, but you will likely be criticized if essential equipment is later found necessary, but was excluded during Preliminary Design. In the first instance, your initial estimates will be somewhat higher than required. In the latter instance, space may not be available to add the equipment if it is deemed essential in the final design.*

Heuristic: 32 *List your assumptions and present them to management as part of your initial design report.*

1. *Clearly explain what assumptions have been made up front.*
2. *Treat assumptions parametrically. Set up a spreadsheet and adjust the value for each assumed parameter over a likely range and recompute estimated costs as a function of each.*
3. *Rank assumptions in terms of their relative economic importance. The most significant assumptions should be ranked first.*
4. *If the project is accepted, work to define an experimental program to evaluate your assumptions, starting with those assumptions which are at the top of your ranked list. (There is little point in experimentally verifying assumptions if they are not economically important.)*

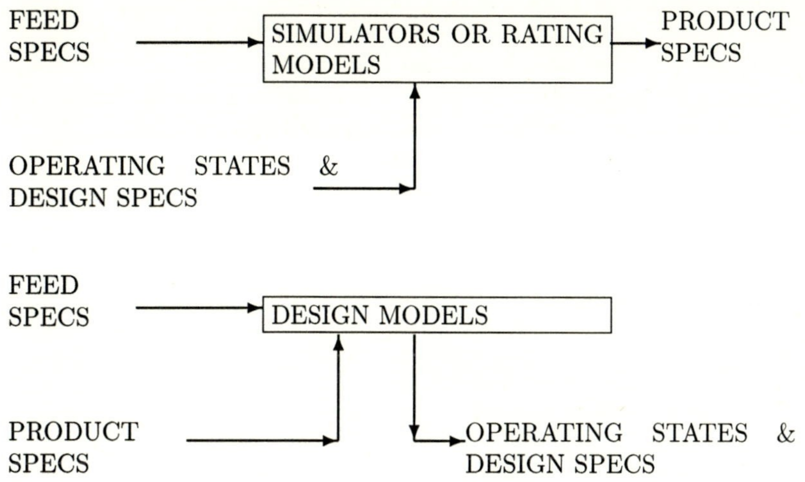

Figure C.1: Simulators, or rating models, differ from design models in the direction of their information flow.

Chapter 1 describes the two general classes of models in use — simulators and design models. These two groups differ primarily in the ways they handle information and variables. As indicated in Figure C.1, simulators compute product compositions and rates as dependent variables while design models compute operating states and specifications as dependent variables. For purposes of Preliminary Design, it is helpful to use design models to estimate and optimize equipment specifications, and then to use simulators to verify the accuracy of the design models. One first step in this procedure is developing an approximate set of material balances, say using a composition node approach.

C.2 Precedence Order

The purpose of material balance approximations as discussed here is to provide reasonable estimates of material and energy requirements for countercurrent separators, or other unit operations, in such a way as to minimize the need for trial-and-error calculations. In some cases, a set of design equations can be solved sequentially without trial-and-error if stream compositions at key points are estimated using approximate methods. Sets of design equations may then be solved sequentially, rather than simultaneously, if the compositions of key streams can be reasonably estimated. The process of solving an equation set in a prescribed sequence is also known as precedence ordering [86].

Suppose we need to solve two equations, say $f_1(x_1, x_2)$ and $f_2(x_1, x_2)$, and both equations are a function of x_1 and x_2. We could solve both equations simultaneously using trial-and-error. On the other hand, we could use precedence ordering to solve these two equations sequentially. We could use equation 1 to compute x_1 as a function of x_2, and then use equation 1 to eliminate x_1 from equation 2.

Example: C.1 *Use precedence ordering to solve the equation set*

$$f_1(x_1, x_2) = 8x_1 - 5x_2^2 + 6 = 0 \quad (C.1)$$
$$f_2(x_1, x_2) = 3x_1 + 7x_2^2 - 10 = 0 \quad (C.2)$$

Solution: C.1 *Rearranging* f_1 *gives* $x_1 = (5x_2^2 - 6)/8$
Eliminating x_1 *from* f_2 *gives* $x_2 = 1.17485$
Substituting into f_1 *gives* $x_1 = 0.11267$.

One approximate method of estimating stream compositions utilizes overall material balance considerations to estimate key system stream flowrates and compositions. This method is called species allocation and several species allocation techniques are presented later in this chapter.

The purpose of species allocation is to estimate material balances and the compositions at particular node points in a flowsheet in order to simplify design calculations by defining a precedence order. For example, the precedence order, or sequence in which equations are solved, to evaluate single countercurrent distillation column can be as follows:

1. Complete a species allocation.
   (a) Choose the design basis.
   (b) Estimate the final product compositions.
   (c) Solve the overall material balance equations.
   (d) Define key stream (or node) compositions.
2. Calculate the vapor and liquid temperatures at each composition node.
3. Calculate the minimum theoretical stages in each tower section.
4. Calculate the minimum vapor rate at the pinch point.
5. Calculate the vapor rate at the pinch point.
6. Calculate the theoretical stages in each section.
7. Calculate the actual stages in each section.
8. Calculate the vapor and liquid flowrates at each node.
9. Size all equipment items.

10. Calculate the utility requirements.

11. Calculate the capital investment and annual utility costs.

Step 1 in this example, the species allocation, is used to estimate stream compositions at key points. It is then possible to complete Steps 2 through 11 sequentially, or in a precedence order as shown, without extensive trial-and-error calculations. By so doing, it becomes possible to adjust the design variables to minimize the cost of the unit operation.

From node composition estimates and specified node pressures, the node temperatures (dew and bubble points) can be computed. Once node temperatures have been computed, then vapor and liquid enthalpies can be estimated. Given relative volatilities and feed fractional vaporization, then the Underwood equations can be used to estimate minimum vapor rates. The Fenske equations can be used to estimate minimum theoretical trays. Gilliland correlations can be used to estimate theoretical trays and efficiency correlations yield actual tray requirements.

Even using this precedence order, some trial-and-error calculations are still required. Specifically, the following calculations are still iterative:

1. Bubble point temperatures

2. Dew point temperatures

3. Feed tray compositions

4. Underwood equation roots

Nonetheless, modeling distillation systems using this short-cut method is much easier than simulating them with equilibrium stage models.

C.3 Precedence Order

A precedence order defines the sequence in which equations may be solved to complete the overall calculations. For a particular set of equations, a number of precedence orders may exist that can be used to solve the equation set without trial-and-error calculations. However, some precedence orders may be more convenient than others and the goal of this section is to describe some simple rules that can be used to study equation sets in a more systematic manner.

Consider an equation f_1 where:

$$f_1 = f_1(x_1, x_2) \tag{C.3}$$

If the variable, x_1, is a given variable or selected as an independent *design* variable, then x_2 may be computed as a dependent variable. In this case, we can show the information flow as:

$$x_1 \rightarrow \boxed{f_1(x_1, x_2)} \rightarrow x_2 \tag{C.4}$$

and we can say that f_1 is used to compute x_2 from a value of x_1.

On the other hand, if we need to solve two equations simultaneously, say:

$$f_1(x_1, x_2) = 0 \tag{C.5}$$
$$f_2(x_1, x_2) = 0 \tag{C.6}$$

then these equations could be solved simultaneously by linearization as previously discussed.

On the other hand, we could set up the precedence order in several ways. For example, we could use the order:

$$x_1^j \rightarrow \boxed{f_1(x_1^j, x_2^j)} \rightarrow x_2^j \tag{C.7}$$
$$x_2^j \rightarrow \boxed{f_2(x_1^{j+1}, x_2^j)} \rightarrow x_1^{j+1} \tag{C.8}$$

where f_1 is used to compute x_2^j from the jth guess of x_1^j and f_2 is used to compute a new value for x_1^{j+1}. If $x_1^{j+1} = x_1^j$, then the correct values for the pair, (x_1, x_2), have been reached. Otherwise, an update strategy is needed. If successive substitution is used for x_1, then:

$$x_1^j = x_1^{j+1} \tag{C.9}$$
$$j = j+1 \tag{C.10}$$

and the iteration is continued until a solution is reached (*i.e.* until $|x_1^{j+1} - x_1^j| < \epsilon$, where ϵ is some arbitrarily small tolerance.

On the other hand, a precedence order could be set up as:

$$x_1^j \rightarrow \boxed{f_2(x_1^j, x_2^j)} \rightarrow x_2^j \tag{C.11}$$
$$x_2^j \rightarrow \boxed{f_1(x_1^{j+1}, x_2^j)} \rightarrow x_1^{j+1} \tag{C.12}$$

In general, these two procedures won't give identical results. Therefore, two different precedence orders exist in this case. Moreover, since both equations involve both variables, we can say that *persistent* information recycle exists in this two equation system.

Several general rules are helpful toward establishing useful information flow schemes (*i.e.* a precedence orders) to solve an equation set.

1. Seek strategies that exploit the modular layout of processes.
2. Seek to minimize persistent information recycle (*i.e.* trial-and-error calculations.
3. Identify the *core* or *dominant* calculational problems.
4. Keep variables of choice as independent design variables (*e.g.* the type of heat exchanger selected).
5. Choose constraints as dependent variables.
6. Note those variables which are fixed by the environment (*e.g.* cooling water T, feed rates, *etc.*)

C.3.1 A Design Variable Selection Algorithm

The design variable selection algorithm is a simple procedure that can be used to analyze a set of equations and choose an appropriate set of design (*i.e.* independent, optimization, or input) variables with which to initiate the calculation sequence. It utilizes the following logic:

1. Construct the structural array showing the equations as rows and the variables as columns. For each equation (row), place a check mark in those columns whose variables appear in the equation.

2. Find a variable which appears in only one equation. Delete that variable and equation.

3. Number the order in which equations and variables are deleted.

4. Return to step 2 until all equations are deleted.

After all equations have been deleted, then:

1. The remaining *undeleted* variables are the design variables.

2. The deletion order is the *reverse* of the computational order (*i.e.* the precedence order).

Example: C.2 *Determine a precedence order for solving the following equation set:*

$$
\begin{aligned}
f_1(x_1, x_2) &= 0 \\
f_2(x_1, x_2, x_3) &= 0 \\
f_3(x_1, x_2, x_3, x_4) &= 0 \\
f_4(x_2, x_3, x_4, x_5) &= 0
\end{aligned}
$$

Solution: C.2 *The degrees of freedom are:*
DF = Variables - Equations = 5 - 4 = 1
Hence there is one independent (design) variable that can be specified. The structural array for this problem is:

	x_1	x_2	x_3	x_4	x_5
f_1	$\surd$	$\surd$			
f_2	$\surd$	$\surd$	$\surd$		
f_3	$\surd$	$\surd$	$\surd$	$\surd$	
f_4		$\surd$	$\surd$	$\surd$	$\surd$

and since x_5 is the only variable appearing in one equation, it and equation f_4 must be deleted first. The deletion order and reduced structural array are then:

1. *Use equation f_4 to calculate x_5.*

	x_1	x_2	x_3	x_4
f_1	√	√		
f_2	√	√	√	
f_3	√	√	√	√

At this point x_4 appears in only one equation. Therefore, delete x_4 and f_3 from the structural array. The deletion order and structural array are then:

1. *Use equation f_4 to calculate x_5.*

2. *Use equation f_3 to calculate x_4.*

	x_1	x_2	x_3
f_1	√	√	
f_2	√	√	√

Now x_3 and f_2 can be deleted. The deletion order and structural array are then:

1. *Use equation f_4 to calculate x_5.*

2. *Use equation f_3 to calculate x_4.*

3. *Use equation f_2 to calculate x_3.*

	x_1	x_2
f_1	√	√

From the reduced structural array, it is clear that either x_1 or x_2 may be selected as the design variable. If x_2 and f_1 are deleted, then the design variable selection algorithm terminates. The precedence order is then:

1. *Use equation f_1 to calculate x_2.*

2. *Use equation f_2 to calculate x_3.*

3. *Use equation f_3 to calculate x_4.*

4. *Use equation f_4 to calculate x_5.*

with x_1 selected as the design variable.

If more than one variable appear in only one equation during any step of the design variable selection algorithm, then either variable may be selected. In such cases, multiple precedence orders may be generated.

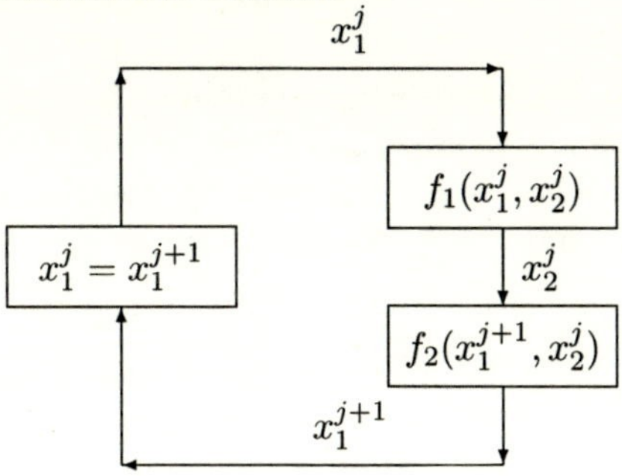

Figure C.2: Solution of two equations by successive substitution of x_1 and the use of only one tear variable, x_1^j.

C.3.2 Persistent Recycle

Whenever the design variable selection algorithm is used to arrive at a reduced structural array where all of the variables appear in two or more equations, then the reduced structural array contains *persistent* information recycle and trial–and–error calculations are required. On the one hand, one may choose to linearize the remaining equations in the reduced structural array and compute the remaining variables simultaneously. On the other hand, one might choose to utilize successive substitution and minimize the number of tear variables that are required to solve the problem iteratively.

For example, if one wishes to solve the equation set:

$$f_1(x_1, x_2) = 0 \tag{C.13}$$
$$f_2(x_1, x_2) = 0 \tag{C.14}$$
$$f_3(x_1, x_2, x_3) = 0 \tag{C.15}$$

then x_3 and f_3 may be deleted leaving the reduced structural array:

	x_1	x_2
f_1	√	√
f_2	√	√

which has persistent recycle. Linearization and the simultaneous solution of two equations will require the use of x_1 and x_2 as iteration variables. However, only one variable (either x_1 or x_2) is required using successive substitution as shown in Figure C.2. While this latter technique is often slower than linearization, it may also be more stable. Therefore, it should always be considered as an alternative whenever:

- General procedures are not required for equation solution.
- Numerical instabilities occur using linearization techniques.
- Only a limited number of linearization variables are required to solve the problem.

If these conditions are not met, then linearization is preferable. On the other hand, if these conditions are met, then a recycle strategy and the selection of *tear* variables should be investigated. Tear variables are simply those variables that are selected as recycle variables (as is x_1 in Figure C.2.) However, the following guidelines should be observed:

1. The choice of recycle (tear) variables affects the numerical stability of the recycle scheme.

2. Always try to minimize the number of torn variables and recycle equations.

3. If a particular set of tear variables is unstable, then choose another set.

4. Delete those equations containing the fewest variables first.

5. One variable in each deleted equation *must be* a design variable.

The rules for applying the recycle variable selection algorithm are as follows:

1. Apply the design variable selection algorithm until it terminates.

2. Define $k = \min[\rho(x_i) - 1]$ where $\rho(x_i)$ is the number of equations in which the variable x_i appears.

3. Define sets of k equations such that when each set is deleted there remains a reduced structural array with at least one variable that appears in only one equation.

4. Delete the equation set.

5. Apply the design variable selection algorithm to the reduced structural array.

6. Try another deletion set if no precedence order results.

7. If the deletion of no set of k equations yields a precedence order, then increment $k = k + 1$ and go to step 2.

Equations may be solved simultaneously or using precedence ordering. In the former cases, the approach is usually to linearize the equation set and then to solve simultaneously by iterating on the linearization variables. In certain cases, however, one may end up dividing by zero. Also, a region of stability exists so that the step sizes must be constrained.

Precedence ordering may eliminate the need to solve some equations simultaneously in favor of a sequential solution method. In such instances, the design variable selection algorithm offers a logical approach to setting up precedence orders. Whenever persistent recycle exists, then a residual equation set may be solved either simultaneously or using successive substitution methods with torn variables.

C.4 Composition Node Estimation Techniques

There are many approaches to short-cut design. The definition and use of composition nodes to devise precedence orders for solving equations that are appropriate for Preliminary Design is one such method.

Complex flowsheets may be represented in a simplified form as collections of composition nodes. A *composition node* is defined as any point in a flowsheet where a feed or product stream exists and one or more composition estimates are needed to approximate the material and energy balances at that point, and to size major equipment items. One node is defined for each stream composition that must be estimated.

The basic philosophy underlying composition node development is the notion that it is often sufficient to generate material balances that approximate the process without solving a set of fundamental equations to make these approximations. By so doing, considerable difficulties may be avoided. For example, trace components can make equation sets stiff[1] and difficult to solve. On the other hand, trace species may have negligible effects on equipment specifications such a condenser duties or tower dimensions. In such cases, they may either be safely ignored altogether or else their compositions may be approximated using simplified methods.

Of course, one cannot define and specify composition nodes in a process without consuming degrees of freedom. In fact, systems based on composition nodes are often overspecified whenever the process involves more than two species. Consequently, composition node models usually don't satisfy all of the fundamental relationships (*e.g.* equilibrium or rate requirements) exactly.

From the design standpoint, is is often sufficient to build a process that either meets or exceeds its design basis requirements. Although excessive overdesign is not desirable, this condition is usually less objectionable than underdesign (*i.e.* failing to meet one or more specifications). Therefore, engineers routinely provide some overdesign. If the initial expense of overdesign is small compared to the cost of equipment replacement from underdesign, this approach is justified.

Consider a chemical process battery limits operation that consists of only two unit operations in series — a mixer followed by a heat exchanger. These units are serially connected as shown in Figure C.3.

Suppose that a chemical reaction occurs in the heat exchanger and that the extent of this chemical reaction is a function of the mixing rate, m, and the residence time, τ_m, in the mixer. In addition, suppose that the extent of reaction depends upon the residence time in the exchanger, τ_e, and the process stream exit temperature, T, from the heat exchanger.

Now this problem may be analyzed by fixing the material balances between the two unit operations and at the heat exchanger outlet. For example, suppose that m and τ_m are constrained so that essentially perfect mixing is always achieved at the mixer outlet (*i.e.* the heat exchanger feed). Also assume that the heat exchanger

[1]Equation sets tend to be stiff whenever the ranges of dependent variables or parameters are large. For example, if some species have mole fractions of about 0.5, but others have mole fractions around 10^{-6}, then the equation set may be stiff. Equations sets with large ranges in their eigenvalues are also said to be stiff.

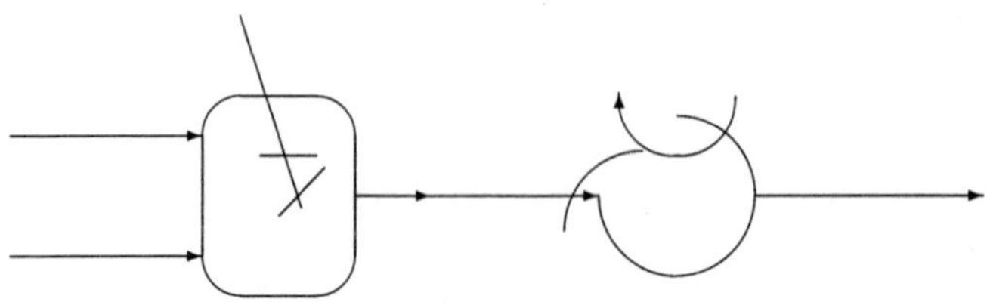

Figure C.3: Example process consisting of a mixer followed by a heat exchanger.

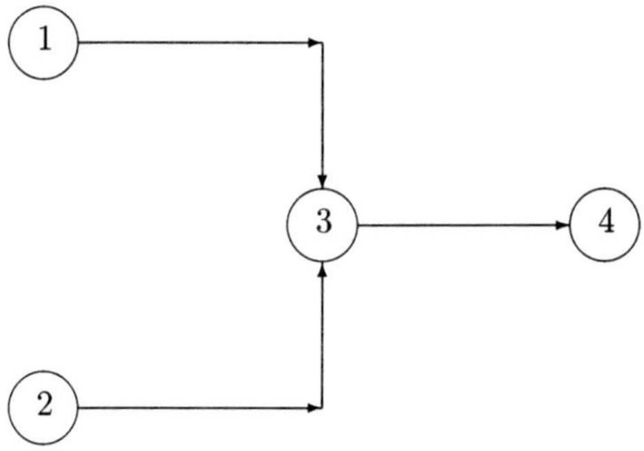

Figure C.4: Composition node representation of the example process.

is always operated so as to achieve the same degree of conversion. In this case, the system can be represented by four composition nodes as shown in Figure C.4.

The use of composition nodes in this case decouples the state optimization and the process synthesis steps because the intermediate stream compositions and their characteristics (m i.e. perfect mixing) are fixed. At this point, however, we are only concerned with understanding how one may strategically assign composition nodes and use them to accurately estimate material flowrates.

C.4.1 Fractionator Product Nodes

Design models exploit precedence ordering whenever possible. Unit operations with countercurrent flow often cause the greatest difficulties, although reactor models can also be challenging. In any event, calculations are simplified whenever some precedence ordering can be achieved.

In the case of countercurrent operations, it is helpful to assume that streams passing each other in a column have the *same* composition. This assumption is not equivalent to those underlying equilibrium stage models, but it is related to them.

Countercurrent processes may be modeled as a set of equilibrium stages. In this case, the vapor and liquid streams leaving a stage are assumed to be in thermodynamic equilibrium. For stage j in Figure C.5, streams V_j and L_j are in equilibrium. For species i at equilibrium:

$$y_{ij} = K_{ij}x_{ij} \tag{C.16}$$

where K_{ij} = the thermodynamic equilibrium constant for species i on stage j
x_{ij} = the mole fraction of species i in the liquid phase exiting stage j
y_{ij} = the mole fraction of species i in the vapor phase exiting stage j

and for the temperatures of the vapor and liquid streams at equilibrium:

$$T_{vj} = T_{Lj} = T_j \tag{C.17}$$

That is, the temperature of the vapor stream, T_{vj}, is equal to the liquid stream temperature, T_{Lj}, since the two streams are at the same pressure, P_j, and are at equilibrium.

On the other hand, the composition node model assumes that L_j and V_j have the *same* composition. Given a node composition and pressure, then a dew point temperature, T_{dj}, and a bubble point temperature, T_{bj}, may be calculated. In cases where $T_{dj} \approx T_{bj}$, it is helpful to assume that the vapor temperature equals T_{dj} and that the liquid temperature equals T_{bj}. Whenever $T_{dj} \gg T_{bj}$, it is better to use an average temperature for the jth stage:

$$T_j \approx \frac{T_{dj} + T_{bj}}{2} \tag{C.18}$$

In either case, the composition node model assumes:

$$y_{ij} = x_{ij} \tag{C.19}$$

and these mole fractions are estimated using species allocation techniques as described in Section C.5.

In reality, countercurrent processes are rate controlled rather than at equilibrium. Thermodynamic equilibrium is achieved only when the system is on total reflux at steady state or if feed and product rates are only infinitesimally small. During normal operations, the process is as far from equilibrium as possible without violating the design specifications, in order to maximize throughput.

For a distillation section producing a top product, P_1, a material balance around the top section yields

$$V_{j+1} = L_j + P_1 \tag{C.20}$$

$$y_{ij+1}V_{j+1} = x_{ij}L_j + y_{i1}P_1 \tag{C.21}$$

if the column utilizes a total condenser and the distillate composition equals y_{i1} for species i. Also, the streams V_{j+1} and L_j are passing each other in the column. If

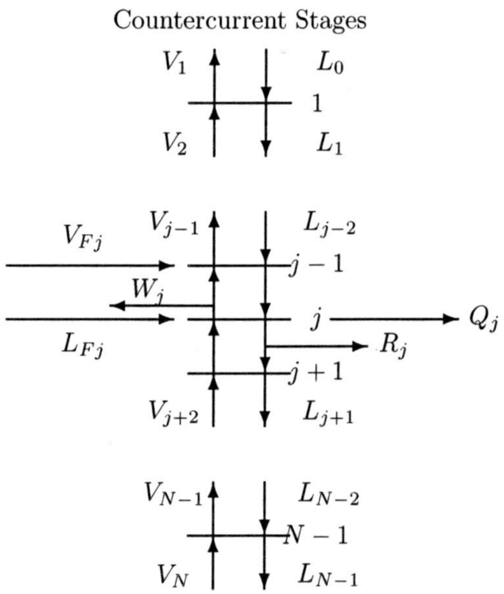

Figure C.5: For equilibrium stage calculations, a total of $N(2NC + 6)$ variables must be specified. NC is the number of species.

these two equations are combined to eliminate L_j, then the operating line equation is:

$$y_{ij+1} = \frac{x_{ij}(V_{j+1} - P_1)}{V_{j+1}} + y_{i1}\left(\frac{P_1}{V_{j+1}}\right) \tag{C.22}$$

Taking the limit of Equation C.22 as the vapor rate becomes infinitely large, then:

$$\lim_{V_{j+1}\to\infty} y_{ij+1} = x_{ij} \tag{C.23}$$

Similarly, taking the limit of Equation C.22 as $P_1 \to 0$ also yields $y_{ij+1} = x_{ij}$ which is the basic assumption of the composition node model. Therefore, the composition node model is exact at infinite reflux or zero distillate rate in the top section (i.e., at total reflux).

Actual operation lies between the composition node and the equilibrium stage assumptions. If we consider compositions, then:

$$\text{For } K_{ij} > 1: \qquad K_{ij}x_{ij} > y_{ij} > x_{ij} \tag{C.24}$$

$$\text{For } K_{ij} = 1: \qquad K_{ij}x_{ij} = y_{ij} = x_{ij} \tag{C.25}$$

$$\text{For } K_{ij} < 1: \qquad K_{ij}x_{ij} < y_{ij} < x_{ij} \tag{C.26}$$

where y_{ij} indicates the actual composition that is achieved in the countercurrent process. For $K_{ij} > 1$, the vapor mole fraction is always less than the equilibrium mole fraction and greater than the composition node assumption (*i.e.* $y_{ij} = x_{ij}$). For $K_{ij} < 1$, the vapor mole fraction is always greater than equilibrium mole fraction and less than the composition node assumption. Therefore, the equilibrium stage and composition node models bound actual behavior.

If we consider the temperatures during actual operation:

$$T_{dj} > T_{vj} > T_j > T_{Lj} > T_{bj} \tag{C.27}$$

where T_{vj} and T_{Lj} are the actual vapor and liquid phase temperatures, respectively. Except in the case of retrograde behavior, the vapor beneath the *jth* stage is always greater than the temperature of the liquid falling from the stage above. While both phases approach the same temperature, the vapor tends to be slightly hotter than the liquid leaving the stage. In any event, these two stream temperatures fall between the dew point and bubble point temperatures, T_{dj} and T_{bj}, respectively. Therefore, the composition node and equilibrium stage models bound the true stage temperature for a given stage pressure.

Countercurrent fractionators (distillation, liquid/liquid extraction, absorption, *etc.*) all may be represented as composition nodes. The emphasis here on fractionators to illustrate the use of composition nodes does not limit the generality of this discussion. In fact, node representation is useful so long as both bulk phases are continuously transported through a contactor. However, node representation must be extended to apply it to unit operations such as fixed bed adsorption where one of the phases is stationary.

A simple distillation tower, receiving a single feed and producing only two products (i.e., a distillate and a bottoms), may be represented with four nodes as shown

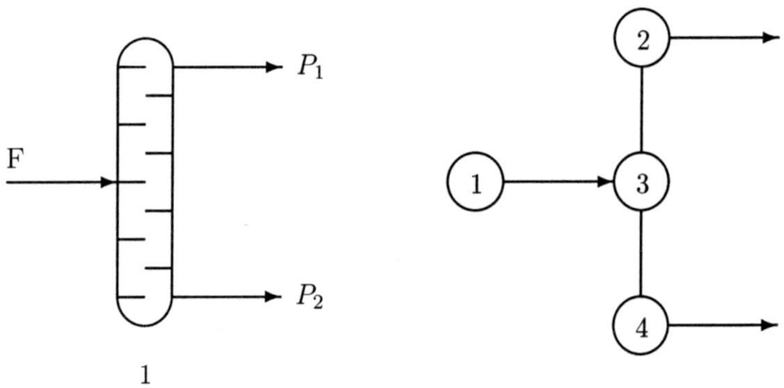

Optimization Variables

Variable	Definition
f	Feed fractional vaporization
P	Tower pressure at node 2
P_F	Feed pressure at node 1
V/V_{min}	Vapor to minimum vapor rate at the pinch point.

Figure C.6: Node representation of a simple tower, design 1, receiving a single feed and producing only a distillate, P_1, and a bottoms product, P_2.

in Figure C.6. The external feed to the column is labeled as node 1. The distillate is node 2. Node 4 represents the bottoms product composition and node 3 is the average composition on the feed tray in the tower. Although Figure C.6 only illustrates a tray column, additional unit operations are implied. For example, the column *subproblem* includes three heat exchangers — the feed heat exchanger, a vapor condenser, and a reboiler. Also, three pumps are involved — the feed pump, the reflux return pump, and the reboiler feed pump. In addition, the condenser is followed by a reflux accumulator drum that is sized to dampen process upsets.

Figure C.6 tabulates four optimization variables which may be used to define those operating conditions that minimize the costs of construction and tower operation. For example, the feed fractional vaporization is one possible optimization variable. It is defined as:

$$f = \frac{H_F - H_L^{sat}}{H_v^{sat} - H_L^{sat}} \tag{C.28}$$

where f	=	the fractional vaporization of the tower feed at node 3
H_F	=	the enthalpy of the tower feed based upon f, H_L^{sat} and H_v^{sat}.
H_L^{sat}	=	the enthalpy of the saturated liquid feed using the composition of node 1 with the T, P at node 3.
H_v^{sat}	=	the enthalpy of a saturated feed using the node 1 composition and the T, P at node 3.

During an optimization, f may be adjusted using the methods discussed in Chapters ?? and ?? to minimize the estimated cost of building and operating the column. Similarly, the V/V_{min} ratio may be adjusted as well as the operating pressure at the top of the column, P, which is defined at node 2.

The feed pressure variable, P_F, is special. For the first tower in a sequence, it is may be fixed as part of the problem statement. As towers are connected into sequences, it is fixed by the conditions of the preceding tower. However, it may float as an optimization variable during the initial optimization of the separate subproblems. Economic interactions between subproblems can be correlated using P_F.

Other fractionator configurations may be modeled using composition nodes. Several are listed in Appendix D.

C.4.2 Feed Tray Nodes

The composition of liquid on the feed tray, (say at node 3 in Figure C.6) can be approximated using the feed line equation [64]:

$$y_i = [(f-1)x_i + z_i]/f \tag{C.29}$$

where f	=	the fractional vaporization of the feed, $(H_F - H_L)/(H_v - H_L)$, an optimization variable.
K_i	=	the vapor/liquid equilibrium constant on the feed tray
x_i	=	mole fraction of species i in the liquid
y_i	=	mole fraction of species i in the vapor
z_i	=	mole fraction of species i in feed (node 1 in Figure C.6

If it is assumed that the liquid and vapor are in equilibrium, then the liquid composition on the feed tray can be calculated as:

$$x_i = z_i/[f(K_i - 1) + 1] \tag{C.30}$$

Experience has shown that this approximation is one of the least accurate for estimating node compositions because it does not consider the effects of internal vapor and liquid streams entering and leaving the feed tray. The major impact of poor feed tray composition estimates is less certainty in the best location for the feed tray.

Although tower costs are relatively insensitive to the feed tray composition estimate, an improved estimation procedure considers the effects of internal vapor and liquid traffic. Writing a balance for the *ith* species around the *jth* feed tray gives

$$x_{i,j} = \frac{x_{i,F}F + x_{i,j-1}L_{j-1} + x_{i,j+1}V_{j+1}}{V_j + L_j} \tag{C.31}$$

where the $j-1$ subscript refers to stream rates and compositions above the feed tray (e.g., at the distillate tray, node 2) and $\jmath+1$ referst to stream rates and compositions below the feed tray (e.g.., at the bottoms tray, node 4).

This improvement in estimating the feed tray composition requires trial-and-error since the internal vapor and liquid rates are not estimated until all composition nodes have been estimated. For the first iteration, the L and V rates can be set equal to the feedrate, F. During subsequent optimization trials (i.e., iterations), they previously computed values may be used.

To use this latter method of estimating feed tray compositions, species allocation is first used to estimate the distillate and bottoms product compositions (in simple towers). Then Equation C.31 is used to estimate the feed tray composition.

C.5 Species Allocation Methods

Species allocation is used to assign species material balances in a realistic manner so that a precedence order can be established to simplify design calculations. This simplification permits the optimization and synthesis of many alternative networks or chemical manufacturing systems.

In some cases, only a partial species allocation is possible and some trial-and-error remains. However, the equation set that must be solved iteratively is reduced compared to the original problem. In other cases, the species allocation results in unacceptable inaccuracies in the design calculations. In all cases, the ability to allocate species is determined largely by ingenuity and insight leading to simplifying assumptions. The example procedures described here are illustrative, but not comprehensive of all possible allocation techniques.

C.5.1 Using Overall Balances

At steady state, the overall material balances around a separation process can be written as:

$$\boldsymbol{f} = \boldsymbol{X}\boldsymbol{p} \tag{C.32}$$

where $\boldsymbol{f}$ = the original feed rate, a column vector with dimension NC, where NC is the number of species to be allocated.

$\boldsymbol{X}$ = the final product composition matrix which has dimensions NC by NC. Each column in this matrix corresponds to the composition of one of the final products.

$\boldsymbol{p}$ = the final product rate, a column vector with dimension NC. Each element in $\boldsymbol{p}$ corresponds to one final product.

Also:

$$F = \sum_{i=1}^{NC} f_i = \sum_{i=1}^{NC} p_i \tag{C.33}$$

where $\boldsymbol{f}^t$ = $[f_1, f_2, \ldots, f_{NC}]$ is a row vector, the transpose of $\boldsymbol{f}$

$\boldsymbol{p}^t$ = $[p_1, p_2, \ldots, f_{NC}]$ is a row vector, transpose of $\boldsymbol{p}$

If the feed vector $\boldsymbol{f}$ is specified and the final product composition matrix $\boldsymbol{X}$ can be estimated, then the final product rates, $\boldsymbol{p}$, can be calculated from:

$$\boldsymbol{p} = \boldsymbol{X}^{-1}\mathbf{f} \tag{C.34}$$

where $\boldsymbol{X}^{-1}$ is the inverse[2] of $\boldsymbol{X}$.

Example: C.3 *A process produces 3 product streams p_1, p_2, and p_3. The composition of p_1 is (0.85, 0.10, 0.05). The composition of p_2 is (0.04, 0.92, 0.04) and the composition of p_3 is (0.01, 0.04, 0.95). The feedrate is (850,500,300) of moles of species A, B, and C, respectively. Write an expression for the final product composition matrix, $\boldsymbol{X}$. Complete the overall species allocation for this system by estimating $\boldsymbol{p}$.*

Solution: C.3 *A series of overall and component mass balances can solve this species allocation problem.*

Feed $= F = 850 + 500 + 300 = p_1 + p_2 + p_3 = 1650$

$$\boldsymbol{f} = \boldsymbol{X}\boldsymbol{p}$$

$$\begin{bmatrix} 850 \\ 500 \\ 300 \end{bmatrix} = \begin{bmatrix} 0.85 & 0.04 & 0.01 \\ 0.10 & 0.92 & 0.04 \\ 0.05 & 0.04 & 0.95 \end{bmatrix} \begin{bmatrix} p_1 \\ p_2 \\ p_3 \end{bmatrix}$$

Solving these equations simultaneously yields:

$$\boldsymbol{p} = \begin{bmatrix} p_1 \\ p_2 \\ p_3 \end{bmatrix} = \begin{bmatrix} 977.0 \\ 426.6 \\ 246.4 \end{bmatrix}$$

[2]The inverse only exists if a matrix is square and if its determinant is not equal to zero. See Appendix ?? for more details.

where the product rates are in lb-moles/hr.

Once the final product rate vector $\boldsymbol{p}$ has been estimated, then all possible intermediate streams can be calculated simply by summation in the direction reverse to material flow (*i.e.* starting with the last products). That is

$$F(ij) = \sum_{k=i}^{j} p_k \qquad j > i \tag{C.35}$$

and the intermediate stream compositions can be estimated from

$$X'(i, ij) = \frac{\sum_{k=i}^{j} X(i,k)p_k}{F(ij)} \qquad j > i \tag{C.36}$$

where $F(ij)$ = the intermediate flowrate for a stream producing final products i through j

p_k = the kth final product

$X(i, ij)$ = the element in the final product composition matrix $\boldsymbol{X}$ corresponding to the ith species and producing final products i through j.

$X'(i, ij)$ = the composition for species i of the intermediate stream producing the final products i through j.

Example: C.4 *Given a three component feed that is to be separated by simple towers with a flowrate of 100 lb-mole/hr and a composition of (0.3, 0.4, 0.3), find the flowrates of streams p_1, p_2, and p_3. Also compute the interconnecting stream flows for simple towers. The final product composition array is:*

$$\mathbf{X} = \begin{bmatrix} 0.95 & 0.02 & 0.02 \\ 0.03 & 0.95 & 0.03 \\ 0.02 & 0.03 & 0.95 \end{bmatrix}$$

Solution: C.4 *For species 1:*

$$f_1 = X(1,1)p_1 + X(1,2)p_2 + X(1,3)p_3$$

For species 2:

$$f_2 = X(2,1)p_1 + X(2,2)p_2 + X(2,3)p_3$$

For species 3:

$$f_3 = X(3,1)p_1 + X(3,2)p_2 + X(3,3)p_3$$

In matrix notation:

$$\begin{bmatrix} 30 \\ 40 \\ 30 \end{bmatrix} = \begin{bmatrix} 0.95 & 0.02 & 0.02 \\ 0.03 & 0.95 & 0.03 \\ 0.02 & 0.03 & 0.95 \end{bmatrix} \begin{bmatrix} p_1 \\ p_2 \\ p_3 \end{bmatrix}$$

and $p_1 = 30.1$ *lb-mole/hr,* $p_2 = 40.2$ *lb-mole/hr, and* $p_3 = 29.7$ *lb-mole/hr. The interconnecting streams for simple distillation towers are then:*
$F(12) = p_1 + p_2 = 30.1 + 40.2 = 70.3$
$F(23) = p_2 + p_3 = 40.2 + 29.7 = 69.9$
The composition of $F(12)$ *is:*
$X'(1,12) = [0.95(30.1) + 0.02(40.2)]/70.3 = 0.418$
$X'(2,12) = [0.03(30.1) + 0.95(40.2)]/70.3 = 0.556$
$X'(3,12) = [0.02(30.1) + 0.03(40.2)]/70.3 = 0.026$
The composition of $F(23)$ *is:*
$X'(1,23) = [0.02(40.2) + 0.02(29.7)]/69.9 = 0.020$
$X'(2,23) = [0.95(40.2) + 0.03(29.7)]/69.9 = 0.559$
$X'(3,23) = [0.03(40.2) + 0.95(29.7)]/69.9 = 0.421$

After generating composition estimates, they are inserted into the appropriate nodes. To model a Design 1 tower shown in Figure C.6, composition estimates based on overall balances are used in nodes 1, 2, and 4. To model a tower producing p_1 and p_2, the $X'(i,12)$ compositions are used in node 1 while the p_1 and p_2 compositions are used in nodes 2 and 4 respectively. To model a tower processing the original feed to produce p_1 as distillate, the original feed compositions $[X'(i,123) = f_i/F(123)]$ are inserted into node 1. The estimates for p_1 are inserted into node 2 and the estimates for $X'(i,23)$ are used for node 4.

C.5.2 Fewer Products than Species

If one wishes to model a system that involves NC species and NP products, where $NP < NC$, then a species allocation technique based upon the overall material balances alone will not work. In this case, the $\boldsymbol{X}$ matrix is no longer symmetric and its inverse doesn't exist. Additional relationships must be used to fix the degrees of freedom ($DF = NC - NP > 0$).

One approach uses the Fenske equation. In this case, a new matrix is defined that consists of the applicable material balance relationships and the Fenske equation to specify additional linear relationships.

In modeling a simple tower as in Figures C.6 or D.1 ($NP = 2$) where $NC > NP$, then the Fenske equation is applied to the overheads and bottoms component flowrates, instead of stream compositions. That is:

$$\frac{d_i}{b_i} = \alpha_{ir}^N \frac{d_r}{b_r} \tag{C.37}$$

where b_i = the bottoms flowrate for the ith species
d_i = the distillate flowrate for the ith species
α_{ir} = the relative volatility of the ith species to a reference species r ($\alpha_{ir} = K_i/K_r$)

To use the Fenske equation in this case, an augmented matrix $\boldsymbol{A}$ is generated by writing the material balance equations for NC species in the form:

$$d_i + b_i = f_i \tag{C.38}$$

To generate an additional NC equations

1. Fix one of the distillate to bottoms ratios, say that of the reference species, where $g \equiv d_r/b_r$.

2. Assume a number N of theoretical stages to act as a dummy variable to specify the degree of separation.

3. Require all other distillate to bottoms ratios (d_i/b_i) to distribute according to the Fenske equation:

$$\frac{d_i}{b_i} = \alpha_{ir}^N g \tag{C.39}$$

4. Calculate $2NC$ d_i and b_i values to complete the species allocation.

This approach results in the matrix equations:

$$\boldsymbol{A\phi} = \boldsymbol{\gamma} \tag{C.40}$$

which can be written as:

$$\begin{bmatrix} 1 & & & 1 & & \\ & 1 & & & 1 & \\ & & \vdots & & & \vdots \\ & & 1 & & & 1 \\ 1 & & & -\alpha_{1r}^N g & & \\ & 1 & & & -\alpha_{2r}^N g & \\ & & \vdots & & & \vdots \\ & & 1 & & & -g \end{bmatrix} \begin{bmatrix} d_1 \\ d_2 \\ \vdots \\ d_{NC} \\ b_1 \\ b_2 \\ \vdots \\ b_{NC} \end{bmatrix} = \begin{bmatrix} f_1 \\ f_2 \\ \vdots \\ f_{NC} \\ 0 \\ 0 \\ \vdots \\ 0 \end{bmatrix} \tag{C.41}$$

Example: C.5 *A ternary mixture must be distilled so that the distillate to bottoms ratio for the heavy species is 1 to 20 on a molar basis. The middle to heavy species volatility ratio is 1.5. The light to heavy species volatility ratio is 1.8. The separator should be designed as 5 theoretical stages between the distillate and bottoms. The light, middle, and heavy species feed rates are 50, 30, and 40 lb-moles/hr. Based on this information, estimate the distillate and bottoms compositions and flowrates.*

Solution: C.5 *For the heavy species,* $d_3/b_3 = g = 1/20 = 0.05$. *Six equations must be solved. They are:*

$$\begin{aligned} d_1 + b_1 &= 50 \\ d_2 + b_2 &= 30 \\ d_3 + b_3 &= 40 \\ d_1 - 1.8^5(0.05)b_1 &= 0 \\ d_2 - 1.5^5(0.05)b_2 &= 0 \\ d_3 - 0.05b_3 &= 0 \end{aligned}$$

These six equations may be solved simultaneously to obtain the d_i and the b_i values. Alternatively, three sets of two equations may be solved in precedence order as:

$$\begin{aligned} d_1 + b_1 &= 50 \\ d_1 - 1.8^5(0.05)b_1 &= 0 \end{aligned}$$

$$\begin{aligned} d_2 + b_2 &= 30 \\ d_2 - 1.5^5(0.05)b_2 &= 0 \end{aligned}$$

$$\begin{aligned} d_3 + b_3 &= 40 \\ d_3 - 0.05b_3 &= 0 \end{aligned}$$

Where the substitutions:

$$\begin{aligned} d_i/b_i &= g_r\alpha_{ir}^N \\ b_i &= f_i/(d_i/b_i + 1) \\ d_i &= b_i(d_i/b_i) \end{aligned}$$

establish the precedence order. For this problem:
$(d_1, d_2, d_3) = (24.29, 8.256, 1.90)$
$(b_1, b_2, b_3) = (25.71, 21.74, 38.09)$
and the distillate rate is 34.45 lb-mole/hr. The bottoms rate is 85.55 lb-mole/hr. The distillate composition is (0.71, 0.24, 0.06) and the bottoms composition is (0.30, 0.25, 0.45).

C.5.3 Distributed Species

The Fenske equation can also be used to specify internal nodes 2 and 4 in Design 5 (Figure D.4) where the overall material balances must be satisfied, but one wishes to optimize on the distillate and bottoms products compositions. In this case, the variables F_2/F_1 and N are optimization variables that can be used with the overall material balances to estimate internal node compositions.

In this instance, the relative flows of the various species are estimated relative to the total flow ratio F_2/F_3 in Figure D.4 using the Fenske form:

$$\frac{b_i}{d_i} = \left(\frac{F_3}{F_2}\right)\alpha_i^{-N} \tag{C.42}$$

However, the overall material balances must be satisfied:

$$F_2 = \sum_{i=1}^{NC} d_i \tag{C.43}$$

$$F_3 = \sum_{i=1}^{NC} b_i \tag{C.44}$$

If Equation C.42 is used to generate estimates of b_i/d_i, then Equations C.43 and C.44 will not be generally satisfied. However, a multiplier, θ, can be specified such that:

$$d_i = \frac{f_i}{1 + \theta\,(b_i/d_i)} \tag{C.45}$$

which is monotonic in θ for fixed estimates of b_i/d_i from Equation C.42. Equation C.45 is stable with respect to Newton-Raphson convergence using θ as the convergence variable since it is monotonic in θ. When a correct value for θ has been found such that Equations C.43 and C.44 are satisfied for an estimated set of b_i/d_i ratios, then distillate compositions can be estimated from:

$$X(i,2) = \frac{f_i}{[1 + \theta(b_i/d_i)]F_2} \tag{C.46}$$

and bottoms compositions estimated from:

$$X(i,4) = \frac{f_i}{[1 + d_i/(\theta b_i)]F_2} \tag{C.47}$$

while satisfying the constraint

$$F_1 = F_2 + F_3 \tag{C.48}$$

where b_i = the bottoms flowrate for species i exiting node 4 in Design 5
d_i = the distillate flowrate for species i exiting node 2 in Design 5
F_1 = initial feedrate to first tower in Design 5
F_2 = distillate rate from first tower in Design 5
F_3 = bottoms rate from first tower in Design 5
f_i = feedrate of species i in F_1 in Design 5
θ = dummy convergence variable

Example: C.6 *A distillation tower is operated such that its distillate to feed ratio is 0.55 on a molar basis. The binary feed is 66 mol% A. The relative volatility of A to B is 1.35. At total reflux the tower has the equivalent of 12 theoretical stages. Use this information to estimate the distillate and bottoms compositions.*

Solution: C.6 *Find θ by trial-and-error. Basis: one mole of feed. Initially set $\theta = 1$.*
$d_a/b_a\theta = \theta(D/F)\alpha^N$, $d_b/b_b\theta = \theta(D/F)$, $b_i = f_i/(\theta d_i/b_i + 1)$, $d_i = f_i - b_i$
$(d_a/b_a)\theta = 0.55 \times 1 \times 1.35^{12} = 20.1543$, $b_a = f_a/(\theta(d_a/b_a) + 1 = 0.03119932$
$(d_b/b_b)\theta = 0.55 \times 1$, $b_b\theta = f_b/(\theta(d_b/b_b) + 1) = 0.21935$
Using the Solver function in an Excel spreadsheet, with the constraint $\theta \geq 0.0$, yields $\theta = 0.182588$, $b_a = 0.141$, $b_b = 0.3089$, and $B = 0.4499$ as desired. The estimated distillate composition is (0.692, 0.308).

C.6 Vapor/Liquid Rates and Duties

The Fenske-Underwood-Gilliland method can be used to estimate the minimum theoretical stages in a column, the minimum vapor requirements, and the theoretical stages.

Once composition nodes have been estimated, the column pressure has been specified, and the vapor rate at the pinch is determined using Underwood, it is then possible to estimate the vapor and liquid enthalpies computed at the dew and bubble point temperatures, respectively. The overall material and energy balances for each section can then be combined to compute the vapor and liquid flowrates in each section. Thus, even if the Underwood method, which assumes constant molal overflow, is used to estimate the minimum vapor rate at the pinch tray, this assumption can be dropped as the overall energy and material balances are used around the various node pairs. These details are briefly described below.

C.6.1 Minimum Vapor Rates

The Underwood [109, 110] equations are derived in the appendix. Here their use is summarized.

First, $NC-1$ roots, λ_k, are calculated from the equation:

$$f = \sum_{i=1}^{NC} \frac{x_{F,i}\alpha_{F,i}}{\alpha_{F,i}-\lambda_k} \qquad k = 1, 2, \ldots, NC-1 \tag{C.49}$$

where f = the fractional vaporization of the tower feed closest to the section in question
$x_{F,i}$ = the mole fraction of the ith component on the feed tray
$\alpha_{F,i}$ = the relative volatility[3] of the ith component to some reference component at the feed tray conditions.

Each root, λ_k, lies between the relative volatilities of components k and $k+1$.

These relative volatilities are based on conditions at the feed tray and the method assumes that the relative volatilities are constant in the pinch zone and throughout the tower. Subsequent calculations may refine this assumption and permit adjustments to the relative volatilities throughout the tower. However, the accuracy of this approximation decreases as the temperature drop up the tower increases.

The fractional vaporization variable, f, is defined as:

$$f = \frac{H_F - H_L^{sat}}{H_V^{sat} - H_L^{sat}} \tag{C.50}$$

where f = the feed fractional vaporization
H_F = the enthalpy of the feed mixture entering the tower
H_V^{sat} = the enthalpy of a saturated vapor mixture at the same composition and pressure as the feed
H_L^{sat} = the enthalpy of a saturated liquid mixture at the same composition and pressure as the feed

Also, if $f < 0$, then the column feed is a subcooled liquid. If $f = 0$, then the feed is a saturated liquid. On the other hand, if $1 > f > 0$, then the feed is partially vaporized and if $f = 1$, then the feed is a saturated vapor. If $f > 1$, the the column feed is a superheated gas mixture.

Having determined the set of Underwood roots, then a minimum vapor rate, $V^u_{n,k}$, may be calculated for each tower section n based on root λ_k by the equation:

$$V^u_{n,k} = \sum_{\substack{j=m \\ m<n}}^{n} \left[W_j \sum_{i=1}^{NC} x_{i,j} \left(\frac{\alpha_{f,i}}{\alpha_{f,i} - \lambda_k} \right) \right] \tag{C.51}$$

where W_j is positive if it represents a product flow rate and negative for feed rates. The compositions $x_{i,j}$ refer either to the product or feeds, but the α and λ values always refer to the feed tray conditions that were initially referenced. Section m is the top section in the tower.

This equation is simply a generalization of the Underwood minimum reflux equation to allow its use for complex towers.

There will be $NC - 1$ Underwood vapor rates for each section with one corresponding to each root of the Underwood equation. The *minimum* vapor rate for a given section, V_{min} is the *maximum* vapor rate in the set for each section:

$$V_{min,n} = \max_k \left\{ V^u_{n,k} \right\} \tag{C.52}$$

C.6.2 Pinch Points

The pinch point in a column is that location at which the vapor rate first equals the minimum rate for a section. Let the vapor rate immediately above a feed point be designated as V_n^+ and as V_n^- immediately below. Now the minimum vapor required immediately above the feed point is determined as:

$$V^+_{min,n} = \max_k \left\{ V^{u+}_{n,k} \right\} \tag{C.53}$$

while the minimum vapor required immediately below the feed point is determined as:

$$V^-_{min,n} = \max_k \left\{ V^{u-}_{n,k} \right\} \tag{C.54}$$

and the pinch point at that feed point is given by:

$$V_{min,n} = \max \left\{ V^-_{min,n}, V^+_{min,n} \right\} \tag{C.55}$$

Usually, the pinch point (either above or below a feed tray) is defined by the section having the largest value overall for the minimum vapor rate.

Except when azeotropic products are produced or the mixture is highly nonideal (*e.g.* ethanol and water), the pinch point in a column occurs either immediately above or below a column feed depending upon the values of the Underwood roots, the relative volatilities at that feed, and the product flow rates W_j.

However, changes in the enthalpies of liquid and vapor traffic at different locations in the column can also shift the location of the pinch point. In order to account for such changes, θ_n values can be defined for each section of a column

$$\theta_n = \frac{V_n}{V_{min,n}} \tag{C.56}$$

where V_n is the estimated vapor traffic using the combined overall energy and material balances. The pinch is more generally identified as:

$$\theta_{min} = \min_n \{\theta_n\} \tag{C.57}$$

and it should always be true that $\theta_{min} \geq \theta$ during each iteration of an optimization study where θ is the independent optimization variable. In cases where $\theta_{min} < \theta$, then the vapor rates in the column must be adjusted upwards during that iteration.

C.6.3 Node Vapor and Liquid Rates

The composition node design method permits two vapor and liquid overflow rates to be defined for each node in a tower. A general representation of countercurrent nodes appears as Figure C.7. These rates are calculated at points just above and just below each node. However, only one composition is defined for a node, so it must be assumed that the physical properties are the same for the vapor immediately above and below a node and likewise for the liquids.

Given the vapor rate at some point in the tower (usually given at the pinch point relative to the minimum vapor rate), then the vapor rates and liquid overflows at all other points in the tower can be calculated. First, the liquid overflow at the same point can be determined from a mass balance over the upper sections of the tower. For example, if the vapor rate is specified at a point just below node n, then:

$$L_n^- = V_n^- - \sum_{j=N_t}^{n} W_j \tag{C.58}$$

where L_n^- = liquid molar flowrate just below node n
N_t = number of the top node in the countercurrent column (*i.e.* the condenser tray)
V_n^- = vapor flowrate just below node n
W_j = either a feed or product flowrate. For feeds, $W_j < 0$. For products, $W_j > 0$. W_j is defined similarly in Section C.6.1.

On the other hand, if the vapor rate is known at a point just above node n, then the mass balance equation is

$$L_n^+ = V_n^+ + \sum_{j=n}^{N_b} W_j \tag{C.59}$$

where L_n^+ = liquid molar flowrate just above node n
N_b = number of the bottom node in the countercurrent column (*i.e.* the reboiler tray)
V_n^+ = vapor flowrate just above node n
W_j = either a feed or product flowrate. For feeds, $W_j < 0$. For products, $W_j > 0$. W_j is defined similarly in Section C.6.1.

As soon as the overall vapor and liquid flowrates passing each other are known at

Countercurrent Nodes

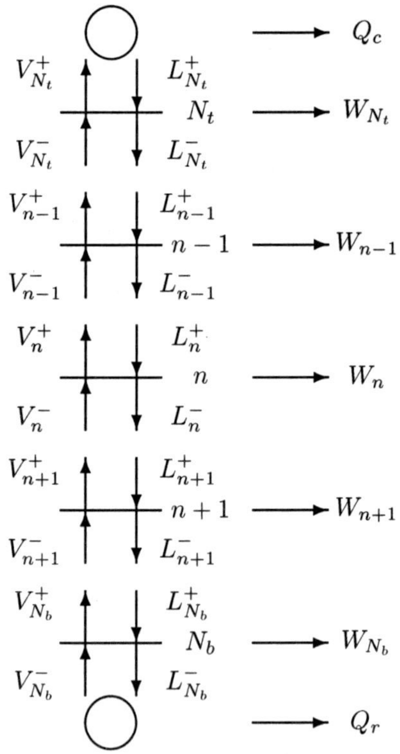

Figure C.7: For countercurrent node calculations, a total of $(N_b - N_t + 1)(NC + 2) + 1$ variables must be specified.

one point in a tower, the rates above and below all nodes in the tower can be computed by combining the overall energy and material balances. The overall mass and energy balance equations relating the rates around an adiabatic node are:

$$L_n^- + V_n^+ + W_n = L_n^+ + V_n^- \quad (C.60)$$

$$L_n^- H_{Ln} + V_n^+ H_{vn} + H_{W_n} W_n = L_n^+ H_{Ln} + V_n^- H_{vn} \quad (C.61)$$

where H_{vn} = the molar enthalpy of vapor at node n
H_{Ln} = the molar enthalpy of liquid at node n
H_{W_n} = the molar enthalpy of feed or product at node n. For a liquid draw, $H_{W_n} = H_{Ln}$. For a vapor draw, $H_{W_n} = H_{vn}$. For a feed, $H_{W_n} = H_{Fn}$.

Similarly, the flow rates between nodes in the same tower are related by the overall mass and energy balance equations since there is no mass or heat interchange with the surroundings between two nodes:

$$L_{n+1}^+ + V_n^- = L_n^- + V_{n+1}^+ \quad (C.62)$$

$$L_{n+1}^+ H_{Ln+1} + V_n^- H_{vn} = L_n^- H_{Ln} + V_{n+1}^+ H_{vn+1} \quad (C.63)$$

C.6.4 Reboiler and Condenser Duties

After computing all vapor and liquid flowrates in a tower, both above and below each node, the duties may be computed. For a condenser:

$$Q_c = (H_{vN_t} - H_{LN_t})\, V_{N_t}^+ \quad (C.64)$$

and for the reboiler:

$$Q_r = (H_{vN_b} - H_{LN_b})\, V_{N_b}^- \quad (C.65)$$

where Q_c = condener duty, Btu/h
Q_r = reboiler duty, Btu/h

Now these equations can be used to calculate the liquid and vapor flows above and below each node in a tower if the vapor rate at a single point, the flow rates of all feed and product streams, and the enthalpies of liquid and vapor at each node are known. Thus, they provide a means for dropping the constant molal overflow assumptions. For many towers, the changes in latent heats of vaporization and sensible enthalpies are significant and this technique (*i.e.* combining overall energy and material balances) allows for corrections in vapor and liquid flowrates that are very important.

Of course, variables must be specified to complete these calculations. In particular, NC compositions, node pressures P_n, and node products (or feeds) W_n must be specified for each node. If there are $(N_b - N_t + 1)$ nodes in a tower, then $(N_b - N_t + 1)(NC + 2)$ of these variables must be fixed. Also, one additional optimization variable, $\theta = V/V_{min}$ at the pinch point must be specified. Therefore, the calculations can be completed after a total of $(N_b - N_t + 1)(NC + 2) + 1$ variables are fixed as information inputs.

Example: C.7 *A distillation tower is operating such that the liquid overflow rate just above the feed point is 685.3 lb-mole/hr. If the overhead product rate is 286.3 lb-mole/hr, calculate the vapor rate just above the feed.*
Solution: C.7 *Write a balance around the top section of the tower:*
$V_3^+ = L_3^+ + P_1 = 286.3 + 685.3 = 971.6$ *lb-mole/hr*

Example: C.8 *A simple distillation tower operates with a vapor rate of 1250 lb-mole/hr just above the feed point. The vapor enthalpy at that point is 21,550 Btu/lb-mole. The overhead product rate is 810 lb-mole/hr. At the reflux tray (node 2 in Figure C.6), the vapor and liquid enthalpies are 18,700 and 1,600 Btu/lb-mole, respectively. The liquid enthalpy just above the feed tray is 2,990 Btu/lb-mole. The feed rate is 12,000 lb-mole/hr with an enthalpy of 1,100 Btu/lb-mole of feed. The vapor and liquid enthalpies at the bottom (node 4 in Figure C.6) are 81,500 and 11,000 Btu/lb-mole, respectively.*

What is the overflow rate just above the feed tray? What are the vapor and overflow rates at the top of the tower? Estimate the condenser duty. What is the vapor rate just below the feed tray? What are the vapor and overflow rates at the bottom of the tower? What is the reboiler duty?
Solution: C.8 *Use the overall material balance around the top section (nodes 2 and 3) to estimate the overflow rate just above the feed tray.*
$L_3 = V_3 - P_1 = 1250 - 810 = 440$ *lb-mole/hr*

Combine the overall energy and material balances around nodes 2 and 3 to estimate the vapor and liquid rates at node 2.

$$\begin{aligned} L_2 &= V_2 - P_1 \\ V_2 H_{v2} + L_3 H_{L3} &= L_2 H_{L2} + V_3 H_{v3} \\ V_2 &= (V_3 H_{v3} - P_1 H_{L2} - L_3 H_{L3})/(H_{v2} - H_{L2}) \end{aligned}$$

First compute $V_2 = 1422.56$ *lb-mole/hr. Then compute* $L_2 = 612.57$ *lb-mole/hr.*

Estimate the condenser duty based upon the vapor and liquid rates and enthalpies at node 2.
$Q_c = V_2(H_{v2} - H_{L2}) = 2.43 \times 10^7$ *Btu/hr.*

Use the overall tower balances to compute the bottoms product rate.
$P_2 = F - P_1 = 11,190$ *lb-mole/hr.*

Combine the overall energy and material balances around the feed tray (node 3) and the overall material balance around the bottom section (nodes 3 and 4) to estimate the vapor rate just below the feed tray (node 3).

$$\begin{aligned} L_3^- &= V_3^- + P_2 \\ L_3^+ + F + V_3^- &= V_3^+ + L_3^- \\ H_{L3} L_3^+ + H_F F + H_{v3} V_3^- &= H_{v3} V_3^+ + H_{L3}(V_3^- + P_2) \\ (H_{v3} - H_{L3}) V_3^- &= H_{v3} V_3^+ + H_{L3} P_2 - H_{L3} L_3^+ - H_F F \end{aligned}$$

and $V_3^- = 2471.98$ *lb-mole/hr. Therefore,* $L_3^- = 13,661.98$ *lb-mole/hr.*

Write the overall energy and material balances around nodes 3 and 4 to estimate the vapor and liquid rates at node 4.

$$\begin{aligned} L_3^- &= V_3^- + P_2 \\ L_4 &= V_4 + P_2 \\ (H_{v4} - H_{L4})V_4 &= H_{v3}V_3^- + H_{L4}P_2 - H_{L3}L_3^- \end{aligned}$$

From which we obtain $V_4 = 1,922.15$ *lb-mole/hr and* $L_4 = 13,112.15$ *lb-mole/hr.*

Use the vapor rate and enthalpies at node 4 to estimate the reboiler duty:
$Q_r = V_4(H_{v4} - H_{L4}) = 1.355 \times 10^8$ *Btu/hr*

Use the overall energy balances around the tower to check the reboiler duty:
$Q_r = Q_c - H_F F + P_1 H_{L2} + P_2 H_{L4} = 1.355 \times 10^8$ *Btu/hr*

C.7 Summary

The composition node method provides a means of rapidly designing flowsheets containing countercurrent flows with a minimum amount of trial-and-error calculations. While the operating conditions for complex distillation arrangements cannot be determined with the accuracy of equilibrium stage calculations, the speed of composition node models makes them attractive for use in Preliminary Design and cost optimization. On the other hand, the optimization of complex fractionators, which may have eight or more degrees of freedom, is practically impossible using conventional equilibrium stage models.

The composition node design method permits the use of precedence ordering rather than trial-and-error to solve design equations for countercurrent systems. The key to accurate design is a realistic assignment of the composition at each node. This assignment is one of the most difficult aspects of applying the methodology, but several techniques such as overall material balances and the Fenske equation are surprisingly useful and accurate.

Experience shows that the ability to correct the constant molal overflow assumption by combining the overall energy and material balances at each node is a key advantage of these design models. This adjustment makes the calculations much more accurate than McCabe-Thiele diagrams, for example, so that the differences between composition node and equilibrium stage predictions are often insignificant.

C.8 Problems

3.1 In your own words, compare and contrast rating and design models. How are they similar? How do they differ?

3.2 What are the advantages and disadvantages of using a design model for process optimization rather than a rating model?

3.3 Determine a precedence order for the following set of linearly independent equations:

$$
\begin{aligned}
f_1(x_1, x_2, x_3, x_5, x_7) &= 0 \\
f_2(x_2, x_3, x_4, x_5, x_7) &= 0 \\
f_3(x_1, x_4, x_5, x_6) &= 0 \\
f_4(x_4, x_5, x_6) &= 0 \\
f_5(x_2, x_3, x_5, x_6, x_7, x_8) &= 0 \\
f_6(x_6, x_7, x_8) &= 0 \\
f_7(x_5, x_7, x_9) &= 0
\end{aligned}
$$

How many degrees of freedom exist in this system?

3.4 Consider the structural array shown below. There are 9 variables and 7 equations.

	x_1	x_2	x_3	x_4	x_5	x_6	x_7	x_8	x_9
f_1	√			√	√	√			
f_2		√		√			√		
f_3	√		√				√	√	
f_4		√		√		√			
f_5	√			√				√	√
f_6		√	√			√			
f_7	√			√					

Determine a precedence order for solving these equations and answer the following questions. Which variables may be selected as design variables? Which variables may be selected as tear (or recycle) variables if they are needed? If persistent recycle exists, then which equation(s) should be deleted?

Determine the precedence order and then draw it. Show each equation as a numbered circle and each variable as a labeled arrow or edge.

3.5 A heat exchanger is to be analyzed using the following equation set:

$$Q = \dot{m}_1 C_{p_1}(T_2 - T_1) \tag{C.66}$$

$$Q = \dot{m}_2 C_{p_2}(T_3 - T_4) \tag{C.67}$$

$$U = f(\dot{m}_1, \dot{m}_2) \tag{C.68}$$

$$\ln\frac{T_3 - T_4}{T_2 - T_1} = \frac{(T_3 - T_4) - (T_2 - T_1)}{\Delta T_{\text{ln}}} \tag{C.69}$$

$$Q = UA\Delta T_{\text{ln}} \tag{C.70}$$

A = the heat transfer area

Q = the heat transfer duty

$\dot{m}_i$ = the ith stream flowrate

C_{p_i} = the ith stream heat capacity (treat these as constants, not variables)

U = the overall heat transfer coefficient

ΔT_{ln} = the log mean temperature driving force

T_i = the ith stream temperature

Set up the structural array for this problem. Assume that the two stream heat capacities are constants, rather than variables (i.e. don't include them in the structural array). Determine whether or not a precedence order exists. If possible, select a design variable set that uses your precedence order and draw your precedence order. Represent each equation with a numbered circle and each variable as a line(edge) connecting the various circles.

3.6 Develop a precedence order for the following set of equations. Identify the design variables and draw out the information flow diagram (i.e., equations are numbers in circles, variables are lines connecting the circles).

$$f_1(X_1, X_2, X_3) = 0 \tag{C.71}$$
$$f_2(X_1, X_4, X_6) = 0 \tag{C.72}$$
$$f_3(X_2, X_5, X_6) = 0 \tag{C.73}$$
$$f_4(X_4, X_5, X_7) = 0 \tag{C.74}$$
$$f_5(X_2, X_3, X_5) = 0 \tag{C.75}$$

3.7 A distillation tower is operating such that the vapor rate just above the feed point is 703.569 lb-mole/hr. If the overhead product rate is 304.11 lb-mole/hr, then what is the liquid overflow rate in lb-mole/hr just above the feed point?

3.8 A distillation tower is operating such that the vapor rate just above the feed point is 703.569 lb-mole/hr and the enthalpy at that point is 15456.4 Btu/lb-mole. The overhead product rate is 304.11 lb-mole/hr and the liquid enthalpy just above the feed point is 4138.45 Btu/lb-mole. At the top of the tower, the vapor and liquid enthalpies are 8928.23 and 1669.16 Btu/lb-mole respectively. What are that flowrates of the liquid just above the feed tray and the vapor and liquid streams at the top of the tower?

3.9 A simple distillation tower is being modeled using three composition nodes 2, 3, and 4 which refer to conditions at the tower overheads, feed tray, and bottoms products, respectively. At node 2, the estimated vapor and liquid enthalpies are 4113.88 and 1908.61 Btu/lb-mole, respectively. At the feed tray the vapor and liquid enthalpies are 4263.35 and 2126.17 Btu/lb-mole. At node 4 the vapor and liquid enthalpies are 5264.58 and 3127.61 Btu/lb-mole. The tower pinch point occurs just above the feed tray where the vapor rate has been set to a value of 16055.3 lb-mole/hr. The tower feedrate is 3600.6 lb-mole/hr with an enthalpy of 3058.08 Btu/lb-mole. The overhead product rate is 3193.6 lb-mole/hr. What is the bottoms rate from the tower?

What are the vapor rates at the top and bottom of the tower? What are the overflow rates at the top and bottom of the tower? What is the condenser duty? What is the reboiler duty? What are the vapor and liquid flowrates just below the feed tray?

3.10 The composition node design model makes an approximation which says that the vapor and liquid streams passing each other in a distillation tower have exactly the same composition. Under what kind of operating conditions is this approximation exact?

3.11 Perform a species allocation for a ternary mixture that is to be split into two final product streams. In this case, the overhead to bottom flow ratio for the least volatile component is 1 to 25. The relative volatility of the middle component to the heavy component is 1.5 and the volatility of the light component to the heavy component is 1.8. Assume that the tower contains ten theoretical stages and calculate the $\boldsymbol{p}$ vector if the feed rate vector is $\boldsymbol{f}^t = [40, 80, 100]$ lb-mole/hr.

3.12 If the fractional vaporization of a feed to a distillation tower is equal to 1.5, is the feed subcooled, at its bubble point, partially vaporized, at its dew point, or a super-heated gas?

3.13 What does it mean for a distillation tower to have a pinch point? How many theoretical trays are required at a pinch point?

3.14 A bubble point liquid at 675 psia with a specific enthalpy of 4260 BTU/lb-mole is throttled to a reduced pressure of 350 psia. At the lower pressure, the vapor and liquid in equilibrium have specific enthalpies of 6200 and 2050 BTU/lb-mole respectively. What fraction of the liquid feed to the throttle is flashed to vapor during this step?

3.15 A feed stream consists of 250 lb-moles/hr of species A and 150 lb-moles/hr of species B. If it is separated into two product streams, p_a and p_b, such that p_a is 83% A and p_b is 92% B, then what are the flowrates of the two product streams?

3.16 A process produces three product streams, p_a, p_b, and p_c. The composition of p_a is (0.82, 0.11, 0.07). The composition of p_b (0.05, 0.90, 0.05). The composition of p_c is (0.02, 0.07, 0.91). The feedrate is (800, 500, 200) moles of species A, B, and C, respectively. Complete the species allocation for this system including the final product flowrates and the rates and compositions of the intermediate streams: AB, BC, and AC. (The last stream is a candidate system using liquid/liquid extraction where B is extractable, but A and C are not extractable.)

3.17 Estimate the Underwood root for a simple distillation tower separating a binary mixture. The relative volatility of species A to that of species B is 2.3 and essentially constant throughout the column. The feed is introduced half way between the distillate and bottoms product draws. The column feed is

60 mole% species A and the column feed is a saturated liquid. The distillate rate is 300 lb-mole/h and the distillate is 85 mole% species A.

3.18 Estimate the minimum vapor rate for a simple distillation column in its rectification section. The Underwood root is $\lambda = 1.04761$. The relative volatility between species A and B is constant and $\alpha = 1.1$. The binary feed to the column is 50 mole% of species A. The column feedrate is 1000 lb-mole/h. The distillate rate is 500 lb-mole/h and the distillate is 90 mole% species A.

3.19 A simple distillation tower is used to partially separate a mixture of three species. Using the two Underwood roots, the minimum vapor in the rectifying section is either 206.7 or 315.8 lb-mole/h. Similarly, the minimum vapor in the stripping section is either 310.1 or 411.2 lb-mole/h. Based on this information, what is the minimum vapor rate for the tower? Is the pinch above or below the feed tray?

3.20 A distillation tower is used to separate a mixture of species A, B, and C which have constant K-values equal to 1.2, 1.01, and 0.9, respectively. The required overhead product rate (p_a) is 30 lb-moles/hr with 99% purity. The required p_c rate is 25 lb-mole/hr with 99% purity. If Design 6 is used to produce all three final products and p_b can contain no more than 1% C, then what is the minimum vapor rate in this tower based upon the purity? If Design 7 is used to produce all three final products and the middle product can contain no more than 2% A, then what is the minimum vapor rate in the tower based on this purity specification?

3.21 A distillation tower is operated such that its distillate to feed mole ratio is 0.3. The feed composition vector is (0.7, 0.3) in mole fraction units and the relative volatility of species A to species B is constant with a value of 1.2. At total reflux, the tower design is equivalent to six theoretical stages between distillate and the bottoms. Based on this information, estimate the distillate and bottoms compositions.

3.22 A simple distillation tower will be used to partially separate a binary mixture. The tower feedrate is 100 lb-mole/hr of species A and 50 lb-moles/hr of species B. The relative volatility of A to B is 1.5 and essentially constant throughout the tower. The tower is equivalent to 10 theoretical stages based upon the distillate and bottoms compositions. If the total distillate rate is 60 lb-mole/hr, and the distillate to bottoms flowrate ratio is 0.1 for the heavy component (species B), then estimate the distillate and bottoms compositions.

3.23 Consider the distillation of a ternary mixture. The distillate to bottoms molar flow ratio (d/b) is 1/25 for the heavy species. The middle/heavy species relative volatility is 1.2/1; the light/heavy relative volatility is 1.5/1. The separator uses 8 theoretical stages. The light/middle/heavy molar feedrates are: 40/30/50 moles/unit time. What is the total distillate flowrate?

3.24 In problem 23, what is the estimated distillate composition?

3.25 A feed stream consists of 490 lb-mole/hr of A and 170 lb-mole/hr of B. It must be separated so that the distillate is 95 mole% A and the bottoms is 75 mole% B. What is the total distillate molar flowrate?

3.26 A distillation tower is operating at steady state. It has a vapor rate of 2990 lb-mole/h just above the feed tray where the vapor enthalpy is 8430 Btu/lb-mole and the liquid overflow enthalpy is 8352 Btu/lb-mole. The distillate product rate is 1380 lb-mole/h and the feed rate is 5400 lb-mole/h. The feed enthalpy is 8250 Btu/lb-mole. At the bottom of the tower, the vapor and liquid phase enthalpies are 9175 and 8870 Btu/lb-mole, respectively.

Based on this information, answer the following questions. What is the overflow rate just above the feed tray? What are the vapor and overflow rates at the top of the column? What is the condenser duty? What are the vapor and overflow rates just below the feed tray? What are the vapor and overflow rates at the bottom of the tower? What is the estimated reboiler duty?

3.27 In your own words, explain how species allocation and composition node concepts may be used to estimate vapor and liquid phase enthalpies at key points in distillation towers. What is the advantage of this modeling approach compared to simulation?

3.28 A continuous distillation column receives a single feed at a rate of 500 lb-mole/h. This feed is 60 mol% species A, 30% B and 10% species C. It is a saturated liquid as it enters the column. The tower distillate is 97 mol% A, 2% B and 1% C. The bottoms product is 2% species A. The relative volatilities are constant throughout the tower with values of 1.3, 1.1, and 1.0 for A, B, and C, respectively.

Based on this information, answer the following questions. What are the bottoms and distillate rates? What is the bottoms composition? What are the minimum vapor requirements in the rectifying and stripping sections of the column? What are the minimum theoretical tray requirements? What are the keys for the column?

3.29 A binary mixture must be distilled. Estimate the bottoms composition for a continuous, simple column using the following information. The relative volatility is constant at 1.1. The distillate/bottoms molar flow ratio for the less volatile species must be 1/15. When at steady state, the column achieves exactly 25 theoretical stages between the distillate and bottoms. It has a total condenser. The feed rate is 70 and 80 moles/hr of the more and less volatile species, respectively.

3.30 A species allocation for a separation facility is needed. The overall steady state design basis is:

$$\boldsymbol{f} = \boldsymbol{X}\boldsymbol{p}$$

and numerically:

$$\begin{bmatrix} 720 \\ 480 \\ 380 \end{bmatrix} = \begin{bmatrix} 0.90 & 0.10 & 0.02 \\ 0.05 & 0.80 & 0.03 \\ 0.05 & 0.10 & 0.95 \end{bmatrix} \begin{bmatrix} 732.9 \\ 542.8 \\ 304.3 \end{bmatrix}$$

where $\boldsymbol{p}^t = (p_1, p_2, p_3)$. Based on this information, what are the *species* rates for a feed stream producing products p_1 and p_2?

3.31 A binary mixture is separated by distillation. Estimate the distillate and bottoms product rates and compositions from the following information. For the less volatile species, the component distillate to bottoms flow ratio is 1 to 10. The relative volatility, α, is 1.2. The distillation column operates with 20 theoretical stages. The light and heavy species feed rates to the column are 50 and 80 mol/s, respectively.

Is trial-and-error required to answer these questions or can precedence ordering be used?

Appendix D

Composition Node Design Models

While composition nodes are primarily used to establish material balances, they may also be used to estimate special requirements. For example, Figure D.1 illustrates a simple tower with a single mid-reboiler located between the feed and bottoms product trays. In order to optimize a column with a mid-reboiler, one must estimate a composition near the mid-reboiler where a liquid draw is removed and a vapor returned from the mid-reboiler. The local composition in this case is estimated using node 4 in Figure D.1. Adding a mid-reboiler also adds two additional optimization variables — the location of the mid-reboiler and its duty.

Figure D.2 illustrates a complex tower combination consisting of an upstream tower with three sections and a downstream tower with a single rectification[1] section. In this design, the reboiler for the upstream tower provides vapor for both towers, but both towers have their own condensers. Therefore, the downstream tower receives a vapor from the upstream tower and returns a liquid. At this point, the two towers are said to be *thermally-coupled* between nodes.

Two complex towers may also be configured as shown in Figure D.3. In this case, the upstream tower contains three sections and the downstream tower consists of a single stripping[2] section. This design only has one condenser at node 2 which provides reflux for both columns. However, both columns have reboilers at nodes 5 and 7. This design is also thermally-coupled between nodes 3 and 6.

If two towers are connected together as shown in Figure D.4, then nine nodes are required to describe the columns. In this case, both columns have condensers at nodes 2 and 5. They have reboilers at nodes 4 and 9. Nodes 5, 7, and 9 indicate the three product compositions and nodes 2 and 4 indicate the intermediate products, F_2 and F_3 which are produced by the upstream tower.

The upstream tower in Figure D.4 is *simple* in that it only produces two products and includes only two sections. The downstream tower is *complex*. It includes four

[1] In a rectification section, the feed is recovered from the top of the section.

[2] A stripping section produces a product from the bottom of the section.

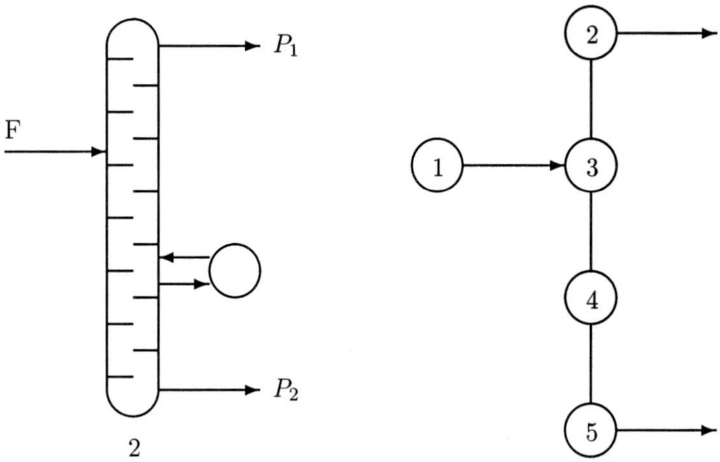

Optimization Variables

Variable	Definition
f	Feed fractional vaporization
P	Tower pressure at node 2
P_F	Feed pressure at node 1
q	Fraction of reboiler duty carried by mid-reboiler.
N	Location of mid-reboiler.
V/V_{min}	Vapor to minimum vapor rate at the pinch point.

Figure D.1: Node representation of a simple tower with a mid-reboiler, design 2.

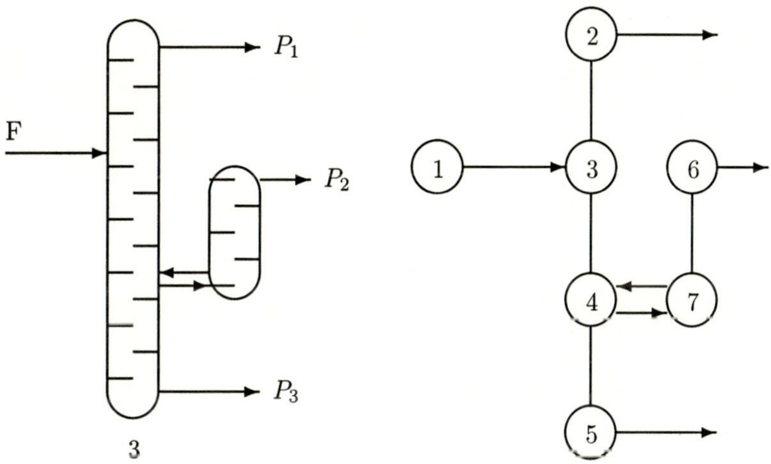

Optimization Variables

Variable	Definition
f	Feed fractional vaporization
P	Tower pressure at node 2
P_F	Feed pressure at node 1
V/V_{min1}	Vapor to minimum vapor rate at the pinch point in the upstream tower.
V/V_{min2}	Vapor to minimum vapor rate at the pinch point in the downstream tower.

Figure D.2: Node representation of a complex tower combination with a downstream rectifier, design 3.

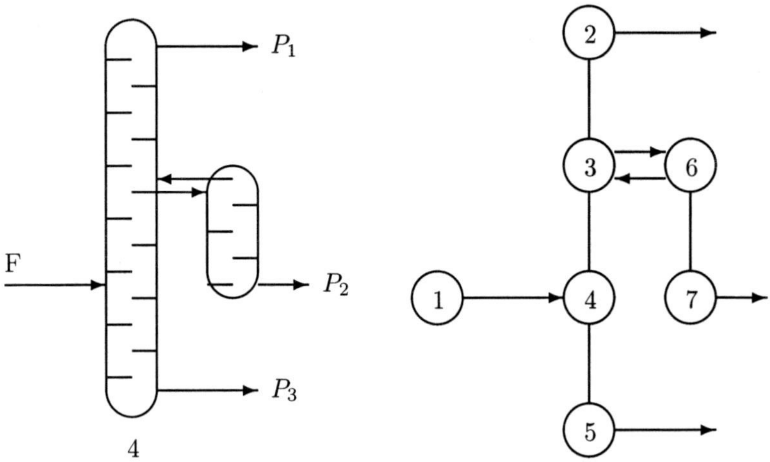

Optimization Variables

Variable	Definition
f	Feed fractional vaporization
P	Tower pressure at node 2
P_F	Feed pressure at node 1
V/V_{min1}	Vapor to minimum vapor rate at the pinch point in the upstream tower.
V/V_{min2}	Vapor to minimum vapor rate at the pinch point in the downstream tower.

Figure D.3: Node representation of a complex tower combination with a downstream stripper, design 4.

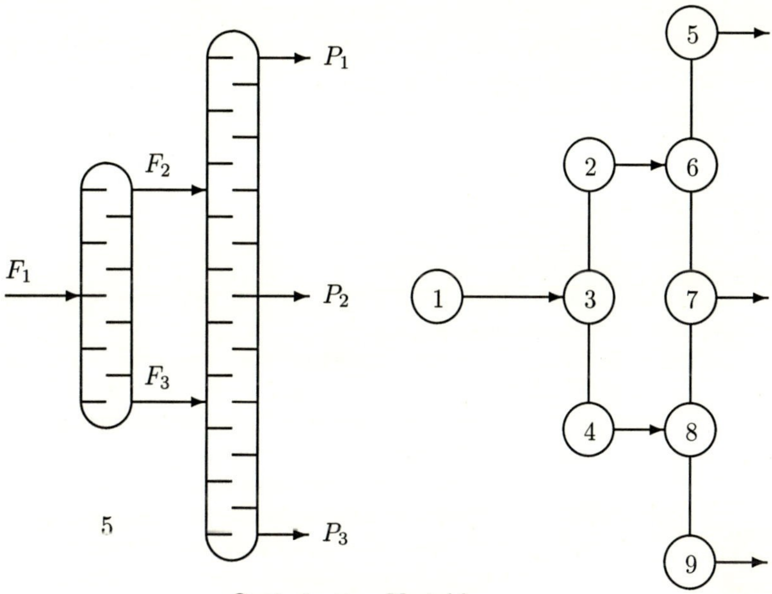

Optimization Variables

Variable	Definition
f_1	Feed 1 fractional vaporization
f_2	Feed 2 fractional vaporization
f_3	Feed 3 fractional vaporization
P_{F_1}	Feed 1 pressure at node 1
P_2	Tower 1 pressure at node 2
P_5	Tower 2 pressure at node 5
V/V_{min1}	Vapor to minimum vapor rate at the pinch point in the upstream tower.
V/V_{min2}	Vapor to minimum vapor rate at the pinch point in the downstream tower.
F_2/F_1	Distillate to Feed 1 ratio in upstream tower.
N_1	Degree of separation in upstream tower.

Figure D.4: Node representation of a complex tower combination of two towers, design 5.

sections and is represented by six nodes. This design also has additional flexibility which is indicated by the optimization variables that are tabulated in the figure.

The designs illustrated in Figures D.5 and D.6 are complex towers with three sections each that produce three products using a single tower. Design 6 in Figure D.5 produces a distillate, P_1, a bottoms product, P_3, and an intermediate product, P_2, that is removed as a lower vapor draw beneath the feed tray and the bottom of the column. On the other hand, Design 7 recovers P_2 as a liquid side draw above the feed tray. Generally these designs are only attractive whenever the bottoms or distillate product rates are small. One of the more popular applications of Design 7 occurs whenever a small amount of noncondensible gas (*e.g.* hydrogen or methane) must be removed from a feed mixture and P_2 is the major distillate product.

Figures C.6 through D.6 illustrate how countercurrent processes may be depicted using composition nodes. The nodes, however, are not all equivalent. For example, the compositions at some nodes vary during an optimization study. At other nodes the compositions are always constant. By definition, compositions vary at *internal* nodes. On the other hand, the compositions are invariant at *external* nodes. Both types of nodes are useful in design models, but the calculational procedures for estimating their compositions differ with the structure.

Feed streams are always designated using a composition node, but product streams are not. Otherwise, the representation would be unnecessarily redundant since a product stream is either a feed to another unit operation or else part of the design basis. In the former case, a separate node is used to describe those products which are also feeds. In the latter case, the design basis products terminate as external nodes and no additional representation is required.

D.1 Additional Considerations for Complex Designs

The general flow of information in the composition node design method as previously outlined holds for complex Designs 3-7 with a few additions. Most differences are related to the greater number of design variables needed for some complex designs and the effect of these variables on the composition of several internal nodes.

D.1.1 Designs 3 and 4

Complex separator Designs 3 and 4 (see Figures D.2 and D.3) are two tower sequences producing three product streams from a single feed. The upstream tower in Design 3 produces a lower vapor side stream which is rectified in the second column to give the middle product with the reflux liquid being returned to the upstream column. Design 4 is similar, except that the second tower produces the middle product by stripping the more volatile components from an upper liquid sidedraw.

Since the composition node design model requires specification of the composition of all nodes near the beginning of the calculation sequence, it is necessary to fix the composition at nodes 4 and 7 in Design 3 and nodes 3 and 6 in Design

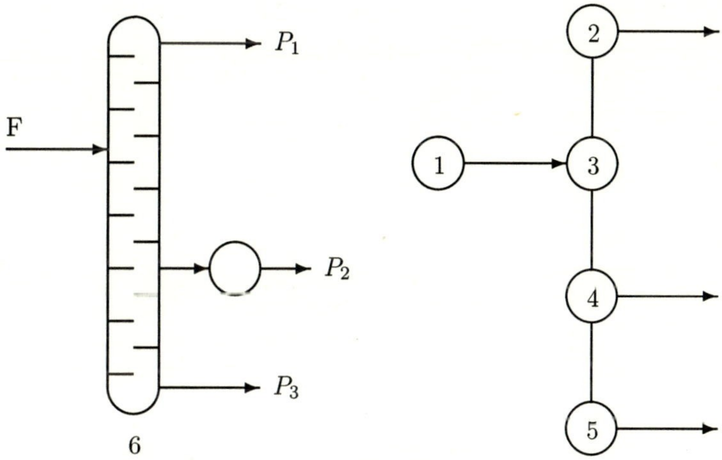

Optimization Variables

Variable	Definition
f	Feed fractional vaporization
P	Tower pressure at node 2
P_F	Feed pressure at node 1
V/V_{min}	Vapor to minimum vapor rate at the pinch point.

Figure D.5: Node representation of a complex tower with a lower vapor sidestream, design 6.

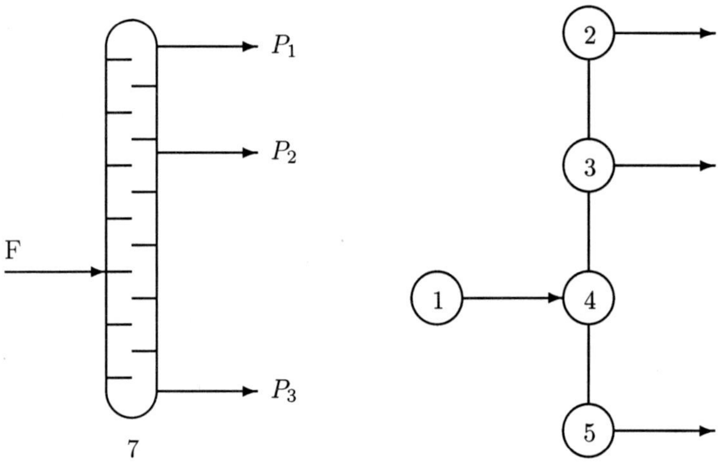

Optimization Variables

Variable	Definition
f	Feed fractional vaporization
P	Tower pressure at node 2
P_F	Feed pressure at node 1
V/V_{min}	Vapor to minimum vapor rate at the pinch point.

Figure D.6: Node representation of a complex tower with an upper liquid side-stream, design 7.

4. It is assumed that most of the major components of the overhead product will have been stripped from the liquid by the time it reaches node 4 in Design 3. As a rough estimate, therefore, it is reasonable to set node 4 equal to the composition of the mixture which would remain after removing the overhead product from the original feed. Also, since the node model makes no distinction between the liquid and vapor compositions, both the liquid sidestream and vapor returned from the rectifying column (node 7) are assumed be at the node 4 composition. This last *no enrichment* assumption is pessimistic since the vapor returning must be enriched in those heavy component(s) that are not removed with the middle product. Ignoring the enrichment results in some overdesign of the main column.

Similar reasoning applied to Design 4 indicates that the composition of nodes 3 and 6 can be estimated as the composition of the original feed minus the bottom product.

The exchange of liquid and vapor streams provides thermal coupling between the two towers of these complex separators. For example, the reboiler of the downstream tower of Design 4 generates vapor to satisfy its own requirements and also increases the vapor rate in the upper section of the first column. Likewise, the condenser of the upstream tower in Design 4 provides the reflux requirements for both itself and the sidestream stripper. Thus, the second tower allows a high purity middle product to be withdrawn without the excessive vapor rate in the main column which would be required by Designs 6 or 7.

Thermal coupling between the two columns also dictates that the minimum vapor requirement for the sidestream stripper be determined first as this quantity sets the flow rates of the sidedraw and return streams. The internal rates in the upstream column can then be calculated in the usual manner.

D.1.2 Design 5

Complex separators like Design 5 offer more flexibility than any of the other designs considered here. This distillation system consists of a simple tower prefractionator whose products feed a second column that produces a liquid sidestream in addition to the overhead and bottom products. Design 5 has eight degrees of freedom that are specified as optimization variables (see Figure D.4). Most of the flexibility in the operation of this design results from the ability to control the compositions and flow rates of the feed introduced into the main tower.

Referring to Figure D.4, nodes 2 and 4 (the top and bottom of the prefractionator) are both internal. Their compositions are calculated based on the amount of the feed taken overhead and the degree of separation specified in this tower. The procedure used to fix these compositions is a slight modification of the iterative, theta method proposed by Holland [44, 45].

After the node 2 and 4 compositions are specified, the design of both towers can proceed in the usual fashion. However, the vapor rates in the downstream column producing a product directly as a sidestream (at node 7) are subject to another constraint in addition to the Underwood minimum vapor rate. The magnitude of this lower limit on the vapor rate is highly dependent on the sidestream purity specifications and, in the case of Design 5, on the composition and flowrates of the two tower feeds.

When the concentration of contaminants at the point of sidestream withdrawal (node 7) are specified at very low values, and these contaminants are some of the major components of the top or bottom products, then a component balances over the upper or lower sections of the column demand that the internal stream rates at the sidedraw plate be great enough to deliver the required quantity of contaminant to the top or bottom of the tower.

The required vapor rate just above the sidestream node is given by a component balance around the upper half of the column written for a light contaminant assuming equilibrium between the liquid and vapor streams at that point:

$$V_{P_2}^{+} = \frac{P_1(X_{LP_2} - X_{LP_1}) - F_2(X_{LP_S} - X_{LF_2})}{(1 - K_{LP_2})X_{LP_2}} \tag{D.1}$$

where P_1 = the distillate molar flow rate from the downstream tower

F_2 = the feedrate entering above the sidestream

X_L = the mole fraction of the light component for which the balance is written. The subscripts P_1, F_2, or P_2 indicate a value in the the streams P_1, F_2, or P_2, respectively

K_{LP_2} = the vapor/liquid equilibrium constant for the component at the sidestream (P_2) conditions

Note that the usual composition node assumption of equal liquid and vapor compositions (i.e., $K_i = 1.0$) cannot be used here as it would lead to an infinite vapor requirement.

A similar balance written for the heavy components which must pass the sidedraw node to exit in the bottom product gives a lower limit on the vapor rate just below the sidestream:

$$V_{P_2}^{-} = \frac{P_3(X_{HP_3} - X_{HP_2}) - F_3(X_{HF_3} - X_{HP_2})}{(1 - K_{HP_2})X_{HP_2}} \tag{D.2}$$

where P_3 = the bottoms flow rate from the downstream tower

F_3 = the molar feed rate entering below the sidestream from the prefractionator

X_H = the mole fraction of the heavy component with the subsripts P_3, P_2, and F_3 referring to the same products and feeds, respectively

Note that the vapor rate constraints given by Equations D.1 and D.2 can be lowered by sending a high percentage of the light components overhead in the prefractionator and most of heavy components out in the prefractionator bottom product (i.e. F_3). It is this capability that allows Design 5 to be used to produce a relatively high purity middle product with reasonable vapor rates in the main column.

D.1.3 Designs 6 and 7

Designs 6 and 7 are the simplest of the separators which split their feed into three products. Both of these designs are single towers from which a middle product is withdrawn as a sidestream. The only difference between the two is that a lower vapor sidestream is produced by Design 6, while Design 7 generates an upper liquid side product. There are only three operating variables for each of these separators (see Figures D.5-D.6).

Being sidestream towers, Designs 6 and 7 are subject to minimum vapor rate constraints which are similar to Equations D.1 and D.2. Design 6 must obey an equation like Equation D.2, and Design 7 is constrained by an equation like Equation D.1. However, there are no feed terms in these cases and it is not possible to reduce the vapor rate limit for these types of towers by appropriate manipulation of the design variables. For this reason, Designs 6 and 7 are often not attractive in comparison with the other complex fractionators unless a high sidestream contaminant level is tolerable or there is relatively little of the contaminant leaving the tower in the other product. The derivation of the appropriate equations for Designs 6 and 7 is left as an exercise.

It has been recognized for some time that advantages exist from the deliberate sequencing of distillation towers. These advantages appear to depend on the structural differences between the designs. Figures C.6-D.6, for example, show seven possible alternatives. Table 12.1 summarizes the sequence alternatives which increase drastically as a function of the number of feed splits that must be performed.

The idea of thermally coupling two towers, as shown for Designs 3 and 4 in Figures D.2 and D.3, is not new. These towers are thermally coupled where they exchange vapor and liquid streams. Grunberg [34] suggested that thermally coupled designs operate more reversibly in the thermodynamic sense and should be favored for that reason. Stupin [98] examined a thermally coupled version of Design 5 in Figure D.4. His analysis suggested that designs like Design 5 might be more stable toward composition disturbances and require less vapor to operate than Design 1.

Petlyuk, Platonov, and Slavinckii [77] considered the separation of ternary mixtures containing equimolar amounts of overhead and bottom products. The amount of middle product was varied. Using approximate equations taken from Robinson and Gilliland [83], they developed expressions for the minimum vapor flow associated with the different designs. Their objective function was the total vapor flow. On this basis, they found that Design 5 always requires less vapor than either the direct or inverted sequences of simple towers. A thermally coupled version of Design 5 requires even less vapor. The advantage they observed for Design 5 increased proportionately with the amount of middle product in the feed.

There are clear structural differences between the designs. There are also differences in the degrees of freedom with which the respective designs can be optimized. But which design is best? How can these structural differences be used to advantage? What makes one design economically preferred over another? Selection heuristics have been suggested from studies by Lockhart [61], Harbert [38], Rudd Powers and Siirola [85], Rod and Marek [84], Heaven [41], and Nishimura and Hiraizumi [71].

D.2 Problems

4.1 Derive Equations D.1 and D.2 from material balance considerations.

4.2 Derive equations similar to Equations D.1 and D.2 for Designs 6 and 7 from material balance considerations.

4.3 What factors can make Designs 5, 6 and 7 more or less economical compared to the alternatives?

Bibliography

[1] Accreditation Board for Engineering and Technology, New York. *Criteria for Accrediting Programs in Engineering*, 1985.

[2] D. H. Allen and R. C. Page. Revised technique for predesign cost estimating. *Chemical Engineering*, pages 142–150, Mar 3, 1975.

[3] T. H. Arnold, Jr. and C. H. Chilton. New index shows plant cost trends. *Chem Eng*, page 143, Feb 18 1963.

[4] G. T. Austin. The industrially significant organic chemicals–part 1. *Chem Eng*, pages 127–132, Jan 21, 1974.

[5] D. F. Baker. Surge control for multistage centrifugal compressors. *Chem Eng*, page 117, May 31, 1982.

[6] B. Batchelor, M. B. Lasala, M. McDevitt, and E. Peacock. Technical and economic feasibility of ultra-high lime treatment of recycled cooling water. *Res J Water Pollut Control Fed*, 63(7):982–990, 1991.

[7] N. P. Baudat and P. A. Darrow. Power recovery in a closed cycle system. *Chem Eng Prog*, page 68, Feb 1980.

[8] R. E. Bellman. *Dynamic Programming*. Princeton University Press, New Jersey, 1957.

[9] D. M. Benforado, G. Ridlehoover, and M. D. Gores. Pollution prevention: One firm's experience. *Chem Eng*, pages 130–133, Sept 1991.

[10] R. S. Bennett. Recompression evaporation. *Chem Eng Prog*, page 67, July 1978.

[11] M. J. P. Bogart. A gas compressor calculator program. *Chem Eng Prog*, page 66, Aug 1981.

[12] C. T. Breuer. Retrofitting coal-fired boilers: Economics for process plants. *Chem Eng*, page 97, Sept 17 1984.

[13] G. S. Brown. How to predict pressure drop before designing the piping. *Chem Eng*, pages 85–86, Mar 16, 1987.

[14] S. Budavari, editor. *The Merck Index.* Merck & Co., Inc, Rahway, N.J., 11th edition, 1989.

[15] P. Y. Burke. Compressor intercooler and aftercoolers: Predicting off-performance. *Chem Eng*, page 107, Sept 20, 1982.

[16] G. M. Cameron, P. D. Nolan, and K. R. Shaw. The cil process for acid manufacture. *Chem Eng Prog*, pages 47–50, Sept 1978.

[17] J. W. Casten. Mechanical recompression evaporators. *Chem Eng Prog*, page 61, July 1978.

[18] R. H. Cavett. Physical data for distillation calculations - vapor-liquid equilibria. In *Proc 27th Midyear Meeting API Div Refining*, volume 42(III), page 351. American Petroleum Institute Division of Refining, 1962.

[19] H. L. Chatelier. Estimation of firedamp by flammability limits. *Ann. Mines*, (8):388–395, 1891.

[20] C. H. Chilton. Plant cost index points up inflection. *Chem Eng*, page 185, Apr 25 1966.

[21] F. D. Clark and A. Lorenzoni. *Applied Cost Engineering.* Marcel Dekker, Inc., New York, 2nd edition, 1985.

[22] R. W. Colbert. Industrial heat exchange networks. *Chem Eng Prog*, 47, July 1982.

[23] C. F. Colebrook. Turbulent flows in pipes, with particular reference to the transition region between smooth and rough pipe laws. *J Inst Civ Eng*, 11:133–156, 1939.

[24] R. T. Crouch, G. D. Stacy, B. F. Houston, and L. D. Gaines. Allocation of plant energy resources. *Chem Eng Prog*, page 37, Oct 1981.

[25] D. A. Crowl and J. F. Louvar. *chemical Process Safety: Fundamentals with Applications.* Prentice Hall, Englewood Cliffs, N.J., 1990.

[26] M. B. Desai. Preliminary cost estimating of process plants. *Chem Eng*, pages 65–70, July 27, 1981.

[27] W. C. Edmister. *Pet Ref*, page 173, Apr 1958.

[28] M. R. Fenske. Fractionation of straight-run Pennsylvania gasoline. *Ind Eng Chem*, 24:482–485, 1932.

[29] J. G. Ferguson, editor. *Funk & Wagnalls New Comprehensive International Dictionary of the English Language.* J. G. Ferguson Publishing Company, New Jersey, 1982.

[30] D. C. Freshwater. The heat pump in multicomponent distillation. *British Chem Eng*, 6(6):388, June 1961.

[31] V. Ganthapy. Evaluate the performance of waste heat boilers. *Chem Eng*, page 291, (Nov 16, 1981.

[32] J. R. Gaston. Antisurge control schemes for turbocompressors. *Chem Eng*, page 139, Apr 19, 1982.

[33] C. H. Gooding. Estimating flash point and lower explosive limit. *Chem Eng*, page 88, Dec 12, 1983.

[34] J. F. Grunberg. The reversible separation of multicomponent mixtures. In *Advances in Cyrogenic Engineering*, volume 2. New York, 1960.

[35] K. M. Guthrie. *Modern Cost Engineering Techniques*, chapter Data and Techniques for Preliminary Capital Cost Estimating, pages 80–106. McGraw-Hill, New York, 1970.

[36] K. M. Guthrie. *Process Plant Estimating, Evaluation, and Control.* Craftsman Book Company of America, 542 Stevens Avenue, Solana Beach, CA 92075, 1974.

[37] J. Happel and D. G. Jordan. *Chemical Process Economics.* Marcel Dekker, Inc., New York, 2nd edition, 1975.

[38] W. D. Harbert. Which tower goes where? *Pet Ref*, 36(3):169, 1957.

[39] P. W. Hart and J. T. Sommerfeld. Cost estimation of specialty chemicals from laboratory-scale prices. *Cost Eng*, 39(3):31–35, March 1997.

[40] Guidelines for hazard evaluation procedures. American Institute of Chemical Engineers, New York, 1985.

[41] D. L. Heaven. Optimum sequencing of distillation columns in multicomponent fractionation. Master's thesis, School of Chemical Engineering, University of California, Berkeley, 1969.

[42] J. E. Hendry. *Computer Aided Synthesis of Optimal Multicomponent Separation Sequences.* PhD thesis, Chemical Engineeering Department, University of Wisconsin, Madison, WI, 1972.

[43] J. E. Hendry and R. R. Hughes. Generating process flowsheets. *Chem Eng Prog*, 68(6):69, 1972.

[44] C. D. Holland. *Fundamentals of Multicomponent Distillation.* McGraw-Hill, New York, 1981.

[45] C. D. Holland and G. P. Pendon. Solve more distillation problems. *Hydrocarb Proc*, page 148, 1974.

[46] W. B. Hooper. *Chem Eng*, pages 127–129, May 17, 1982.

[47] O. A. Hougen, K. M. Watson, and R. A. Ragatz. *Chemical Process Principles*, volume 2. Wiley, Inc., New York, 2nd edition, 1959.

[48] C. Johnnie and D. Aggarwal. Calculating plant utility costs. *Chem Eng Progr*, (11):84, 1977.

[49] G. Jones. Inflammation limits and their practical application in hazardous industrial operations. *Chem. Rev.*, (1):1–26, 1938.

[50] O. R. Jones and S. K. Farr. Selecting an efficient fluid for recovering power from waste heat. *Chem Eng*, page 57, Dec 28, 1981.

[51] M. Kiley. *National Construction Estimator.* Craftsman Book Company of America, Carlsbad, CA, 37th edition, 1989.

[52] C. J. King, D. W. Gantz, and F. J. Barnes. Systematic evolutionary process synthesis. *I& EC Proc Des Dev*, 11(2):271, 1972.

[53] T. A. Kletz. *Hazop and Hazan: Identifying and Assessing Process Industry Hazards.* Taylor and Francis, Philadelphia, PA, 1999.

[54] H. J. Lang. *Chem Eng*, 54(10):117, 1947.

[55] H. J. Lang. *Chem Eng*, 55(6):112, 1948.

[56] R. P. Lapina. How to use the performance curves to evaluate behavior of centrifugal compressors. *Chem Eng*, page 86, Jan 25, 1982.

[57] B. Linhoff and J. R. Flowers. Synthesis of heat exchange networks — I: Systematic generation of energy optimal networks. *AIChE J*, 24(4):633, 1978.

[58] B. Linhoff and J. R. Flowers. Synthesis of heat exchanger networks — II: Evolutionary generation of networks with various criteria of optimality. *AIChE J*, 24(4):642, 1978.

[59] B. Linhoff and D. W. Townsend. Designing total energy systems. *Chem Eng Prog*, 72, July 1982.

[60] B. Linhoff and J. A. Turner. Heat recovery networks: New insights yield big savings. *Chem Eng*, page 56, Nov 2, 1981.

[61] F. J. Lockhart. Multi-column distillation of natural gasoline. *Pet Ref*, 26(8):104, 1947.

[62] J. Matley. Ce plant cost index — revised. *Chem Eng*, pages 153–156, Apr 19, 1982.

[63] J. Matley. Ce cost indexes set slower pace. *Chem Eng*, pages 75–76, Apr 29, 1985.

[64] W. L. McCabe and J. C. Smith. *Unit Operations of Chemical Engineering.* McGraw-Hill, New York, 2nd edition, 1967.

[65] S. F. Miller. Making the most of bottoms and residuals. In *Solvent Waste Reduction*, chapter 12, pages 97–104. Noyes Data Corporation, Park Ridge, NJ, 1990.

[66] K. Molich. Consider gas turbines for heavy loads. *Chem Eng*, page 79, Aug 25, 1980.

[67] Y. K. Molokanov, T. P. Korablina, N. I. Mazurina, and G. A. Nikiforov. An approximation method for calculating the basic parameters of multicomponent fractionation. *Int Chem Eng*, 12(2):209, 1972.

[68] R. L. Motard and A. Westerberg. *Chemical Process Synthesis.* New York, Aug 1980.

[69] R. F. Neerken. Use steam turbines as process drivers. *Chem Eng*, page 63, Aug 25, 1980.

[70] W. L. Nelson. *Petroleum Refining Engineering.* McGraw-Hill, New York, 4th edition, 1958.

[71] H. Nishimura and Y. Hiraizumi. Optimal system pattern for multicomponent distillation systems. *Int Chem Eng*, 11:188, 1971.

[72] H. R. Null. Heat pumps in distillation. *Chem Eng Prog*, 72(7):58, July 1976.

[73] H. E. O'Connell. *Trans AIChE*, 42:751, 1946.

[74] J. Page. *Conceptual Cost Estimating Manual.* Gulf Publishing Company Book Division, Houston, 1984.

[75] R. H. Perry and C. H. Chilton, editors. *Chemical Engineers' Handbook.* McGraw-Hill, New York, 5th edition, 1973.

[76] M. S. Peters and K. D. Timmerhaus. *Plant Design and Economics for Chemical Engineers.* McGraw-Hill, New York, 4th edition, 1991.

[77] F. B. Petlyuk, V. M. Platonov, and D. M. Slavinckii. Thermodynamically optimal method for separating multicomponent mixtures. *Int Chem Eng*, 5:555, 1965.

[78] T. K. Pho and L. Lapidus. Topics in computer-aided design: Part II. Synthesis of optimal heat exchanger networks by tree searching algorithms. *AIChE J*, 19(6):1182, 1973.

[79] R. N. S. Rathore, K. A. V. Wormer, and G. J. Powers. Synthesis strategies for multicomponent separation systems with energy integration. *AIChE J*, 20(3):491, 1974.

[80] F. Redmill, M. Chudleigh, and J. Catmur. *System Safety: HAZOP and Software HAZOP.* Wiley, New York, 1999.

[81] R. C. Reid, J. M. Prausnitz, and B. E. Poling. *The Properties of Gases and Liquids.* McGraw-Hill, New York, 4th edition, 1987.

[82] R. C. Reid and T. K. Sherwood. *The Properties of Gases and Liquids.* McGraw-Hill, New York, 2nd edition, 1966.

[83] C. S. Robinson and E. R. Gilliland. *Elements of Fractional Distillation.* McGraw-Hill, New York, 4th edition, 1950.

[84] V. Rod and J. Marek. Separation sequences in multicomponent rectification. *Coll Czech Chem Comm*, 24:3240, 1959.

[85] D. F. Rudd, G. J. Powers, and J. J. Siirola. *Process Synthesis.* Prentice Hall, New Jersey, 1973.

[86] D. F. Rudd and C. C. Watson. *Strategy of Process Engineering.* John Wiley & Sons, Inc., New York, 1968.

[87] J. seader and E. J. Henley. *Separation Process Principles.* John Wiley & Sons, Inc., 1998.

[88] T. K. Serghides. Estimate friction factor accurately. *Chem Eng*, pages 63–64, Mar 5, 1984.

[89] M. R. Shelton and I. E. Grossmann. A shortcut procedure for refrigeration systems. *Comput Chem Eng*, 9(6):615–619, 1985.

[90] M. Shoaei. An improved method for design calculations in a multicomponent distillation column. Master's thesis, School of Chemical Engineering, Georgia Institute of Technology, Atlanta, GA, Dec 1983.

[91] M. Shoaei, S. S. Lorensen, and D. W. Tedder. A tree search methodology based on sequential dynamic programming and set analysis. *Comput Chem Eng*, 11(5):469–480, 1987.

[92] M. Shoaei and D. W. Tedder. Design calculations for multicomponent distillation by an improved shortcut method. *Chem Eng Res. Des.*, 65:251–260, May 1987.

[93] J. M. Smith and H. C. V. Ness. *Introduction to Chemical Engineering Thermodynamics.* McGraw-Hill, New York, 4th edition, 1987.

[94] J. Sommerfeld and C. Lenk. Thermodynamics helps you predict selling price. *Chem Eng*, pages 136–138, May 4 1970.

[95] A. Stancell. Personal communication to D.W. Tedder. January 2000.

[96] N. V. Steere, editor. *Handbook of Laboratory Safety.* The Chemical Rubber Company, Cleveland, Ohio, 1967.

[97] R. B. Stein. Estimating personnel for a new plant. *Chem Eng*, page 140, Sept 24 1979.

[98] W. J. Stupin. *The Separation of Multicomponent Mixtures in Thermally-Coupled Distillation Systems.* PhD thesis, Chemical Engineering, University of Southern California, Los Angeles, 1970.

[99] W. Y. Tawfik. *Design of Optimal Fuel Grade Ethanol Recovery System Using Solvent Extraction.* PhD thesis, School of Chemical Engineering, Georgia Institute of Technology, Atlanta, GA, 1986.

[100] W. Y. Tawfik, A. J. Eckles, IV, and D. W. Tedder. Performance correlations for reciprocating plate extraction columns. *Sol Ext Ion Exch*, 1988.

[101] D. W. Tedder. *The Heuristic Synthesis and Topology of Optimal Distillation Networks.* PhD thesis, Chemical Engineering Department, University of Wisconsin, Madison, 1975.

[102] D. W. Tedder. Fractional distillation of C_2/C_3 hydrocarbons at optimal pressure. *U.S. Patent No.*, 4,430,102, Feb 7, 1984.

[103] D. W. Tedder and D. F. Rudd. Parametric studies in industrial distillation: Part I design comparisons. *AIChE J*, 24(2):303, Mar 1978.

[104] G. W. Thompson. In A. Weissberger, editor, *Techniques of Organic Chemistry*, chapter Vol. 1, pt. 1, page 473. Interscience Publishers, Inc., New York, 3rd edition, 1959.

[105] G. Thornburg, J. C. McArdle, J. A. Piccari, and C. H. Byszewski. Recovery and recycle of valuable constitutents in spent pickling acids. *Environ Progress*, 9(4):N10–N11, Nov 1990.

[106] B. D. Tyreus and W. L. Luyben. Two towers cheaper than one? *Hydrocarbon Proc*, 54(7):93, July 1975.

[107] G. D. Ulrich. *A Guide to Chemical Engineering Process Design and Economics.* John Wiley & Sons, New York, 1984.

[108] T. Umeda, A. Hirai, and A. Ichikawa. Optimal design of chemical process by feasible decomposition method. *Chem Eng Sci*, 27:795, 1972.

[109] A. J. V. Underwood. Fractional distillation of multicomponent mixtures. *Chem Eng Prog*, 44(8):603, Aug 1948.

[110] A. J. V. Underwood. *Chem Eng Prog*, 45(10):609, 1949.

[111] U.S. Census Bureau. *Statistical Abstract of the United States 1998*, Sept 29 1998.

[112] J. Vennard. *Handbook of Fluid Dynamics*, pages 3–23. McGraw-Hill, New York, 1961.

[113] J. L. Viola, Jr. Estimate capital costs via a new, shortcut method. *Chem Eng*, pages 80–86, Apr 6, 1981.

[114] M. P. Wagle. Estimate relative volatility quickly. *Chem Eng*, page 85, Apr 29, 1985.

[115] T. J. Ward. Predesign estimating of plant capital costs. *Chem Eng*, pages 121–124, Sept 17, 1984.

[116] J. Wei, T. Russell, and M. Swartzlander. *The Stucture of the Chemical Processing Industries: Function and Economics.* McGraw-Hill Book Company, New York, 1979.

[117] H. Wessel. New graph correlates operating labor for chemical processes. *Chem Eng*, (7):209, July 1952.

[118] G. S. Whittlesey and J. D. Muzzy. Vapor recompression can reduce steam costs. *Chem Eng*, 146, Jan 12 1981.

[119] F. Wolff, D. V. Eyre, and M. Grenier. Oxygen plants: 10 years of development and operation. *Chem Eng Prog*, pages 83–87, July 1979.

[120] N. Yamanouchi and H. Nagasawa. Using lng cold for air separation. *Chem Eng Prog*, pages 78–82, July 1979.

[121] M. Zabetakis. Fire and explosion hazards at temperature and pressure extremes. In *AIChE-Inst. Chem. engr. symp. Ser. 2, Chem. engr. Extreme Cond. Proc. Symp.*, pages 99–104. AIChE, new York, 1965.

[122] F. C. Zevnik and R. L. Buchanan. Generalized correlation of process investment. *Chem Eng Prog*, 59(2):70–77, Feb 1963.

Index

PUB = S - MAT, 32

yield = f.g.(conv × recovery), loss = $f_i(1-g_i)$, 37

APUB = $\frac{1}{2}[S - 2(MAT)] - \frac{1}{4}[TIC]$, 46

debits must equal credits, 53

assets = liabilites + equity, 54

GP = revenue - costs, NP = GP - tax - CR, 60

cash flow = NP + CR, RE = NP - dividends, 61

capital investment definitions, 64

manufacturing & total product cost, 66

net taxable income, depreciation, depletion, amortization, 68

cash flow, net profits, net cost, CR = I/payoff years, 70

column ht = trays (spacing) + 10', 101

M&S equip & plant cost, 96

pressure factors (F_p), 111

F_m for pumps & storage tanks, 110

purchase cost: $E' = E F_m F_p$, 104

tray cost (freight guidelines in footnotes), 102

70/30 rule: process piping cost is 70% of piping cost, 113

F_{bm} and f_t; 115-119

$C_{bm}^{a} = C_p^{cs}[F_{bm}^{cs} + (1 + 0.7 f_t)(F_m F_p - 1)]$, 121

TIC = equip + installation + startup, 91

base equip cost: $E = E_0(S/S_0)^a$, 105-108

cost definitions (E, E', C_t^a, F_{bm}^a, C_{bm}^a), 100

process vessel correlations, 101

tray costs, packing costs, 102

liquid pump costs, 104

elements of bare module cost, TIC = 1.18 BMC
  carbon steel, low P: $BMC_i = (1+f_i^{ind})(E_i + M_i + L_i)$ } 111, see table 4.17-4.21

Species allocation & simple distillation separation heuristics, 253

complex tower heuristics, 258

  $K_i = P_i^{sat}/P$, $\alpha_{i,j} = P_i^{sat}/P_j^{sat}$, pick two highest $\alpha_{i,j}$, ESI = lightest of 2/heaviest

  if one separation is much easier than the rest, it may be considered alone to generate an initial separation using a simple tower

time value of money, discount factor = $1/(1+i_a)^n$, 166

uniform payments, 168

Printed in the United States
41385LVS00002B/1-60

Heat Exchanger Matching

$T_{min} = 20$

	Cp	source	target
1 cold	55	50	520
2 cold	95	70	580
3 hot	100	510-20=490	70-20=50
4 hot	85	540-20=520	90-20=70

580, 520, 490, 70, 50 — temp intervals = 4

interval	T_i	Ranges (55, 95, 100, 85)	ΣF_iCp_i	ΔH_i	$\Sigma \Delta H_i$	$Q_n + \Sigma \Delta H_i$
	580				0	7650
1			-95	$-95(\Delta T) = -5700$		
	520				-5700	1950
2			-55-95+85=-65	-1950		
T_p	490				-7650	0
3			-55-95+100+85=35	14700		
	70				7050	14700
4			-55+100=45	900		
	50				7950	15600

$Q_{min, heating} = 7650$ $Q_{min, cooling} = 15,600$

$mCp_{,cold} > mCp_{,hot}$ $mCp_{,hot} > mCp_{,cold}$

MNU = 3+4=7

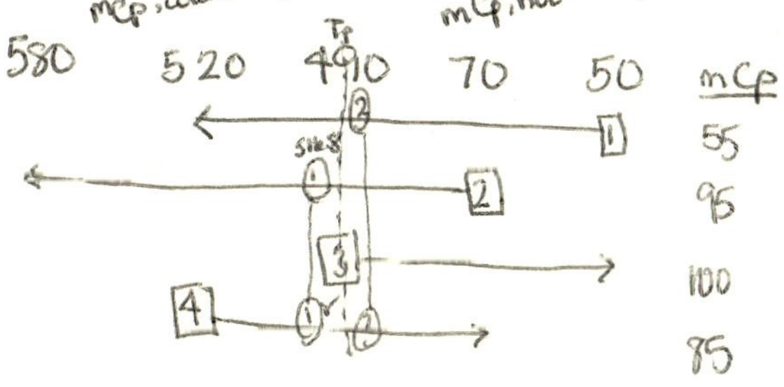

$Q_1 = 85(520-490) = 2550$

$2550 = 95(T_2 - 490) \Rightarrow T_2 = 516.8$

$Q_2 = 55(490-70)$

if $mCp_{hot} \not> mCp_{cold}$, split streams of mCp_{cold}